Time and Ancient Medicine

Time and Ancient Medicine

How Sundials and Water Clocks Changed Medical Science

KASSANDRA J. MILLER

Great Clarendon Street, Oxford, OX2 6DP,
United Kingdom

Oxford University Press is a department of the University of Oxford.
It furthers the University's objective of excellence in research, scholarship,
and education by publishing worldwide. Oxford is a registered trade mark of
Oxford University Press in the UK and in certain other countries

© Kassandra J. Miller 2023

The moral rights of the author have been asserted

All rights reserved. No part of this publication may be reproduced, stored in
a retrieval system, or transmitted, in any form or by any means, without the
prior permission in writing of Oxford University Press, or as expressly permitted
by law, by licence or under terms agreed with the appropriate reprographics
rights organization. Enquiries concerning reproduction outside the scope of the
above should be sent to the Rights Department, Oxford University Press, at the
address above

You must not circulate this work in any other form
and you must impose this same condition on any acquirer

Published in the United States of America by Oxford University Press
198 Madison Avenue, New York, NY 10016, United States of America

British Library Cataloguing in Publication Data
Data available

Library of Congress Control Number: 2023937463

ISBN 978–0–19–888517–7

DOI: 10.1093/oso/9780198885177.001.0001

Printed and bound by
CPI Group (UK) Ltd, Croydon, CR0 4YY

Links to third party websites are provided by Oxford in good faith and
for information only. Oxford disclaims any responsibility for the materials
contained in any third party website referenced in this work.

Acknowledgments

> Time is the most valuable thing a person can spend.
> Theophrastus (DL 5.40.3-4)

It's such a simple thing to spend time—even simpler to waste it. But *making* time for another person is a very different matter; it requires a tremendous amount of generosity, patience, and empathy. I cannot adequately express how grateful I am to everyone who has made time for me and for this project over the years. Your attention and support are priceless gifts, without which we could never have made it to this stage.

This project had its origins in my doctoral dissertation at the University of Chicago, titled A Doctor on the Clock: Hourly Timekeeping and Galen's Scientific Method. Thus, many fervent thanks are due to the friends, colleagues, mentors, students, administrators, and fellowship committees who inspired me to pursue this topic in the first place and provided critical support along the way. Jacqueline Feke, along with my peers within the graduate-student-run reading group on ancient technical writing, opened my eyes to the ways in which a philologist and social historian—with no professional training in medicine, science, or engineering—could nevertheless make meaningful contributions to the study of these domains in antiquity. I am also grateful to Alexander Jones for welcoming me into his seminar on ancient timekeeping at NYU's Institute for the Study of the Ancient World, which helped to spark my enduring fascination with the social history of time, and for being a wonderful mentor ever since. I am deeply indebted to the members of my dissertation committee, Christopher Faraone, Elizabeth Asmis, Jonathan Hall, and Alain Bresson, for their guidance, support, and advocacy, and to Chris especially for his continued mentorship throughout my career. I am also thankful for my peers at UChicago—especially Paul Vadan, Bart van Wassenhove, Erika Jeck, Aimee Genova, Hannah Marcuson, and Emily Wilson—whose friendship has been invaluable, both then and now.

No dissertation can become a monograph without time, resources, and community support. I have been very fortunate to have department chairs—Stacie Raucci at Union College, Lauren Curtis at Bard College, and Kerill O'Neill at Colby College—who have fiercely guarded my research time, even when I was contingent faculty, and who helped me to locate resources to fund and facilitate my work. I am also thankful for Kathy Fox, Leahanna Pelish, and Karin Parsons, departmental administrators whose tireless work enabled me to process reimbursements, organize conferences, print materials, and navigate so many of the logistical challenges involved in bringing a book project to life. I am grateful for the writing workshops and accountability programs, such as those offered by the National Center for Faculty Development & Diversity and FORGE Fuel, that helped me to form and maintain healthy writing habits and to the museums and libraries that facilitated my research. I also owe tremendous debts of gratitude to the frontline and essential workers who helped us all survive the years of pandemic lockdown, during which the bulk of this manuscript was written; to our childcare providers (especially Caroline Stairs, Rebecca Howell, and the staff of Happy Days childcare center); and to our house cleaners (especially Sherry Harper, Carlene, and the team of Pleasant Valley Executive Cleanse & Maintenance Service). Without you, I would not have been able to work at all.

The alchemical processes by which my ideas have germinated, grown, and transmuted were catalyzed by many conferences and conversations with brilliant colleagues and students. This project and I were especially energized by the following conferences: Down to the Hour: Perspectives on Short Time in the Ancient Mediterranean, held at the University of Chicago in 2017; The Day Unit in Antiquity and the Middle Ages, hosted by the Israel Institute for Advanced Studies in 2018; Scientific Traditions in the Ancient Mediterranean and Near East, hosted by NYU's Institute for the Study of the Ancient World in 2019; and Kairos, Krisis, Rhythmos: Time and Time Awareness in Ancient Medicine, hosted by the Einstein Center Chronoi, Berlin, in 2020. I am especially grateful to Jonathan Ben-Dov, Claire Bubb, Philip van der Eijk, Robert Germany, Sarit Kattan Gribetz, Stephan Heilen, Alexander Jones, James Ker, Paul Kosmin, Alexandra von Lieven, Candida Moss, Robert Ritner, Barbara Sattler, Anette Schomberg, Calloway Scott, Peter N. Singer, John Steele, Sacha Stern, Colin Webster, and Anja Wolkenhauer for their inspirational ideas, thoughtful suggestions, and eagerness to collaborate. To Sarah Symons, my co-editor of *Down to the Hour* and now dear friend: you have had a tremendous impact on me, as a researcher, as an educator, and as a person. I can't thank you enough. Finally, I am profoundly grateful to my father, Kenneth Jackson, for curating countless reading lists for me in the philosophy of science and medicine. Those works, and our conversations about them, have shaped my thinking in enduring and deeply generative ways.

At a certain stage, of course, ideas have to be anchored in text, and this text has received the careful attention of many eyes, hands, and minds. I am deeply grateful to the colleagues and family members who volunteered their time to read and respond to drafts of this project—Claire Bubb, Glen Cooper, Deborah Jackson, Kenneth Jackson, Alexander Jones, Paul Keyser, Susan Mattern, Peter N. Singer, Richard Talbert, and Colin Webster—and to the anonymous reviewers of the press. These readers' thoughtful suggestions have improved the book immensely, and any infelicities that remain are, of course, nobody's fault but my own

When this book was out for review, I was fortunate to discover that Peter N. Singer had completed a draft of his monograph, *Time for the Ancients* (De Gruyter, 2022), which also uses the Galenic corpus as a jumping-off point for pursuing questions about time, medicine, and society in ancient Rome, albeit from different perspectives than those adopted here. I am very grateful to Peter for generously sharing and discussing his draft with me and greatly enriching my own thinking. Hearty thanks are also due to my undergraduate research assistant, Sarah Haynes, who meticulously organized this book's primary sources and offered helpful feedback on the work as a whole.

I am especially grateful to Charlotte Loveridge, Jamie Mortimer, and Cathryn Steele of Oxford University Press for seeing the potential in this project and helping to accomplish its final transformation, from manuscript into book. Many thanks are due, as well, to the scholars and institutions who supplied this book's images and, in particular, to my father, Kenneth Jackson, for preparing two original line drawings. It means so much to me to be able to collaborate with you like this and to share your artistic talents with this book's readers.

Lastly—and first and always—I am infinitely grateful for my family, those magical beings who can make any place a home and any moment momentous. Kristin Mahan, Jessica Kwong, Valerie Love, Rachael Richman, and Katie Ryan, thank you for always inspiring me and being there through thick and thin. Heidi, Brian, Max, Julie, and Izzie, I am so honored to be a part of your family and grateful for the love and openness with which you've welcomed me. Mom and Dad, thank you for loving me unconditionally, being my role models and cheerleaders, and always making the time. Jon and Riley, our wonder girl, thank you for helping me to look past the time and savor the moment. I love you beyond measure.

Contents

List of Figures	xi
List of Tables	xiii
Abbreviations	xv
Introduction	1

PART I. CLOCKWORKS

1. Telling Time in the Greco-Roman World	15
Telling Time without Clocks	17
Hours	19
Sundials	23
Water Clocks	27
Conclusion	30
2. Doctors and Clocks: The Emergence of Hourly Timekeeping in Medical Contexts	32
Egyptian and Assyro-Babylonian Precedents	35
Hippocratic Evidence	40
Hellenistic Evidence	45
The Roman Period	48
3. Clocks as Symbols 1: In Galen's Thought	61
Galen's Chronotopic Biography	61
Introducing *Affections and Errors*	63
Galen's Scientific Method	66
Defending Propositions 1a and 1b: Testability	71
Defending Propositions 2a and 2b: Useful Contributions to Intellectual and Civic Communities	75
Defending Propositions 3a and 3b: Progress and Perpetuity	79
Conclusion	83
4. Clocks as Symbols 2: Among the Roman Elite	87
A Very Public Gnomon: Timekeeping and the Imperial Program	87
Clocks and Celestial Geometry: Symbols of Mathematical Ingenuity	93
Clocks and the Educated Elite: Symbols of the *Pepaideumenoi*	96
Clocks and Philosophers: Symbols of the Human Lifespan	100
Clocks in Architectural Writing: Symbols of the Human Body	105
Clocks and Horoscopic Astrology: Symbols of Death and the Afterlife	108
Conclusion: The "Clock-Construction" Lifestyle	111

PART II. HOURS IN ACTION

5. From "Season" to "Hour": Galenic Refinements of
   the Hippocratic *Hōra* — 115
   Galen's Hippocratism — 116
   Galen's *On Critical Days* and the Hippocratic *Epidemics* — 118
   Temporality in Galen's Fever Case Histories — 120
   Hours and Astronomy I: The Period of the Moon — 126
   Hours and Astronomy II: The Periods of Planets — 131
   Conclusion — 134

6. When Is Temporal Exactitude Desirable? — 135
   Insufficient Exactitude: Fevers and Symptoms — 136
   Excessive Exactitude: Fevers and Phlebotomy — 146
   Conclusion — 154

7. "Right Timing" in Sickness and in Health: Hourly
   Timekeeping and *Kairos* in Galen's *On Hygiene* — 156
   Kairos and Galen's Predecessors — 157
   Kairos and Febrile Disease — 161
   Kairos and Health Maintenance — 164
   Kairos and the Elderly — 173
   Conclusion — 178

   Conclusion: From Antiquity to Modernity — 180

Bibliography — 187
Index — 207

List of Figures

A. Portable cylindrical sundial, Tomb of the Physician, Este, Italy. Museo Nazionale Atestino di Este: IG MNA 15397. Photograph by Guido Petruccioli with permission from the Museum Atestino di Este.   16

B. Schematic of sundial hour-lines and day-curves. Modeled after a marble conical dial from Pompeii (Berlin Sundial Collaboration, Ancient Sundials, Dialface ID 174, Napoli, 2014, Edition Topoi, DOI: 10.17171/1-1-2199). Image © Kenneth Jackson.   24

C. Roman mosaic depicting a thinker, possibly Anaximander, with a sundial. Image © GDKE/Rheinisches Landesmuseum Trier, Inv. No. 1907,724, photographer Th. Zühmer.   27

D. Statuette of grieving slave with sundial, Necropolis, Myrina, Greece. National Archaeological Museum, Athens: 5007/D.95. Image © Institute for the Study of the Ancient World, photographer Orestis Kourakis.   51

E. A schematic representation of the Montecitorio obelisk acting as a meridian dial. Image © Kenneth Jackson.   88

F. Roofed spherical sundial with Greek inscriptions. Inv. MA5074. Image © RMN-Grand Palais/Art Resource, NY, photographer Hervé Lewandowski.   94

G. Gem with sitting, reading thinker in front of a sundial. Staatliche Münzsammlung, Munich, Inv. A.197. Reproduced with permission from J. Lang, *Mit Wissen geschmückt? Zur bildlichen Rezeption griechischer Dichter und Denker in der römischen Lebenswelt*, MAR 39 (Wiesbaden 2012), 170, Abb. 165, Kat. Nr. G TypA65, Taf. 21. Image © Köln Digital Archaeology Laboratory, photographer Jörn Lang.   97

H. Muse sarcophagus. Detail of side panel, depicting a seated "philosopher" looking up at a sundial. Image © KHM-Museumsverband.   99

I. Detail of Prometheus sarcophagus. Capitoline Museum, Inv. 638. Image © Shutterstock.   110

J. Mosaic depicting a man looking at a sundial, Daphne. Image © NPL—De Agostini Picture Library/Bridgeman Images.   177

List of Tables

1. The pattern of paroxysms patient experienced in the first five days.   141
2. The paroxysms of Galen's Tertian 1.   141

Abbreviations

Ancient Authors

Aen. Tact.	Aeneas Tacticus
Aët. Amid.	Aëtius of Amida
Libr. med.	Libri medicinales
Alc.	Alciphron
Alex. Aphr.	Alexander of Aphrodisias
In Top.	In Aristotelis Topica commentaria
Arr.	Arrian
Epict.	Epicteti dissertationes
De bell. Afr.	De bello Africo
De bell. Hisp.	De bello Hispaniensi
Anth. Graec.	Anthologia Graeca
Antyl.	Antyllus
Ar.	Aristophanes
Ach.	Acharnenses
Ekk.	Ekklesiazousai
Arist.	Aristotle
Gen. an.	De generatione animalium
Pol.	Politica
Nic. eth.	Ethica Nicomachea
Aristid.	Aelius Aristides
Hier. log.	Hieroi logoi
Artem.	Artemidorus
Oneir.	Oneirocritica
Ath.	Athenaeus
Deipn.	Deipnosophistae
Aul. Gell.	Aulus Gellius
NA	Noctes Atticae
M. Aur.	Marcus Aurelius
Med.	Meditationes
Cael. Aur.	Caelius Aurelianus
Tard. pass.	Tardae passiones
Caes.	Caesar
Gall.	De bella Gallica
Civ.	De bella civile
Callim.	Callimachus
Catull.	Catullus

Cens.   Censorinus
- *DN*   *De die natali*

Cic.   Cicero
- *De nat. deor.*   *De natura deorum*
- *Rep.*   *De republica*
- *Sen.*   *De senectute*
- *Tusc.*   *Tusculanae disputationes*
- *Verr.*   *In Verrem*

Dig.   *Digesta*
Dio Cass.   Dio Cassius
Diod. Sic.   Diodorus Siculus
DL   Diogenes Laertius
Gal.   Galen [see Titles and Abbreviations of Works Attributed to Galen]
- *Hist. Aug. Had.*   *Historia Augusta: Hadrianus*

Hdt.   Herodotus
Hes.   Hesiod
- *WD*   *Opera et dies*

Hippoc.   Hippocrates
- *Aer.*   *De aere aquis et locis*
- *Aph.*   *Aphorismi*
- *De morb.*   *De morbis*
- *Epid.*   *Epidemiae*
- *Int.*   *De internis affectionibus*
- *Nat. hom.*   *De natura hominis*
- *Prisc. med.*   *De prisca medicina*
- *Reg.*   *De diaeta in morbis acutis*
- *Vict.*   *De victu*
- *Hist. Aug.*   *Historia Augusta*
- *Had.*   *Vita Hadriani*

Hor.   Horace
- *Serm.*   *Sermones*

Juv.   Juvenal
- *Sat.*   *Satires*

Luc.   Lucian
- *Hipp.*   *Hippias*

Marcellin.   Marcellinus
- *Puls.*   *De pulsibus*

Mart.   Martial
- *Epig.*   *Epigrammata*

Orib.   Oribasius
- *Coll. med.*   *Collectiones medicae*

Paus.   Pausanias
Pers.   Persius
Petron.   Petronius
- *Sat.*   *Satyricon*

Pl.	Plato
Gorg.	*Gorgias*
Phaedr.	*Phaedrus*
Prot.	*Protagoras*
Rep.	*Respublica*
Ti.	*Timaeus*
Plb.	Polybius
Plin.	Pliny
Ep.	*Epistulae*
NH	*Historia naturalis*
Plut.	Plutarch
Per.	*Pericles*
Philostr.	Philostratus
Im.	*Imagines*
Ptol.	Ptolemy
Anal.	*Analemma*
Sen.	Seneca
Brev.	*De brevitate vitae*
Clem.	*De clementia*
Const.	*De constantia sapientis*
Ep.	*Epistulae*
Ira	*De ira*
Otio	*De otio*
Tranq.	*De tranquillitate animi*
Sext. Emp.	Sextus Empiricus
Math.	*Adversus mathematicos*
Sid. Apoll.	Sidonius Apollinaris
Epist.	*Epistulae*
Simpl.	Simplicius
In Phys.	*In Aristotelis de Physica commentarii*
Sor.	Soranus
Gyn.	*Gynaecia*
Vit. Hipp.	*Vita Hippocratis*
Stob.	Stobaeus
Ecl.	*Ἐκλογαί*
Strab.	Strabo
Geog.	*Geographica*
Suet.	Suetonius
Aug.	*Vita Augustae*
Tib.	*Vita Tiberii*
Vit.	*Vita Vitellii*
Tac.	Tacitus
Ann.	*Annales*
Hist.	*Historiae*
Thess.	Thessalus of Tralles
De virt. herb.	*De virtutibus herbarum*

Ulp.   Ulpian
    Mos. et Rom. legum coll.   *Collatio legum Mosaicarum et Romanarum*
V. Max.   Valerius Maximus
Varr.   Varro
    Ling.   *De lingua Latina*
Verg.   Vergil
    Aen.   *Aeneid*
Vitr.   Vitruvius
    De arch.   *De architectura*
Xen.   Xenophon
    An.   *Anabasis*
    Mem.   *Memorabilia*

Titles and Abbreviations of Works Attributed to Galen

Abbreviation	Latin Title	English Title
AA	De anatomicis administrationibus	On Anatomical Procedures
Adv. Jul.	Adversus Julianum	Against Julian
Adv. Lyc.	Adversus Lycum	Against Lycus
Adv. typ. scr.	Adversus eos qui de typis scripserunt	Against Those Who Have Written on Types (or On Periods)
Aff. pecc. dig.	De propriorum animi cuiusque affectuum dinotione et curatione; De animi cuislibet peccatorum dignotione et curatione	On the Diagnosis and Treatment of the Affections and Errors of the Individual Human Soul (or Affections and Errors)
Alim. fac.	De alimentorum facultatibus	On the Properties of Foodstuffs
Ant.	De antidotis	On Antidotes
[An. ut.]	An animal sit quod in utero geritur	Whether What is in the Womb is an Animal
Ars med.	Ars medica	The Art of Medicine
Art. sang.	An in arteriis natura sanguis contineatur	Whether Blood is Naturally Contained in the Arteries
At. bil.	De atra bile	On Black Bile
Bon. hab.	De bono habitu	On Good Condition
Bon. mal. sac.	De bonis malisque succis	On Good and Bad Juices
CAM	De constitutione artis medicae	On the Composition of the Art of Medicine
[Cath. med. purg.]	Quos, quibus catharticis medicamentis et quando purgare oporteat	Whom to Purge, with What Cleansing Drugs and When
Caus. cont.	De causis contentivis	On Containing Causes
Caus. morb.	De causis morborum	On the Causes of Disease
Caus. proc.	De causis procatarcticis	On Antecedent Causes

Caus. puls.	De causis pulsuum	On the Causes of the Pulse
Caus. resp.	De causis respirationis	On the Causes of Breathing
Caus. symp.	De causis symptomatum	On the Causes of Symptoms
Comp. med. gen.	De compositione medicamentorum per genera	On the Composition of Medications According to Kind
Comp. med. loc.	De compositione medicamentorum secundum locos	On the Composition of Medications According to Places
Cons.	De consuetudinibus	On Habits
Cris.	De crisibus	On Crises
Cur. rat. ven. sect.	De curandi ratione per venae sectionem	On Treatment by Venesection
[Def. med.]	Definitiones medicae	Medical Definitions
Dem.	De demonstration	On Demonstration
Di. dec.	De diebus decretoriis	On Critical Days
Di. Hipp. morb. ac.	De diaeta Hippocratis in morbus acutis	On Hippocrates' Regimen in Acute Diseases
Diff. feb.	De differentiis febrium	On the Distinct Types of Fever
Diff. morb.	De differentiis morborum	On the Distinct Types of Disease
Diff. puls.	De differentiis pulsuum	On the Distinct Types of Pulse
Diff. resp.	De difficultate respirationis	On Difficulty in Breathing
Dig. insomn.	De dignotione ex insomniis	On Diagnosis by Dreams
Dig. puls.	De dignoscendis pulsibus	On Diagnosing Pulses
Elem.	De elementis ex Hippocrate	On the Elements According to Hippocrates
Fasc.	De fasciis	On Bandages
Foet. form.	De foetuum formation	On the Formation of the Fetus
Gal. fasc.	Ex Galeni	From Galen's Notes
Gloss.	Glossarium	Hippocratic Glossary
[Hipp. Alim.]	In Hippocratis De alimento	Commentary on Hippocrates' Nutrition
Hipp. Aph.	In Hippocratis Aphorismos	Commentary on Hippocrates' Aphorisms
Hipp. Art.	In Hippocratis De articulis	Commentary on Hippocrates' On Joints
Hipp. AWP	In Hippocratis De aere aquis et locis	Commentary on Hippocrates' Airs Waters Places
Hipp. Com.	De comate secundum Hippocratem	Coma According to Hippocrates
Hipp. Epid. 1	In Hippocratis Epidemiarum 1	Commentary on Hippocrates' Epidemics 1
Hipp. Epid. 2	In Hippocratis Epidemiarum 2	Commentary on Hippocrates' Epidemics 2
Hipp. Epid. 3	In Hippocratis Epidemiarum 3	Commentary on Hippocrates' Epidemics 3

Hipp. Epid. 4	*In Hippocratis Epdemiarum* 4	*Commentary on Hippocrates' Epidemics 4*
Hipp. Fract.	*In Hippocratis De fracturis*	*Commentary on Hippocrates' Fractures*
[*Hipp. Hum.*]	*In Hippocratis De humoribus*	*Commentary on Hippocrates' Humors*
[*Hipp. Ius*]	*In Hippocratis Ius*	*Commentary on Hippocrates' Oath*
Hipp. Off. med.	*In Hippocratis De officina medici*	*Commentary on Hippocrates' Surgery*
Hipp. Prog.	*In Hippocratis Prognosticum*	*Commentary on Hippocrates' Prognostic*
Hipp. Prorrh.	*In Hippocratis Prorrheticon*	*Commentary on Hippocrates' Prorrhetic*
[*Hipp. Sept.*]	*In Hippocratis De septimanis*	*Commentary on Hippocrates' Sevens*
Hipp. Vict.	*In Hippocratis De victu*	*Commentary on Hippocrates' Regimen for Health*
[*Hipp. Vict. morb.*]	*De victu ratione in morbis acutis ex Hippocratis sententia*	*On Regimen in Acute Diseases According to Hippocrates*
[*Hist. phil.*]	*Historia philosophica*	*History of Philosophy*
HNH	*In Hippocratis De natura hominis*	*Commentary on Hippocrates' Human Nature*
HRCIS	*De hirundinibus, revulsione, cucurbitula, incisione et scarificatione*	*On Leeches, Revulsion, Cupping, Incision, and Scarification*
[*Hum.*]	*De humoribus*	*On Humors*
HVA	*In Hippocratis De victu in morbis acutis*	*Commentary on Hippocrates' Regimen in Acute Diseases*
Inaeq. int.	*De inaequali intemperie*	*Uneven Bad Mixture*
Ind.	*De indolentia*	*On Avoiding Distress*
Inst. log.	*Institutio logica*	*Introduction to Logic*
Inst. od.	*De instrumento odoratus*	*The Organ of Smell*
[*Int.*]	*Introductio seu medicus*	*Introduction*
Lib. prop.	*De libris propriis*	*My Own Books*
Loc. aff.	*De locis affectis*	*On Affected Places*
Marc.	*De marcore*	*On Marasmus*
Med. exp.	*De experientia medica*	*On Medical Experience*
[*Mel.*]	*De melancholia*	*On Melancholy*
MM	*De methodo medendi*	*On the Method of Healing*
MMG	*De methodo medendi ad Glauconem*	*On the Method of Healing, for Glaucon*
Mor.	*De moribus*	*On Character Traits*
Morb. temp.	*De morborum temporibus*	*On Opportune Moments in Disease*
Mot. dub.	*De motibus dubiis*	*On Problematical Movements*
Mot. musc.	*De motu musculorum*	*On the Movement of Muscles*

Musc. diss.	*De musculorum dissectione*	*On the Anatomy of the Muscles*
Nat. fac.	*De facultatibus naturalibus*	*On Natural Faculties*
Nerv. diss.	*De nervorum dissectione*	*On the Anatomy of the Nerves*
Nom. med.	*De nominibus medicis*	*On Medical Terminology*
Opt. const. corp.	*De optima corporis constitutione*	*On the Best Constitution of our Bodies*
Opt. doct.	*De optima doctrina*	*On the Best Method of Teaching*
Opt. med.	*Quod optimus medicus sit quoque philosophus*	*That the Best Doctor is also a Philosopher*
Opt. med. cogn.	*De optimo medico cognoscendo*	*On Discovering the Best Physician*
[*Opt. sect.*]	*De optima secta*	*On the Best Sect*
Ord. lib. prop.	*De ordine librorum propriorum*	*On the Order of My Own Books*
Oss.	*De ossibus ad tirones*	*On Bones for Beginners*
Part. art. med.	*De partibus artis medicae*	*On the Parts of the Art of Medicine*
Part. hom. diff.	*De partium homoeomerium differentia*	*On the Difference Between Uniform Parts*
Parv. pil.	*De parvae pilae exercitio*	*On Exercise with the Small Ball*
PHP	*De placitis Hippocratis et Platonis*	*On the Opinions of Hippocrates and Plato*
Plat. Tim.	*In Platonis Timaeum*	*Commentary on the Medical Statements in the* Timaeus
Plat. Tim. comp.	*Timaei Platonis compendium*	*Compendium of Plato's* Timaeus
Plen.	*De plenitudine*	*On Fulness*
[*Pond. mens.*]	*De ponderibus et mensuris*	*On Weights and Measures*
Praec.	*De praecognitione ad Epigenem*	*On Prognosis, for Epigenes*
Praen.	*De praenotione*	*On Prognosis*
Praes. puls.	*De praesagitione ex pulsibus*	*On Prognosis from the Pulse*
[*Praes. ver. exp.*]	*De praesagitione vera et experta*	*On True and Expert Prognosis*
[*Prog. dec.*]	*Prognostica de decubitu ex mathematica scientia*	*Prognosis Based on the Hour When a Patient Goes to Bed Based on the Science of Astrology*
Prolaps.	*De humero iis modis prolapso quos Hippocrates non vidit*	*On Dislocations Unseen by Hippocrates*
Prop. plac.	*De propriis placitis*	*On My Own Opinions*
Protr.	*Protrepticus*	*An Exhortation to Study the Arts*
Ptis.	*De ptisana*	*On Barley Gruel*
Puer. epil.	*Puero epileptico consilium*	*Advice for an Epileptic Boy*
Puls.	*De pulsibus ad tirones*	*On Pulses for Beginners*
[*Puls. Ant.*]	*De pulsibus ad Antonium*	*On Pulses, for Antonius*
Purg. med. fac.	*De purgantium medicamentorum facultatibus*	*On the Property of Purgatives*
QAM	*Quod animi mores corporis temperamenta sequantur*	*That the Soul's Behavior Depends on Bodily Mixtures*
[*Qual. incorp.*]	*De qualitatibus incorporeis*	*On Non-corporeal Qualities*

[Rem.]	De remediis parabilibus	On Readily Available Remedies
[Ren. aff.]	De renum affectibus	On Affections of the Kidneys
San. tu.	De sanitate tuenda	On Hygiene
Sect.	De sectis ad eos qui introducuntur	On Sects for Beginners
Sem.	De semine	On Semen
Sept. part.	De septimestri partu	On Seven-month Children
Sim. morb.	Quomodo simulantes morbum deprehendendi	How to Detect Malingerers
SMT	De simplicium medicamentorum facultatibus	On the Properties of Simple Drugs
Soph.	De sophismatibus penes dictionem	On Linguistic Sophisms
Sub. nat. fac.	De substantia facultatum naturalium	On the Substance of the Natural Faculties
Subf. emp.	Subfiguratio empirica	A Sketch of Empiricism
[Suc.]	De succedaneis	On Substitute Drugs
Symp. diff.	De symptomatum differentiis	On the Distinct Types of Symptom
Syn. puls.	Synopsis de pulsibus	A Synopsis of the Pulse
Temp.	De temperamentis	On Mixtures
[Ther. Pamph.]	De theriaca ad Pamphilanum	On Theriac, for Pamphilianus
[Ther. Pis.]	De theriaca ad Pisonem	On Theriac, for Piso
Thras.	Thrasybulus sive Itrum medicinae sit aut gymnasticae hygiene	Thrasybulus or Whether Hygiene Belongs to Medicine or Physical Training
Tot. morb. temp.	De totius morbi temporibus	On Opportune Moments in Diseases as a Whole
Trem. palp.	De tremore, palpitatione, convulsione, et rigore	On Tremor, Spasm, Convulsion, and Shivering
Tum. pr. nat.	De tumoribus praeter naturam	On Unnatural Swellings
Typ.	De typis	On Types
UP	De usu partium	On the Use of Parts of the Body
[Ur.]	De urinis	On Urines
[Ur. comp.]	De urinis compendium	A Synopsis of Urines
[Ur. comp. Gal.]	De urinis ex Hippocrate, Galeno	A Synopsis of Urines, according to Hippocrates, Galen
Us. puls.	De usu pulsuum	On the Use of the Pulse
Us. resp.	De usu respirationis	On the Use of Breathing
Ut. diss.	De uteri dissection	On the Anatomy of the Womb
[Ven.]	De venereis	On Sexual Activity
Ven. art. diss.	De venarum arteriarumque dissectione	On the Anatomy of Veins and Arteries
[Ven. sect.]	De venae sectione	On Venesection
Ven. sect. Er.	De venae sectione adversus Erasistratum	On Venesection, Against Erasistratus

Ven. sect. Er. Rom.	*De venae sectione adversus Erasistrateos Romae degentes*	*On Venesection, Against the Erasistrateans in Rome*
Vict. att.	*De victu attenuante*	*On the Thinning Diet*
[*Virt. cent.*]	*De vertutibus centaureae*	*On the Properties of Centaury*
Voc.	*De voce*	*On the Voice*

Reference Works and Editions

AE	1889–2016. *L'Année épigraphique*. Paris: Presses Universitaires de France.
ÄM	Ägyptisches Museum und Papyrussammlung, Berlin, museum siglum.
BM	British Museum, London, museum siglum.
CIL	1862–. *Corpus Inscriptionum Latinarum*. Berlin: De Gruyter et al.
DB	De Boer, W. 1937. *Galeni de animi cuiuslibet affectuum et peccatorum dignotione et curatione*. Corpus Medicorum Graecorum 5.4.1.1. Leipzig: Teubner.
Gardner	Gardner, P. 1882. *The Types of Greek Coins*. Cambridge: Cambridge University Press.
Gibbs	Gibbs, S. L. 1976. *Greek and Roman Sundials*. New Haven: Yale University Press.
I.Aeg. Thrace	Loukopoulou, L. D., et al. 2005. *Epigraphes tēs Thrakēs tou Aigaiou: metaxy tōn potamōn Nestou kai Hevrou (nomoi Xanthēs, Rhodopēs kai Hevrou)*. Athens: Ethnikon Hidryma Ereunōn, Kentron Hellēnikēs kai Rōmaikēs Archaiotētos.
ID	Durrbach, F., et al. 1926–72. *Inscriptions de Délos, I–VII*. Paris: H. Champion.
IEph	Wankel, H., et al. 1979–84. *Die Inschriften von Ephesos*, IGSK 11–17. Bonn: R. Habelt.
IG	1903–. *Inscriptiones Graecae*. Berlin: De Gruyter et al.
IGUR	Moretti, L. 1968–90. *Inscriptiones Graecae Urbis Romae*. Rome: Istituto italiano per la storia antica.
ILS	Dessau, H. 1892–1916. *Inscriptiones Latinae Selectae*. Berlin: Weidmann.
IPriene	Blümel, W., and R. Merkelbach. 2014. *Die Inschriften von Priene*. Bonn: Habelt.
K	Kühn, K. G. 1821–33. *Claudii Galeni Opera Omnia*. Leipzig: C. Knobloch
L	Littré, E. 1839–61. *Oeuvres completes d'Hippocrate*. Paris: Baillière.
LBAT	Pinches, T. G., and J. N. Strassmaier, eds. 1955. *Late Babylonian Astronomical and Related Texts*. Providence, RI: Brown University Press.
LSJ	Liddell, Scott, and Jones Greek-English Lexicon online
MANN	Museo Archeologico Nazionale di Napoli, museum siglum

MDAI(A)	1876–1914, 1952–. *Mitteilungen des deutschen archäologischen Instituts, Athenische Abteilung.* Berlin: De Gruyter et al.
P. Oxy.	The Oxyrhynchus Papyri
P. Coll. Youtie	A. E. Hanson, ed. 1976. *Collectanea Papyrologica: Texts Published in Honor of H.C. Youtie.* Bonn: Habelt.
SEG	1923–. *Supplementum Epigraphicum Graecum.* Brill: Leiden.
SNG von Aulock	Von Aulock, H. 1957–67. *Sylloge Nummorum Graecorum.* Berlin: Mann.
TLG	*Thesaurus Linguae Graecae* online database

Introduction

Imagine a future in which every prescription carries a time stamp, and even an over-the-counter remedy such as Paracetamol (Tylenol), profoundly rhythmic in its liver toxicity, sports a label with timing instructions. Next to your blood type, your medical records could include your chronotype, which your work hours also reflect. Junk food would carry a warning about late-night consumption.[1]

This vision of the future, offered by science writer J. Gamble in 2016, reflects the fact that Western biomedical professionals have become keenly interested in how human bodies perform differently at different times of day and night. This field of medical research, known as chronobiology, investigates how molecular "clocks" orchestrate the rhythms and cycles within us. The "settings" of these molecular clocks vary from person to person, leading some people, for example, to be described as "early birds" or "night owls," or to experience varying degrees of jet lag when adjusting to a new time zone. Molecular clocks can also affect the outcomes of medical treatments. For instance, the rate at which a patient's body processes a medication will, according to this new research, vary depending on that patient's "chronotype" and on the exact time of day at which the medicine was taken. Gamble's vision of the future, projecting outward from current trends in chronobiology, proposes that one day people might be as familiar with their chronotypes as they are with their blood types, and that knowledge of one's chronotype might be used to create medical, dietary, fitness, and work regimens that are precisely timed and specially tailored.

The idea of using high-precision timekeeping instruments to synchronize human behaviors with natural rhythms and periodicities may strike us as exceedingly modern, perhaps bordering on the realm of science fiction. This book, however, argues that many current chronobiological ideas are not entirely new. In fact, ancient physicians hotly debated the role of "clocks"—whether those inside us or those external to us, such as celestial bodies or sundials—within medical theory and practice. Both ancient and modern debates on this topic intersect with a constellation of other issues: to what extent should medicalized time be expressed in numerical rather than qualitative terms? To what extent must

---

[1] Gamble 2016, 9.

medicalized time be standardized? And what degrees of precision and accuracy are medically useful or justified?[2]

This book chronicles how, in the Greek- and Latin-speaking worlds, healers' answers to these questions changed over time and from context to context. In the fifth and fourth centuries BCE, the Hippocratics developed practices (potentially adapted from Assyro-Babylonian and Pharaonic Egyptian approaches to predictive healing) that relied on tracking the temporal trajectories of diseases, and these practices coincided with a burgeoning societal interest in timekeeping and calendrics. It is in this period, the late Classical and early Hellenistic, that sundials, water clocks, and references to numbered hours first appear in our textual and archaeological records. Yet, at the time, these tools remained relatively rare, and there was little consensus among healers about whether and how to employ them. A timekeeping revolution, which originated in the middle Hellenistic period and solidified under the Roman Empire, saw a dramatic increase in the ubiquity and standardization of timekeeping devices. This technological revolution, in turn, helped to generate new ways to use time in medical diagnosis, prognosis, and therapeutics. The temporal parameters of this book were dictated by the availability of evidence for hourly timekeeping. Thus, while the book considers material from circa 400 BCE–300 CE produced by a range of authors writing within many regions that came under Roman rule, it focuses primarily on material from major urban centers during the Roman Imperial period.[3]

The ancient Mediterranean did not have anything like our International Bureau of Weights and Measures, which has declared the "second" to be the base temporal unit for the scientific world and has defined this unit very precisely as "the duration of 9,192,631,770 periods of the radiation corresponding to the transition between the two hyperfine levels of the ground state of the caesium-133 atom."[4] However, as the expanding Roman Empire brought together under a single administration numerous regions and polities with different cultures, languages, and practices of measurement, the need for some degree of standardization was increasingly felt. The Imperial-period physician Galen of Pergamon, for example, often complained that it was difficult to translate pharmacological material from one cultural context to another because there was insufficient standardization in the names of plants and units of measurement appropriate for mixing and dosing

---

[2] A groundbreaking work on the sociology of time (with an emphasis on schedules) in American hospitals is Zerubavel 1979.

[3] This concentration of evidence in Roman Imperial urban centers is no doubt partially attributable to biases in preservation and excavation. Nevertheless, it is clear that, with the advent of sundials, water clocks, and hourly timekeeping frameworks in the Hellenistic period, we begin to see a significant change in how certain kinds of time are marked and measured, a change that reaches its acme under the Roman Empire.

[4] Taylor and Thompson 2006, 113, section 2.1.1.3. These standard seconds are marked using "atomic clocks," which deviate from true by less than 1×10^{-15} seconds over a given month.

medications.[5] Great strides in standardization were made over the course of the Roman period, but adoption was differential: standardization was not effected equally in all parts of the Empire, or in all social, cultural, and professional contexts.

By the height of the Roman Imperial period, residents of most urban centers had access to monumental, domestic, or even portable sundials and water clocks which, between them, could offer continuous numerical timekeeping throughout the day and night and could, in some cases, even measure time down to the half-hour. Yet, as we in the twenty-first century know well, being surrounded by precise timekeeping devices does not mean that we rely on them for timing every activity, or that we always employ the greatest level of available precision when talking about time. The digital timekeepers on most of our phones, computers, and watches can indicate the time down to a fraction of a second. Yet instead of saying, for example, that we arrived for an appointment at 10:27:49 a.m., we will typically round up and say that we arrived at 10:30. Likewise, if we are told that a medication is meant to be taken two hours after eating, we tend to interpret this as a general guideline with some temporal leeway. Moreover, there are occasions when we abandon "clock time" and numerical precision all together and time our activities using qualitative indicators. For instance, while some of us decide when to break for lunch by using a clock (e.g., because our company gives us a predetermined lunch hour, or because we are coordinating a date with someone else), many of us decide when to eat based on cues from our bodies, such as whether our stomachs are growling or we can feel ourselves getting "hangry."[6] We might also decide when to eat lunch based on the sequence of our activities, perhaps telling ourselves that we can take a break to eat *after* we have completed a certain project or finished writing that email. Thus, even in the clock-dense environments of the modern day, different people make different choices within different contexts about the degree to which quantitative temporal precision is relevant. This book seeks to demonstrate that the same principle holds true for Greco-Roman antiquity and investigates how various personal, social, and cultural factors led individual physicians to mark and measure time in different ways.

Increasingly, but in different ways, ancient Greek physicians chose to mark and measure time *quantitatively*, using devices like sundials and water clocks that could measure time in numbered hours. To many ancient physicians, numbers and mathematics offered the tantalizing possibility of describing, predicting, and thereby controlling patient outcomes with precision. Yet this attitude was not shared across the board. Many ancient physicians also argued that, since medicine

---

[5] See, e.g., *Alim. fac.* 6.628 K.
[6] This fun portmanteau (added to the *Oxford English Dictionary* in 2018) describes the irritability that often attends hypoglycemia.

deals with humans—who have diverse bodies, habits, and preferences and are embedded in different social frameworks—"real-world" medicine should be considered an inexact science, one that does not readily conform to abstract, mathematical models. This debate is, itself, intertwined with a number of other issues, such as the availability of hourly timekeeping devices (based, for instance, on one's geographical location and social class), one's ideological orientation toward numbers and mathematics, and the degree to which the subject matter at hand lends itself to numerical, temporal measurement.

The most dominant voice in this debate belongs to Galen of Pergamon (129–216 CE). Galen was born and raised on the west coast of Asia Minor and then traveled throughout the Mediterranean to study with a variety of philosophical and medical teachers before ultimately settling in Rome. There he became the personal physician to two Roman emperors, Marcus Aurelius and Commodus.[7] As S. Mattern points out in her 2013 biography of Galen, "the most modern edition of his corpus runs to 22 volumes, including about 150 titles, making up one-eighth of all the classical Greek literature that survives."[8] In addition to producing a vast amount of textual material, Galen also wrote on a wide range of topics. He considered himself an authority not only on medical practice, but also on pharmacology, philosophy, linguistics, and basic astronomy and mathematics. As a result, his writings allow the modern scholar to see how these disciplines intersected and mutually influenced one another within Galen's thought. Additionally, since Galen engages both regularly and critically with the works of other theorists and practitioners, his corpus is a rich source for (his own interpretations of) the views of his contemporary and historical rivals. The present book proposes that Galen's texts—although they construct the evidence according to his own biases—capture a lively debate surrounding the medical use of timekeeping within his own and earlier periods.

This book explores how individual physicians, like Galen and his interlocutors, decided when to indicate and measure medically significant time using numbered hours—rather than by referencing, for instance, the sequence of a patient's activities or the cues generated by a patient's body. It considers how these doctors' decisions were affected by the availability and relative accuracy of sundials and water clocks in their environments. It also asks how their decisions about when and why to employ these tools were affected by their own personal values, biases, and ideological commitments, as well as by broader social and intellectual trends, such as the increasing popularity of astronomy and astrology. Finally, this book investigates how the ways in which physicians chose to talk about time could, in turn, help them to articulate their ideological commitments and support their stances within a network of active debates. It focuses on four kinds of debates,

---

[7] For Galen's biography, see below, "Galen's Chronotopic Biography" and Chapter 3, n. 1.
[8] Mattern 2013, 3.

broadly speaking: (a) debates over specific questions, such as how to define the length of fetal gestation or the periods of irregular intermittent fevers; (b) debates over proper scientific methodology, such as those that raged between the so-called Rationalist, Empiricist, Methodist, and Pneumatist schools of medical thought during the Roman period; (c) debates over the status of medicine as a formal art or *tekhnē*, which are closely intertwined with discussions about the value of mathematics and whether medicine can and should be considered an exact science (like astronomy and geometry); and finally, (d) debates over the relative status of Hippocrates of Cos and other "founding fathers" of medicine. The case studies that anchor each of this book's chapters investigate how these various kinds of debates intersected, interacted, and informed the thinking of individual physicians.

The medical environment in which physicians like Galen wrote and practiced was very different from the present. While there were certainly recognized schools of medical *thought*, there were no formal institutions for medical education or accreditation. In a world in which, technically speaking, anyone could lay claim to medical expertise, practicing healers and medical writers were under constant pressure to demonstrate their own merits and discredit the claims of their rivals. The backing of high-profile patrons or the brand name of a particular teacher or medical sect could provide some defense against this cutthroat environment.[9] Nevertheless, one could often find several rival healers crowded around the sickbeds of wealthy individuals, with each physician trying to persuade the patient to choose the course of treatment that he (for these physicians were overwhelmingly male) personally recommended.[10] Aspiring students of medicine were not required to commit exclusively to one school of thought, but were free to hop from teacher to teacher as they pleased. By the Roman period, this had created a market for introductory medical "textbooks" or "handbooks" that advertised the author's expertise but omitted enough critical information that the would-be student would have to come and pay for lessons in person if he wanted to unlock the true secrets of the art.[11]

The medical marketplaces of Greco-Roman antiquity were vast, diverse, and limited neither to male practitioners nor to those who explained health and disease by reference to natural, physical causes. According to their writings, male doctors (*iatroi* in Greek; *medici* in Latin) often consulted with male gymnastics instructors, as well as with "midwives" (*maiai* in Greek; *obstetrices* in Latin) and generalist "female physicians" (*iatrinai* in Greek; *medicae* in Latin).

---

[9] Galen is not shy about his successes curing such eminent philosophers as Glaucon (*Loc. aff.* 8.361–6 K) and Eudemus (*Praen.* 14.605–19 K).
[10] *Diff. puls.* 8.511–14 K and *Dig. puls.* 8.900–16 K. For more on the competitive nature of ancient medicine, see Barras et al. 1995, p. vii; Hankinson 2008a; Lloyd 2008; Mattern 2008, 69–97. On the agonistic aspect of Galen's public dissections, see Gleason 2007.
[11] Barton 1994, 156.

While these last seem to have claimed similar competencies to their male counterparts, they are far less frequently attested, and there remains much scholarly debate over their medical purviews, social status, and demographic distribution and whether or not any of them produced medical treatises or compilations in their own names.[12] There was also a wide range of healers, of both genders, who offered metaphysical explanations and solutions for human pathologies.[13] Temple priests, as well as private oracle-mongers, amulet-makers, root-cutters, and itinerant ritual specialists promised to preserve health and cure disease by appealing—often through prayers, rites, and numinous objects—to deities, daimons, and other supernatural entities. To draw a sharp line, as has often been done, between the "rational" and "irrational" within Greco-Roman medicine is not appropriate.[14] Many recent analyses have shown that so-called "irrational" medical practices are often rigorously logical and empirical, as long as one accepts their underlying axioms. Conversely, Greek "rational" medicine often appeals to intuition and to principles like sympathy that are today strongly associated with "magical" thinking.[15]

Because of this book's commitment to following the evidence for clocks and hourly timekeeping within Greco-Roman medicine, the voices of elite, male physicians dominate the discussion. These healers, unlike many of their female and lower class counterparts, had the good fortune not only to learn how to read and write, but also to have their writings copied and passed down to us over the millennia. Wherever possible, I endeavor to incorporate evidence from documentary papyri, stone inscriptions, artistic products, and archaeological remains—as well as literary genres, like comedy—that opens windows onto the experiences of healers and patients who were female, enslaved, and/or of low socioeconomic status.[16] This book also engages in only limited ways with the temporality of public and private religious healing because our evidence for clock- and hour-use within these contexts is scarce.

By offering case studies of how Galen of Pergamon, as well as his contemporaries and interlocutors, chose to engage with hourly timekeeping, this book aims to shed greater light on the diverse and multi-faceted ways in which individuals interact with technologies. It illustrates how modes of interaction are not simply determined by a technology's availability, accuracy, and ease of use, but by its cultural and intellectual cachet and by the claims it enables a user to make about

---

[12] Some of our ancient, male writers cite medical works by "Metrodora" and "Cleopatra," but it is not clear whether these are pseudonyms and, if indeed they are, whether the authors they mask were female or male. On this debate, and on female healers in the Greco-Roman world more generally, see esp. Flemming 2007b, 2013.

[13] On the diversity and pluralism of Greco-Roman medical marketplaces, see esp. Nutton 2013; Totelin and Flemming 2020.

[14] See, e.g., Longrigg 1993, 2013.

[15] A succinct overview of these critiques is offered at Horstmanshoff et al. 2004, 3–7.

[16] On gendered temporalities in Roman Egypt, see Remijsen 2023.

him- or herself. This book also explores some of the antecedents to modern debates about the relative value of quantitative versus qualitative forms of knowledge and about the level of precision desirable in particular social or professional contexts.

Ultimately, though, this book also seeks to revise a neat and tenacious narrative about the origin, development, and impact of clock technology within the Western world. Though this narrative was reproduced most frequently among twentieth-century historians of the early modern period, it is still influential today.[17] According to this narrative, the social history of clocks only begins in the medieval period—specifically, in the fourteenth century CE, at the moment when the mechanical clock first appears in the historical record. The narrative then progresses, often in a teleological manner, to the present day and a Western world whose political, economic, social, and psychological structures are dominated by awareness of the clock. According to this account, increasing access to ever more accurate clocks fostered a new form of temporal awareness among fifteenth–seventeenth-century Europeans. The development of this sensibility was considered a *sine qua non* of the nineteenth- and twentieth-century Industrial Revolutions, prompting the historian L. Mumford to famously declare that "the clock, not the steam-engine, is the key-machine of the modern industrial age."[18] In a prominent article, the social historian E. P. Thompson put this sentiment another way: "Without time-discipline, we could not have the insistent energies of industrial man."[19]

The problems with this tenacious narrative are myriad and, fortunately, scholarship of the past twenty years has begun to confront them.[20] I would like to draw attention, however, to two of the narrative's underlying assumptions which should provoke the student of ancient history. The first assumption is that an investigation into the significance of clocks and hours should only begin with the medieval period. This overlooks the numerous attestations of shadow- and water-based clocks in ancient Mesopotamia, Egypt, Greece, Rome, and China. While more scholars have begun to acknowledge that these deserve mention in the history of western horology, that mention has typically been brief. Of G. Dohrn-van Rossum's much-cited *History of the Hour*, for example, a mere twelve pages are devoted to the subject of clocks in pre-medieval periods.[21] This comparative neglect of the ancient evidence is especially surprising in light of a second assumption embedded in the traditional narrative: namely, that increased exposure to clocks and the temporal unit of the "hour" can have a transformative effect

---

[17] Some notable champions of this narrative include Thompson 1967; Landes 1983; Jenzen and Glasemann 1989; Dohrn-van Rossum 1996.
[18] Mumford 1934, 14.
[19] Thompson 1967, 93.
[20] See Sauter 2007, 685–6, for discussion. For additional criticism of Thompson 1967, see Glennie and Thrift 1996.
[21] Dohrn-van Rossum 1996.

upon cultural institutions and modes of thought. If this assumption is valid, how can the former assumption, that the social history of clocks only begins in the fourteenth century, be successfully defended? Why has so little attention been paid to the socially and intellectually transformative potential of clocks in antiquity?

The study of ancient timepieces has traditionally been the preserve, not of social historians, but of historians of technology, concerned primarily with questions of clock design, classification, and accuracy.[22] Studies on such topics are of critical importance to classicists and social historians of the ancient world, since they allow us to better understand the prevalence and properties of ancient clocks, as well as the materials, technical knowledge, and other resources required to produce and use them.[23] However, social historians active in the twentieth century exhibited little interest in ancient clocks. This is surprising given the growing enthusiasm among social historians for discussing the political, economic, and cultic roles of other timekeeping technologies, such as calendars and chronological systems.[24]

Fortunately, the tide is beginning to turn. Since the year 2000, works by scholars such as R. Hannah, S. Remijsen, A. Wolkenhauer, A. Jones, J. Bonnin, S. Heilen, R. Talbert, and J. Ker have shed light on the social history of clocks and hours in various Greco-Roman cultural contexts,[25] while research by scholars such as S. Symons, A. von Lieven, A. Schomberg, J. Steele, S. Kattan Gribetz, S. Stern, and J. Ben-Dov have expanded our knowledge of how clocks and hours functioned in ancient Egyptian, Assyrian, and Jewish contexts.[26] In 2020, the first edited volume on the comparative social history of hourly timekeeping in Egypt, Mesopotamia, Greece, and Rome came out with Brill,[27] while initiatives such as

---

[22] The *Journal for the History of Astronomy*, for instance, has published numerous articles that attempt to classify various clock types or to reconstruct their design and operation. Examples include Arnaldi and Schaldach 1997; Evans 1999; Catamo et al. 2000. See also Turner 1989 in *History of Science*. The latest sundial finds are more likely to receive in-depth treatment in the *British Sundial Society Bulletin*, a non-academic publication for enthusiasts, than in journals of ancient history or classical archaeology. See, e.g., Symons 1998; Bonnin 2010b; Bonnin and Savoie 2013; Schaldach and Feustel 2013.

[23] The major catalogues are Price 1969 (portable sundials); Buchner 1976a (more portable sundials); Gibbs 1976 (stone sundials); Schaldach 1998 (Roman sundials); Schaldach 2006 (Greek sundials); Bonnin 2015 (all sundials then extant, including iconographic representations of dials); Graßhoff et al. 2015 (a running database of extant sundials); Talbert 2017 (Roman portable sundials).

[24] Feeney's *Caesar's Calendar* (2007) has been particularly influential, highlighting Rome's strategic use of calendars to construct its past, establish synchronicities, and integrate its own temporal frameworks with those of foreign communities. Another significant contribution has been Rüpke's *The Roman Calendar from Numa to Constantine* (2011), which offers a long-ranging diachronic analysis and seeks to situate Roman calendars within their institutional and societal contexts. See also Samuel 1972. Other scholars have taken a comparative approach, setting Roman calendars alongside those of other ancient cultures (esp. Mesopotamia and Egypt; see, e.g., Stern 2012) or those of more recent communities (esp. within western Europe and the United States).

[25] Particularly Remijsen 2007; Hannah 2009; Wolkenhauer 2011; Bonnin 2015; A. Jones 2016; Talbert 2017; K. J. Miller 2018; K. J. Miller and Symons 2020; Heilen 2020; Remijsen 2021; Ker 2023.

[26] See esp. the contributions in Ben-Dov and Doering 2017; K. J. Miller and Symons 2020.

[27] K. J. Miller and Symons 2020.

the Einstein Center Berlin's multi-faceted and multi-year Chronoi project attest to the fact that this is a promising and fruitful area for future research.[28]

The present book is divided into two parts. Part I, "Clockworks," establishes the sociocultural and theoretical background of clock use: first, in Greek and Roman thought more generally; then, in Greco-Roman medicine specifically; and finally, within Galen's own scientific methodology. Chapter 1, "Telling Time in the Roman Empire," introduces readers to the temporal landscape of the Imperial period and the key concepts with which this book engages. The chapter reviews the archaeological evidence for sundials and water clocks and discusses both their capabilities and their changing prevalence over time. It also recounts how the Greek term *hōra* evolved to mean "hour" (as well as "season" and a general span of "time"), and it distinguishes between the two different kinds of hour (seasonal and equinoctial) that ancient timekeeping tools could measure. The aim of this chapter is to help readers better understand the temporal frameworks and lexicons that were available to physicians like Galen and, thus, to better appreciate the temporal choices that these physicians faced.

The second chapter, "Doctors and Clocks: The Emergence of Hourly Timekeeping in Medical Contexts," reviews the textual and archaeological evidence for hourly timekeeping within *medical* contexts specifically, during the periods leading up to that of Galen—i.e., in the late Classical, Hellenistic, and early Roman periods. While acknowledging the challenges presented by the scant and problematic evidence, this chapter charts the development of a growing interest in precise medical timekeeping and the concept of *kairos* (the "right or opportune moment" to perform an intervention). The chapter then sketches the temporal landscape of the Roman Imperial period, reviewing the methods for hourly timekeeping that would have been available to Galen and his contemporaries in order to set the scene for the series of Galenic case studies that follow.

Chapter 3, "Clocks as Symbols 1: In Galen's Thought," focuses on the roles of sundials and water clocks within a single Galenic text, *On the Diagnosis and Treatment of the Affections and Errors of the Individual Human Soul* (henceforth, *Affections and Errors*). Surprisingly, this treatise on ethical psychology features an extended discussion of sundial and water-clock construction that highlights the differences between Galen's method of scientific inquiry and the methods used by members of contemporary philosophical schools. The chapter demonstrates that Galen associated clock design with the positive concepts of verifiability, clarity, concord, utility, and long-term scientific progress, while associating sectarian philosophies with the opposite. I suggest that Galen chose to use clock-construction as his central example because of a series of personal ideological commitments, including his belief in long-term scientific progress, his dream of

---

[28] In 2017, the Israel Institute for Advanced Studies also hosted a multi-day conference on "The Day Unit in Antiquity" (HUJI 2017).

achieving clear communication among scientists, and his insistence that proper scientific method involves a balance between logic and empiricism.

The fourth chapter, "Clocks as Symbols 2: Among the Roman Elite," contextualizes Galen's attitude toward clocks by investigating the semiotic fields of sundials and water clocks under the Roman Empire. This chapter shows how—across media such as sarcophagi, gemstones, friezes, and obelisks—sundials had become symbols connoting a range of concepts, including the human lifespan, immortality, the idea of the "scientific thinker," Greek education, and Roman Imperial power. This chapter therefore reveals the kinds of social capital available to medical writers and practitioners who associated themselves with clock technology and allows us to revisit with fresh eyes Galen's appeal to this technology in *Affections and Errors*. Ultimately, I propose that Galen reconfigures the common clock tropes of his day in order to promote his own scientific method as the best way to live one's life.

Part II of the book, "Hours in Action," examines a series of Galenic treatises as case studies to discover how he and his rival physicians incorporated hourly timekeeping into their medical practice. Chapter 5, "From 'Season' to 'Hour': Galenic Refinements of the Hippocratic *Hōra*," examines how Galen incorporates references to hourly timekeeping into his defense and refinement of Hippocratic critical-day schemes, which were designed to help physicians anticipate turning points in febrile diseases. This chapter proposes that Galen, in both his fever case histories and his astrological explanations for critical days, used hourly timekeeping to support two interrelated claims: that he alone was the true successor of Hippocrates, and that, in keeping with Hippocratic injunctions, his own approach to medicine integrated current astronomical knowledge and practices.

In Chapter 5, we see Galen aiming for ever greater temporal precision and may come away with the impression that he would prefer doctors to keep their eyes on the clock at all times. Chapter 6, however, encourages us to refine that notion. This chapter, "When is Temporal Exactitude Desirable?", argues that Galen actually seeks to present himself as occupying a reasonable middle stance in an ongoing debate over the role of numerical precision in medical timekeeping. By examining a series of passages drawn from Galen's works on fevers and bloodletting—as well as his magnum opus, *On the Method of Healing*—it becomes evident that, while Galen accuses some physicians of being insufficiently precise in their numerical timekeeping, he also accuses others of being *overly* precise and programmatic. This chapter, like the preceding one, highlights the fact that, even in times and places where clocks were easily accessible, medical authors made different choices about when and how to employ hourly timekeeping. It reveals that certain medical concepts and treatments were seen as being particularly ripe for debates over numerical temporal precision.

The seventh and final chapter, "'Right Timing' in Sickness and in Health: Hourly Timekeeping and *Kairos* in Galen's *On Hygiene*," explores more deeply how Galen's interest in hourly timekeeping relates to his understanding of "right timing." The primary case study for this discussion is Galen's *On Hygiene*, a treatise in which he expounds upon the nature of health and strategies for maintaining it. This chapter demonstrates how, for Galen, hours can be used both to calculate and to articulate individual *kairoi*, and that the window of timely or "kairotic" action varies in aperture depending on whether a patient is sick, healthy, or simply aged. Ultimately, we see that here too Galen's use of hours is motivated by his desire to practice good scientific method and to build upon the theories of his Classical and Hellenistic predecessors. However, the chapter also establishes important connections between Galen's approach to hourly timekeeping and his personal theories of health and disease.

A short conclusion follows some of these themes into twentieth- and twenty-first-century Western biomedicine. It further investigates the connections between Imperial-period debates about medical timekeeping and present-day discussions surrounding chronobiology, precision medicine, and qualitative health research to highlight certain continuities in medical deliberations across millennia.

A central goal of this book is to offer the first sustained treatment of the social history of hourly timekeeping within Greco-Roman medicine.[29] While the book focuses primarily on the writings of Galen, it uses these writings as a lens through which to explore the larger valences of clock technology within Roman Imperial society, and to explore how ideas about the proper use of clock technology in medicine evolved out of earlier frameworks. These ideas and debates have retained their relevance in the present day and may encourage us to more greatly appreciate and more critically examine the clock technologies in our own hospitals, doctors' offices, and homes.

---

[29] Earlier treatments of this topic have been article- or chapter-length: e.g., Miller 2018; Miller 2020; Singer 2022, 1-32. Singer 2022 offers the first broad overview of how Imperial-period physicians like Galen engaged with time more generally (not only with hours, but also with seasons, months, lifespans, intellectual genealogies, and intervals of time too short to measure).

PART I
CLOCKWORKS

1
Telling Time in the Greco-Roman World

In 1901, excavators in Palazzina Capodaglio, near the Italian town of Este, unearthed from the first-century CE "tomb of the physician" a very peculiar find.[1] In among the surgical instruments, spoons, glass vessels, and other grave goods was a small, hollowed-out cylindrical tube, about 62 mm high and 25 mm in diameter. It was finely crafted out of bone and possessed a removable cap or stopper that had a bronze ring attached to the top and two bronze "wings" affixed to its sides. These could be hinged out at right angles or folded down to fit within the tube itself (Figure A). The excavators, nonplussed, classified this object as an *astuccio* or "case," and for about eighty years, it sat on display in the Museum of Este, largely forgotten.[2] In 1984, however, S. Bonomi proposed that this object was, in fact, a portable sundial: the "wings" were bronze pointers, or "gnomons," that could be flipped out to cast a shadow on one or the other half of the dial's body where faint nets of inscribed hour and date lines could be detected.[3] A more detailed and accessible study of the sundial's form and function was published by M. Arnaldi and K. Schaldach in 1997, and now this small, portable sundial is widely recognized as the sole surviving example of an ancient physician's timepiece.

For scholars of ancient medical timekeeping, the so-called "Este Dial" raises many tantalizing questions. First, did this sundial really belong to the physician buried in this tomb, or was it placed in the grave as an offering by a grieving friend or relative? Second, if the dial did belong to the physician, did he actually use it during his patient visitations, or did he employ it strictly for other purposes—to coordinate social activities, for example, or simply to show off, as a prestige item, to his friends? Third, if the physician did incorporate this clock into his medical practice, in what contexts did he do so? And fourth, how representative might this physician have been? Was it common for ancient physicians to have recourse to sundials, portable or otherwise? And if so, why have we not found them—or other timekeeping tools, like water clocks—in other caches of medical instruments?

By cobbling together scattered archaeological remains and indirect references to sundials and water clocks in our textual sources, we can begin to reconstruct the range of timekeeping options that might have been available to ancient Greek and Roman physicians of different time periods. This chapter will attempt to provide a

---

[1] Some material from this chapter appeared in K. J. Miller 2018.
[2] Arnaldi and Schaldach 1997, p. xxviii.   [3] Bonomi 1984.

Time and Ancient Medicine: How Sundials and Water Clocks Changed Medical Science. Kassandra J. Miller, Oxford University Press. © Kassandra J. Miller 2023. DOI: 10.1093/oso/9780198885177.003.0002

Figure A Portable cylindrical sundial, first century CE, bone, Tomb of the Physician, Este, Italy, h. 6.2 cm; diam. 2.5 cm, Museo Nazionale Atestino di Este: IG MNA 15397. Photograph by Guido Petruccioli with permission from the Museum Atestino di Este.

concise overview of the development of hourly timekeeping strategies in the Greek- and Latin-speaking worlds, as well as a brief survey of the evidence for their antecedents in ancient Egypt and Mesopotamia. The following chapter will then offer a chronological overview of how these developments in hourly timekeeping intersected with medical practice.[4] These historical tours will set the stage for the Imperial-period case studies considered in the succeeding chapters and will help readers to appreciate some of the broader trends in the production and use of these technologies over time.

[4] For more detailed accounts of timekeeping in the Greco-Roman world, see Hannah 2009; A. Jones 2016; Talbert 2017.

Telling Time without Clocks

Inhabitants of the ancient Mediterranean had at their disposal a wide range of strategies for both "time-indication" and "time-reckoning." M. P. Nilsson, who first introduced these terms in 1920, defined "time-indication" as the observation of "concrete phenomena of the heavens and of Nature,"[5] such as the length of time it takes for the sun to move from the horizon to mid-heaven or for a candle to burn down to its nub. The durations of these time-units can vary significantly, both in relation to each other and from instance to instance, from observer to observer. The time between sunrise and midday, for example, may differ observationally depending on whether that day is cloudy or clear, or whether the observer's view of sunrise is impeded by a hill or mountain. It will also differ in absolute terms depending on the time of year, as the earth moves along its orbit. Likewise, the rate at which a candle burns will depend on its composition and size, the length of its wick, and the way in which it is tended. Thus, such time-indications are not naturally standardized, nor do they necessarily come together to produce total, uniform coverage of times within the day or night. Instead, these time-indicators can overlap—yielding periods of what we might call "hyper-indicated time"—or leave gaps—thereby creating stretches of time that is "unindicated." Time-*reckoning*, on the other hand, according to Nilsson, involves the standardization and organization of individual time-indicators into a continuous counting system. In many modern societies, for example, we "reckon" daily time using a seamless system of hours, minutes, and seconds. In this system, each individual hour, minute, or second is identical in length to others of its type, and every moment of day and night can be marked with a precise and accurate temporal tag, such as 14:53:21 Eastern Standard Time.

Methods of indicating "short time" (i.e., periods shorter than a full day) are visible even within our earliest written sources. I divide these methods into four general categories according to the kind of phenomena being observed. The first is time-indication by means of *personal* phenomena—i.e., the activities (conscious or unconscious) of one's own body. This could include, for instance, going to sleep when you "feel tired" or after you "have brushed your teeth." The second is time-indication by means of *social* phenomena. In the Hippocratic *Epidemics*, for example, we see times marked in reference to the opening or filling of the marketplace,[6] and in the present day, we might promise to meet a friend for drinks "after the faculty meeting." The third category indicates time by the positions of *celestial* phenomena (such as the sun, moon, and stars) as they move through the sky. Pertaining to this category are not only references like "at daybreak" and "after sunset," but also strategies that mark the position of a

---

[5] Nilsson 1920, 9.   [6] *Epid.* 7.92, 5.448 L and 5.62, 5.242 L, respectively.

celestial body using a proxy, such as the shadow cast when the sun strikes an object. The Greek comic poet Aristophanes, for instance, informs us that Athenians of the Classical period sometimes used the length and position of their own shadows to determine the right moment to dine.[7] Finally, the fourth category concerns the time it takes for a *terrestrial, non-human entity* to complete an action (which, in practice, often involves the consumption of a "fuel"). For example, the Greek military strategist Aeneas Tacticus describes soldiers using lamps lined with wax to measure out their night watches,[8] and Athenian (and later, Roman) courtrooms often allotted speaking time to prosecutors and defendants on the basis of the time it took for water to drain out of a bowl of designated size.[9] These latter tools are often referred to in our sources as "water clocks" (*klepsydrai* in Greek, *clepsydrae* in Latin), though it is important to note that these functioned more like modern-day egg-timers than modern-day clocks and should not be confused with the water clocks used for continuous time-reckoning that will be discussed shortly.[10]

Beginning sometime during the late Classical or early Hellenistic period, tools for continuous time-reckoning emerged which both standardized and systematized certain of these earlier time-indicators. For example, out of the general practice of using shadows to mark and measure the movement of the sun developed the specific idea of the sundial, which uses the shadow of a gnomon to track the sun's motion along a grid of hour and date lines. Furthermore, from processes like using flowing water to time speeches in the courtroom emerged the idea of using larger, more sophisticated water clocks to keep continuous time throughout the day or night. Both of these time-reckoning tools, the sundial and the water clock, allowed Greeks and Romans to layer over their temporal landscapes a grid of regular, continuous, and numerically counted hours which introduced new ways of thinking about and engaging with units of "short" time.[11] The Greeks and Romans, however, were neither the only nor even the first ancient

---

[7] Ar. *Ekk.* 651–2. For discussion, and the ways in which this practice could be formalized into shadow tables, see Hannah 2020, 323–6.

[8] Aen. Tact. 22.24–5. Cf. Plb. 10.44. By adjusting their lamps' capacities with different amounts of wax, the soldiers could supposedly recalibrate these timers to fit the changing seasonal hours. It is difficult to imagine how anything resembling precision was obtained by this method. Testimonies by Latin authors indicate, however, that lamps were indeed used to time night watches in the military and during work shifts in the mines. See Plin. *NH* 33.96–7 (mines) and Caes. *Gall.* 5.13.3–4 (military).

[9] The excavations at the Athenian agora uncovered one such vessel. See Camp 1986, 111–12.

[10] For more on the terminology associated with water clocks, see Bonnin 2015, 87–98. On the use of clepsydras to time speeches within Greek and Roman law courts, see Allen 1996; Ker 2009; Riggsby 2009.

[11] These developments in short-time reckoning are somewhat analogous to the developments in yearly reckoning that took place under the Seleucid Empire. Kosmin (2018) has explored how the Seleucids parted ways with other Hellenistic kingdoms, which used systems of regnal years, and instead introduced an innovative chronological system (the Seleucid Era count) that identified 311 BCE as "Year 1" and continued to mark each successive year with a successive number. The count continued in an unbroken, linear sequence regardless of who was on the throne and, as Kosmin proposes, facilitated new ways of conceptualizing the past, present, and future.

Mediterranean cultures to develop methods for hourly timekeeping. Nor were their sundials and water clocks constructed or distributed uniformly. Thus, in what follows, I will offer an overview of the terminology, forms, and functions related to Greco-Roman sundials, water clocks, and the concept of the "hour" and briefly discuss their antecedents in other ancient cultures. A review of the timekeeping technologies used in Pharaonic Egypt and Mesopotamia, to which early Greek-speakers were likely exposed, will highlight certain ways in which Greco-Roman clocks and temporal *Gestalten* differed from those of their southern and eastern neighbors. This will help us to appreciate how Greek and Roman daily timekeeping practices were both informed by cultural exchange and very culturally contingent. We will also see how time-reckoning systems—in antiquity as today—never come to replace strategies of time-indication; rather, strategies of both kinds tend to coexist, sometimes in harmony and sometimes in conflict.

Hours

In the modern Western world, we are accustomed to thinking of an "hour" as a temporal unit of fixed length. Regardless of one's location or the season of the year, an hour will always be composed of sixty minutes of sixty seconds each, and twenty-four hours will invariably make up a "day." In the Greek and Roman worlds, however, equinoctial hours (i.e., hours of consistent length) enjoyed a very limited use. They are unattested before the Hellenistic period and, even in later periods, remain the almost exclusive preserve of technical astronomy. In all other contexts it was the seasonal hour that prevailed, a unit produced by dividing the time between a given sunrise and sunset (for daytime hours) or a given sunset and sunrise (for night-time hours) into twelve equal parts. Thus, while the seasonal hours on any particular day were equal to one another, they were unequal to the hours on any other day, since the absolute lengths of daylight and night-time vary throughout the year.[12]

The concept of sub-dividing the day into hour-units first emerged, probably independently, in both ancient Egypt and Mesopotamia around the third millennium BCE or earlier.[13] In Egypt, these hour attestations first appear during the time of the *Pyramid Texts* and are closely associated with the practice of representing celestial movements at night. One of the earliest appearances of the term "hour" (*wnw.t* in Middle Egyptian) is on a diagonal star table painted on a wall of the Osireion at Abydos and dated to the mid-thirteenth century BCE.[14] Such star tables, which

---

[12] On seasonal hours in Greco-Roman antiquity, see, e.g., Neugebauer 1975, 1069; Evans 2005, 277; Hannah 2009, 73–5. On their introduction into Babylonian timekeeping, see Rochberg-Halton 1989.
[13] von Lieven and Schomberg 2020, 58 n. 32.   [14] Symons et al. 2013, 879–80.

were also painted inside coffin lids, offered individuals a way of sub-dividing the night into smaller time-units by charting the sequential rise of selected asterisms known as "decans." The tables that appear within coffin lids may have been intended to help the spirits of the deceased to keep track of time in the eternal night of the underworld. Those on temple walls seem to have been used by *wab*-priests to track the location of the sun during its nightly journey through the underworld, and thereby to ensure that nocturnal rituals were performed at the most auspicious times. Ultimately, seasonal hours were used in Egypt, as well, to keep track of time during the day, as we will see in our discussion of Egyptian sundials and water clocks below. Throughout the Pharaonic periods, though, it is noteworthy that the "hours" were, to our knowledge, almost exclusively linked to ritual praxis, and they were even represented as divinities themselves.[15] This allows for an interesting comparison between hourly timekeeping in Egypt and in Greece and Rome, where clocks are occasionally attested in sanctuaries, but the "hour" itself (unlike other Greco-Roman temporal concepts like the "year," "season," "month," "day," or "night") is never divinely personified; it is always presented as a numerical, man-made construction. This idea, that the Greco-Roman "hour" is a mathematical object, will become especially significant for us when we discuss its conceptual roles in Imperial-period medicine.

Babylonian "hours" are also attested from a very early period, but they differ markedly from the Egyptian *wnw.t*. Seasonal hours (*simanu* in Akkadian) were a late development in ancient Mesopotamia.[16] Instead, most astronomical cuneiform texts from the region discuss time spans in terms of the fixed-length units *bēru*, UŠ, and NINDA. Within this system, a full day (measured from one sunset to the next) was divided into twelve units of fixed, equal length called *bēru*. Thus, a single *bēru* would have been equal to two of our modern hours (hence the term's frequent translation into English as "double hour"). Each *bēru* could be subdivided into thirty UŠ of four minutes each, and in fact, our 360° circle and our habit of measuring arcs and time in degrees both derive from the multiplication of thirty UŠ by twelve *bēru*.[17] An UŠ, in turn, could be divided into sixty NINDA, each with a duration of four seconds.[18] Short time in ancient Mesopotamia could also be described in terms of the three "watches" of the day or night. Though, as J. Steele suggests, "in most contexts the three watches probably represented only broad ranges of time that were only approximately the same length and were not defined or measured precisely, akin to the way that English speakers today may often refer to the period before lunch as the 'morning', even if lunch is not eaten until well after 12 noon."[19] References to seasonal hours,

---

[15] von Lieven and Schomberg 2020, 54–5. [16] See Rochberg-Halton 1989 for discussion.
[17] Neugebauer 1983, 8.   [18] For the remaining discussion, I rely heavily on Steele 2020, 96–8.
[19] Steele 2020, 96–7.

in contrast, can only be securely dated from the second century BCE onward, and they never become common.[20]

Hours do not appear in our Greek sources until much later, and in our Roman sources later still. One of the earliest, rather elliptical references may appear in the fifth century BCE, in Herodotus's *Histories*. Herodotus claims that "the Greeks learned of the *polos* and the *gnomon* and the twelve divisions of the day from the Babylonians."[21] Many scholars have assumed that, with this assertion, Herodotus meant his readers to understand that the Greeks adopted both sundial technology and the concept of seasonal hours directly from ancient Mesopotamia.[22] More recent scholarship has called such a conclusion into question, pointing out that there is insufficient evidence to determine whether Herodotus's "twelve divisions of the day" actually refer to the number of daylight or night-time hours or, rather, to the number of *bēru* or "double hours" into which Babylonians preferred to divide the full (i.e., the twenty-four-hour) day.[23] Thus, Herodotus's *Histories* may not deserve credit for being the first Greek text to refer to Greek seasonal or equinoctial hours. Other candidates for that title have been proposed, such as certain Hippocratic works.[24] In each case, however, the evidence is not entirely secure, as we shall see in the following chapter with regard to the Hippocratic texts. By the mid to late fourth century BCE, however, we begin to see sundials with hour markings, which suggests that the term *hōra* as hour had begun to circulate, as well.

Prior to that, in the Archaic and Classical periods, Greeks primarily used the term *hōra* to mean "season," though it could also designate a "right moment" or a general span of "time."[25] Hesiod's *Hōrai*, for example, which are divine personifications of this concept, are four in number, not twelve, and while modern translators often call these goddesses the "Hours," Hesiod's *Hōrai* actually represent the seasonal subdivisions of the year (Spring, Summer, Autumn, and Winter).[26] The new definition of *hōra* as "hour" never superseded the old; instead, each definition continued to be used alongside the others. Clocks themselves were similarly multivalent. Like the word *hōra*, these devices could represent multiple types of time at once; they marked the hour of the day, like modern clocks, but also tracked the time of year, like modern calendars.[27] Thus, they were tools for

---

[20] The references of which I am aware include a handful of Seleucid-era horoscopes (BM 33018, 35515, 38104, and 41301) and—possibly—a pair of highly fragmentary instruction manuals for the construction of sundials (LBAT 1494 and 1495 [= BM 34719 and 34067] and BM 35010).

[21] Hdt. 2.109.3: πόλον μὲν γὰρ καὶ γνώμονα καὶ τὰ δυώδεκα μέρεα τῆς ἡμέρης παρὰ Βαβυλωνίων ἔμαθον οἱ Ἕλληνες. Translations are my own unless otherwise stated.

[22] See, e.g., Hannah 2009, 73.   [23] See Rochberg-Halton 1989, 165.

[24] *Epid.* 4.12, 5.150 L, and *Int.* 7.238.22–3. See Langholf 1973.

[25] See LSJ "ὥρα" and Langholf 1973. For discussion of other terms relating to clocks and hours, see Rehm 1913 (*horologium*); Thalheim 1921 (*klepsydra*); Schaldach 2006, 4; Hannah 2009, 68–115; Sontheimer 2011.

[26] Hes. *WD* 75.

[27] In Gibbs's view, the design of the sundial, in particular, "gives the impression that the Greeks had embodied within it a model of solar motion" (1976, 42).

measuring both kinds of *hōra*, the season and the seasonal or equinoctial hour,[28] and are often referred to in our sources with words like the Latin *horologium* (literally, a "*hora*-reckoner") or, in Greek, a *hōroskopeion* (a "*hōra*-watcher").[29]

By the third century BCE, however, the concept of the seasonal hour seems to have spread within the ancient Mediterranean more widely. Papyrus records indicate, for example, that the Ptolemaic postal system time-stamped letters to the numbered hour (a practice that would later be employed by the Roman army, as well).[30] Meanwhile, also in the third century BCE, the hour makes its first appearance in Rome when, according to later written sources like Pliny's *Natural History*, the first clocks were brought to Rome as spoils from Greek-speaking communities.[31] In Latin, as in Greek, the term *hora* was multivalent. In some contexts, particularly in poetry, it typically referred to a season of the year or to a general span of time. In other genres of writing, however, the term is regularly paired with an ordinal number, to indicate the hour of the day or night, or with a cardinal number, to indicate a duration in hours. In our extant writings from Republican Rome, the term *hora* as "hour" appears especially often in descriptions of military operations. Livy, Caesar, and the unknown authors of *On the African War* and *On the Spanish War*, for instance, often structure their narratives of battles and campaign movements using numbered hours.[32] As this trend persists through the Imperial period, it may be relevant for our purposes that Galen, the physician whose extant writings are most concerned with numbered hours, had experience accompanying Roman armies into the field.

By the time the Roman comic poet Plautus was writing, in the late third and early second century BCE, it seems that hourly timekeeping was a sufficiently common practice—at least in major urban centers like Rome and Athens—for him to poke fun at it on stage. The later author Aulus Gellius quotes the following Plautine fragment, which he attributes to a play called *The Boeotian Women*:[33]

---

[28] For overviews on the history of Greek clocks and conception of the hour, see Bilfinger 1886, 1888; Brauneiser 1944; Drachmann 1963; Price 1969; Brumbaugh 1975; Bowen and Goldstein 1988; Hannah 2008; 2009; Remijsen 2021. There is an extensive body of literature on Greek understandings of time in general. A few examples include Aveni 1989; Darbo-Peschanski 2000; Sorabji 2006.

[29] Though ὡρολόγιον and ὡροσκοπεῖον (Greek) and *horologium* (Latin) are some of the most common terms for sundials and water clocks, our literary and epigraphic sources employ a wide range. On the inconsistency of this terminology, see Bonnin 2010a, 185; 2015, 73–98.

[30] Remijsen 2007.

[31] Plin. *NH* 7.212–215. For discussion of this passage, see Wolkenhauer 2011, 86–91; 2020, 217–24; A. Jones 2020, 137–43.

[32] See Liv. 22.1.6, 35.1.1, and many more examples; Caes. *Gall.* 26 and *Civ.* 1.64, 80, 82; *De bell. Afr.* 19, 61, 69, 70, 78; *De bell. Hisp.* 27.

[33] Aul. Gell. *NA* 3.3, 4: *ut illum di perdant, primus qui horas repperit, / quique adeo primus statuit hic solarium! / qui mihi conminuit misero articulatim diem. / Nam me puero venter erat solarium / multo omnium istorum optimum et verissimum: / ubi is te monebat, esses, nisi cum nihil erat. / Nunc etiam quod est, non estur, nisi soli libet; / itaque adeo iam oppletum oppidum est solariis, / maior pars populi aridi reptant fame.*

> May the gods damn that man who first figured out
> how to tell the hours, and first set up a sundial here
> to chop up my days into sorry little pieces.
> For when I was a boy, my belly was my sundial
> by far the best and truest of them all:
> whenever it commands, you eat—
> except when there's nothing to be had.
> Now, though, even the food that's there
> can't be eaten unless the sun allows.
> These days, the town is so stuffed with sundials
> that most people creep along shriveled up with hunger.

The Greek title of this play suggests that it was inspired by an earlier Greek comedy, perhaps one written by Menander, Theophilus, Antiphanes, Diphilos, or another New Comic poet.[34] Hence, by the mid-Hellenistic period it seems to have been the case that, at least in certain parts of the Greek- and Latin-speaking worlds, the concept of the hour and the tools for hourly timekeeping were fairly commonplace. This passage also suggests that, by this time, some Greeks and Romans were beginning to wrestle with tensions familiar to us in the modern day, such as those between "body time" and "clock time," or between viewing the clock as a useful tool and as an unyielding master. But how did tools for hourly timekeeping work?[35] Where, exactly, could they be found? And how did they compare to their Egyptian and Babylonian predecessors?

Sundials

A sundial, in general, is composed of a shadow-caster (the "gnomon") and a designated shadow-casting surface (Figure B). The surfaces of Greek sundials are usually inscribed with a net of lines consisting of a sequence of concentric semi-ellipses (which mark the solstices and equinoxes) and a series of hour lines radiating out from the gnomon. As the sun traces its path across the sky, one can tell the hour by the lateral movement of the gnomon's shadow across the net of lines and the time of year by the shadow's length. In order to create a sundial that would be accurate for a given location, a designer had to know the latitude of that location and the tilt of the ecliptic in degrees.[36] Happily, this also means that a

---

[34] On this motif, see Gratwick 1979; Wolkenhauer 2011, 126–7.
[35] Overviews can be found at Mills 1996; Bonnin 2012a; 2012b.
[36] Greeks used a value of 24°, a bit higher than the actual 23.4°. Ancient technical treatments of sundial construction suggest that they were usually designed using an *analemma* (a mathematical diagram for generating nets of hour and date lines) such as those offered by Vitr. 9.7 and Ptol. *Anal.* (a short treatise entirely devoted to the process of generating a plane diagram that maps out the location of the sun).

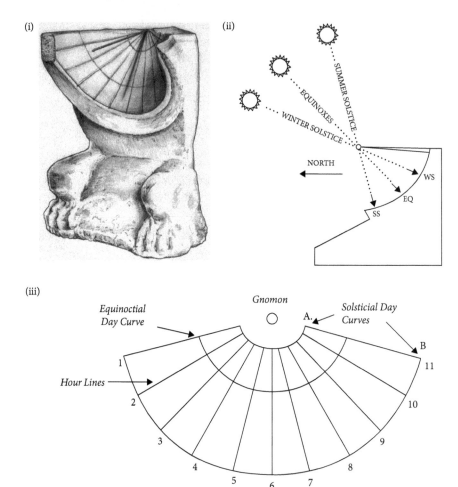

A. Winter Solstice on north facing dials, Summer Solstice on south facing dials
B. Summer Solstice on north facing dials, Winter Solstice on south facing dials

Figure B Schematic of sundial hour-lines and day-curves. Sundial modeled after a marble conical dial from Pompeii (Berlin Sundial Collaboration, Ancient Sundials, Dialface ID 174, Napoli, 2014, Edition Topoi, DOI: 10.17171/1-1-2199). Image © Kenneth Jackson.

modern historian can often reconstruct the latitude for which a given sundial was designed based on the dial's gnomon length and the value for the ecliptic commonly used at the time. This has allowed modern scholars such as S. Gibbs, K. Schaldach, J. Bonnin, and A. Jones to determine which of our extant sundials were tailored to the same geographical locations in which they were found—and which were significantly displaced.[37]

---

[37] Gibbs 1976; Schaldach 2006; Bonnin 2015. Displacements are discussed by A. Jones (2020).

Well before sundials began to be used in Greece and Rome, they were employed in Egypt. The earliest secure example is a stone L-shaped sundial from the time of Tuthmosis III, around 1500 BCE.[38] This same design continues to be attested into the early centuries CE, and seems to have been unique to Egypt.[39] These dials are "L-shaped" because each consists of a horizontal, rectangular shadow-casting surface and a vertical, rectangular gnomon. The shadow-casting surface bears markings for the hours, though how exactly the tool was used is still a matter of debate.[40] Egyptians also produced at least two other kinds of dials unique to the region: the "semi-circular dial" and the "sloping sundial." Like the L-shaped dials, dials of these types were small enough to be portable and, in order to function properly, required the operator to make sure that the dials were level and oriented correctly with respect to the sun. While semi-circular sundials may be attested from Pharaonic Egypt through to the Roman Imperial period,[41] our only examples of datable "sloping sundials" derive from the Hellenistic period, a time of increased cultural contact and exchange with Greeks and Romans.[42] While we have no evidence that Greeks and Romans adopted or adapted the aforementioned dial types (L-shaped, semi-circular, and sloped), from the Hellenistic period onward a particularly Greco-Roman style of sundial, the "concave dial," begins to appear in Egyptian contexts. Babylonian sundials may also have borne the stamp of Greco-Roman influence.[43]

While, as mentioned above, the remains of a few Hellenistic-period handbooks seem to describe the construction of sundials, no actual sundials or sundial components from ancient Mesopotamia have been identified. On the basis of this very limited evidence, we can speculate that Babylonian sundials would have been planar, i.e., with shadow-casting surfaces that are flat.[44] This would dovetail with what we know of Babylonian astronomy, which seems to have been more arithmetical than geometrical, and did not work with concepts like the celestial "sphere." It seems, then, that one of the things that set early Greek horology apart from its Egyptian and Babylonian predecessors was its interest in making sundials that reproduced a *geometrical* view of the cosmos. Greco-Roman "spherical,"

---

[38] ÄM 14744. ÄM 19743, dated to the period between 1000 and 600 BCE, is of a similar type and is also mostly complete. Two further component parts, ÄM 14573 and N 781, both dated to ca.1380 BCE, have also been attributed to L-shaped sundials, but for problems with this identification, see Symons 2020, 16.

[39] Papyrus EA10673, which dates to the first–third century CE, includes an illustration of an L-shaped sundial. Image available at Griffith and Petrie 1889, plate XV; Leitz 2014, 489–91 (Fragments 59–63) and plate 116.

[40] For various interpretations, see Borchardt 1910, 12–14; Hoffmann 2016; Symons 2016; 2020, 17.

[41] The beginning of this date range depends on whether the example from the Valley of the Kings can be securely dated to the reign of Seti II (c.1200 BCE) as Bickel and Gautschy (2014) assert. For problems with this assertion, see Symons 2020, 20–1.

[42] The two datable examples are SS Qantara and SS Fitz E.GA.4596.1943 in McMaster's Ancient Egyptian Astronomy Database (Symons et al. 2013). The sloping sundial also appears as a hieroglyph during this time (Daumas 1995).

[43] See Steele 2020, 117.   [44] Rochberg-Halton 1989, 162–5.

"hemispherical," and "conical" sundials each attempted, through shadow-casting surfaces of different curvatures, to map the arc of the heavens onto a dial face.

At the time of writing (in 2020 CE), over 600 Greco-Roman sundials and fragments thereof have been identified.[45] Some of these, like the Egyptian examples discussed above, were portable, and could even be calibrated for multiple latitudes. Most, however, would have been fixed in place, being either monumental or sized for the home and garden. The diversity in their designs is substantial. Though the vast majority of fixed dials are marble or limestone, local tufas and other porous stones were sometimes used, and portable dials could appear in bronze or bone.[46] While spherical[47] and conical[48] dials are the most common, we also have examples of hemispherical,[49] cylindrical,[50] and vertical and horizontal planar dials,[51] as well as dials that have multiple faces[52] or are shaped like globes.[53] The whimsicality and technical sophistication of some of these pieces suggests that they were valuable not only (or perhaps even primarily) as useful tools, but also as display pieces that could amuse guests and enhance their owners' prestige.[54]

Amidst all of this variety, though, a few trends can be detected. Chronologically speaking, there is a marked decrease over time in the quality and accuracy of sundials. Concern for their calendrical (as opposed to horological) functions likewise declines. While most Hellenistic Greek clocks are finely worked and carefully calibrated, the numerous dials of Roman-period Pompeii, for example, are much cheaper and of shoddier workmanship. By late antiquity, many clocks are downright imprecise, and their date lines are either strictly ornamental or missing entirely. This suggests a larger cultural shift in the valuation of precision and accuracy in hourly timekeeping. There is little correlation between chronology and style, although J. Bonnin has noted that the *pelecinum* dial (which is shaped like a double-headed axe) comes to dominate the iconographic representations of sundials in late antiquity (Figure C).[55] Nor are there particularly strong geographical trends with regard to style. The eastern half of the Roman Empire seems to have had a predilection for conical dials while the west favored spherical, but variety is present everywhere. Each of the primary production centers (including, at different times, Athens, Kos, Rhodes, Delos, Pompeii, and Aquileia) seems to have had its preferences and minor signature features, but these do not constitute formal, regional styles.

---

[45] Graßhoff et al. 2015.
[46] On portable dials, see Price 1969; Buchner 1976a; Arnaldi and Schaldach 1997; Talbert 2017.
[47] E.g., Gibbs 1976, 1001–1005G and Schaldach 2006, nos. 9, 27, 34, 40.
[48] E.g., Gibbs 1976, 3001G–3054G and Schaldach 2006, nos. 2–5, 10–12, 14–20, 24–6, 28–31, 35, 37, 39.
[49] E.g., Gibbs 1976, 1069G–1075 and Schaldach 2006, nos. 8, 13, 32.
[50] E.g., Schaldach 2006, nos. 36, 41.
[51] Vertical, e.g., Gibbs 1976, 5001–9 and Schaldach 2006, nos. 1, 22. Horizontal, e.g., Gibbs 1976, 4001G.
[52] E.g., Gibbs 1976, 7001G, 7002G, and Schaldach 2006, no. 33.
[53] For the Prosymna globe, see Schaldach and Feustel 2013. For the Matellica globe, see Evans 2005.
[54] A. Jones 2016, 25.
[55] Bonnin 2013. Although Hannah does not propose a chronological sequence of sundial types, he sees formal connections between them (2009, 95).

Figure C Roman mosaic depicting a thinker, possibly Anaximander, with a *pelecinum* sundial. Image © GDKE/Rheinisches Landesmuseum Trier, Inv. No. 1907,724, photographer Th. Zühmer.

Water Clocks

In contrast to sundials, very few water clocks from any of the cultures under consideration here have survived to the present day.[56] In general, water clocks seem to have come in two basic types and a range of sizes. The earliest kind was the "outflow clock," which operated analogously to the courtroom *klepsydrai* mentioned earlier. An outflow clock is a vessel perforated at the bottom and fitted

---

[56] For an overviews of Greek water clock history and technology, see Schmidt 1906; Landels 1979; Lewis 2000; Hannah 2009, 97–115. One of the earliest references to water clocks in Greek textual sources appears at Ar. *Ach.* 693–4.

with a spout through which water can drain. The operator pours water into the vessel up to the fill-line and can then check the time by periodically measuring the level of the draining water against a vertical hour scale (the first hour at the top, the twelfth at the bottom).[57] This method can be problematic, however, because the change in water pressure as the vessel empties can prevent the outflow rate from remaining uniform. This difficulty was not fully surmounted, in any ancient Mediterranean culture, until the invention of the "inflow clock," credited to Ctesibius in the third century BCE. In the case of inflow clocks, water drains from a supply vessel into a smaller vessel whose first hour mark is near the bottom, the twelfth at the top.[58] Some later Greco-Roman water clocks also sported bells, whistles, and/or mechanical clockwork to make the pronouncement of the hour a more notable experience.[59]

The Egyptian and Babylonian predecessors to Greco-Roman water clocks did not, to our knowledge, feature auditory cues or mechanical displays. They are attested, however, very early on and have several innovative features. The nobleman Amenemhet claims to have invented the Egyptian water clock (*dbḥ*) in the first half of the sixteenth century BCE,[60] although the earliest surviving example, found in the temple of Ammon at Karnak, dates to c.1375 BCE. The exterior of this tapered outflow clock, shaped rather like a flower pot (▽), is decorated with astronomical scenes, while the interior is inscribed with twelve scales that mark the night-time hours during each month of the year.[61] Although water clocks, by nature, are best suited to measuring equinoctial hours, the inclusion of twelve separate hour scales allowed this clock to approximate the shift in hour length throughout the year and, thereby, to do a decent job of measuring seasonal hours. The shape of this outflow clock, too, was an innovative solution to a technical problem. A. Schomberg puts it this way:[62]

> As Borchardt was the first to note, the calculation of the behavior of the water level in an outflow clock and the clock's discharge rate is based on the application of "Torricelli's theorem." Applied to the outflow clock, this means that as the water level within a cylindrical vessel decreases (and thereby, the water pressure inside the vessel, as well), the speed of the outflow also slows down. The water would not flow out continuously over a period of 12 hours, but would flow faster at the beginning and then increasingly slower, ending in a trickle. By using the shape of a truncated cone instead of a cylinder, the Egyptians successfully

---

[57] In order to keep track of seasonal hours, these clocks needed multiple hour scales (usually one for each month of the year).
[58] Landels 1979, 193–4. In more sophisticated inflow clocks, this process may have been regulated by means of a float.
[59] Cf. Luc. *Hipp.* 8.　　[60] Cenotaph of Seti I; see Frankfort 1933.
[61] Cotterell et al. 1986, 36–7; Schaldach 2006, 14; von Lieven and Schomberg 2020, 58–79.
[62] von Lieven and Schomberg 2020, 68–9.

developed an outflow clock even before the fifteenth century BCE that counteracted this crucial drawback inherent in the outflow clock in comparison to the inflow clock.

While, from the Hellenistic period onward, the inflow clock seems to have achieved ascendancy in Egypt, these "flower pot" outflow clocks continued to be used. Fragmentary examples from the Hellenistic period suggest that Greeks and Romans of that time (especially those who lived in or visited the multicultural metropolis of Alexandria) would have been exposed to this technology, and indeed, we know of a few Egyptian or Egyptian-style water clocks that were imported to Rome under the Empire.[63] Thus, while Egyptian "hours" and sundials likely influenced Greeks and Romans on the conceptual level, in the area of design, it may have been the Egyptian water clock that left the greatest mark on Greco-Roman horology.

Babylonian water clocks, too, made a great impression on Greco-Roman timekeeping, though in a different way.[64] While examples of actual Babylonian water clocks, like examples of actual Babylonian sundials, have not been identified in the archaeological record, the textual record attests to the use of *dibdibbu*, which seem to have been inflow water clocks.[65] How exactly Babylonian time-units were measured is still, in part, a mystery. However, prior to the Hellenistic period, we encounter some cuneiform texts that convert the length of a watch or a duration in *bēru* ("double hours") into a certain number of "minas," a unit of weight that corresponds to the amount of water drained out of or into a water clock. Babylonian water clocks, analogously to Egyptian hourly timekeepers, seem to have been used primarily within ritual contexts, most specifically to track and predict celestial events that could be considered omens. Toward this end, Babylonian priests compiled extensive tables of astronomical data, which included the precise times (in Babylonian equinoctial hours) at which celestial bodies rose, set, and attained other significant positions.[66]

We do not know precisely when or how Greek-speakers began to engage with Babylonian astronomical data and methods, but we do know that Hipparchus, who was active in the second century BCE, had access to Babylonian lunar eclipse reports that ranged from the mid-eighth century BCE to at least the early fourth century BCE.[67] Toomer has suggested that Hipparchus himself was instrumental in copying sections from the Babylonian archives and conveying them to Alexandria,

---

[63] von Lieven and Schomberg 2020, 76–81.
[64] On Babylonian water clocks, see Brown et al. 1999; Fermor and Steele 2000.
[65] J. Steele notes that the term *dibdibbu* "may have its origin in the sound of dripping water" (2016, 50).
[66] These are found in *The Astronomical Diaries and Related Texts*, vols. I–IV (Sachs and Hunger 1988, 1989, 1996, 2000).
[67] For an overview of the many, essential contributions of Babylonian to Greek astronomy, see A. Jones 1991; Rochberg 2020b.

though this cannot be proven.[68] While Hipparchus's own lunar system borrows heavily from Babylonian System B lunar theory, he was but one of many Greek astronomers and astrologers who, over the coming centuries, would be inspired by Babylonian table-making and adapt Babylonian predictive systems to his own purposes. Through such channels, whatever their shape, Greek astronomers became acquainted with Babylonian equinoctial hours and adapted the concept to suit their own temporal frameworks.[69]

In the Greco-Roman world, the foundations of monumental water clocks have been uncovered at Oropos, Samos, Pergamon, Ephesus, Priene, and Athens (both in the so-called Tower of the Winds and near the Rectangular Peribolos), and written accounts attest to their presence in Rome, as well.[70] These clocks were situated in public spaces, such as marketplaces or theaters, and installed in large, roofed structures where they could operate unaffected by nightfall or inclement weather. A smaller example, exquisitely wrought in bronze, has been associated with a temple of the healing god Borvo-Apollo in Roman Gaul, and it is likely that we possess fragments of other small water clocks that have been mistakenly classed as regular vessels. They must have been more common than our sparse material evidence suggests, because we have many written testimonies of their use—whether direct (e.g., that someone owned or saw a water clock) or indirect (e.g., the specification of an hour of the night, when sundials could not function).

Conclusion

Our brief overview of the Egyptian and Babylonian antecedents to Greco-Roman *hōrai/horae,* sundials, and water clocks should suggest that, around the same time that Greek horological tools came on the scene, a complex process of cultural borrowing, adaptation, and innovation was at work. This overview also suggests that hubs for cross-cultural intellectual and mercantile exchange, like Hellenistic Alexandria, may have played important roles in exposing Greeks and Romans (particularly those among the intellectual elite) to preexisting modes of hourly timekeeping, and in turn, exposing Egyptians and Babylonians to the timekeeping adaptations and innovations introduced by Greeks and Romans. We know that many prominent Hellenistic physicians, such as Herophilus and Erasistratus, were affiliated with or visited the multicultural, interdisciplinary "think tank" housed within the Library of Alexandria, and they may have been early adopters of clock technology within the domain of medicine. However, as we will see in the next

---

[68] Toomer 1988, 357–60.
[69] For more on the "set of diverse but interacting astronomies" that make up Hellenistic astronomy as a whole, see Bowen and Rochberg 2020, 3.
[70] E.g., Plin. *NH.* 7.191ff.

chapter, our textual evidence for medical timekeeping in the Classical and Hellenistic periods is sparse and problematic.

In fact, the interpretation of our Greco-Roman horological data overall is fraught with challenges. To begin with, few of our extant sundials and water clocks are well provenienced (or, for that matter, provenanced), which makes it difficult, if not impossible, to reconstruct their exact locations, degrees of visibility, and potential functions. The clocks themselves tell us little, since only a small proportion bear inscriptions or non-generic decorations.[71] In order to assemble these enigmatic data into a fuller picture of how clocks were actually used and understood, particularly within medical contexts, it is imperative that we also consult our epigraphic and literary sources and explore how individual authors engaged with this technology. We will begin to do so in the next chapter.[72]

---

[71] Most ancient sundials are plain or decorated with standard architectural motifs: rosettes, acanthus leaves, lion's-feet bases.

[72] The literary evidence for clocks and hours clusters within particular genres. These include technical works by astronomers, engineers, and mathematicians (e.g., Posidonius, Ctesibius, Heron, Geminus, Ptolemy); erudite compendia (e.g., by Pliny, Vitruvius, Athenaeus); ethnographic works (e.g., by Strabo, Pausanias, Plutarch); military narratives (e.g., by Xenophon, Polybius); comedies (e.g., by Attic Middle Comedians, Plautus, Alciphron); and medical treatises, such as the ones explored in this book.

2

Doctors and Clocks

The Emergence of Hourly Timekeeping in Medical Contexts

The famed Roman senator and soldier Cato the Elder is not typically thought of as a physician. However, his treatise *On Agriculture*, written in the mid-second century BCE, contains a wealth of advice on how to treat ailments at home using plants from one's own fields and gardens. In a particularly encomiastic chapter on cabbage, Cato offers the following instructions for clearing out the digestive tract:[1]

> Take four pounds of very smooth cabbage leaves, make them into three equal bunches and tie them together. Set a pot of water on the fire, and when it begins to boil sink one bunch *for a short time*, which will stop the boiling. When it begins again, sink the bunch briefly *while you count to five* and remove. Do the same with the second and third bunches, then throw the three together and macerate. *After macerating*, squeeze through a cloth about a *hemina* of the juice into an earthen cup; add a lump of salt the size of a pea, and enough crushed cumin to give it an odor, and let the cup stand in the air *through a calm night*. *Before taking a dose* of this, one should take a hot bath, drink honey-water, and go to bed fasting. *Early the next morning*, he should drink the juice and walk about *for four hours*, attending to any business he has. *When the desire comes on him and he is seized with nausea*, he should lie down and purge himself; he will evacuate such a quantity of bile and mucus that he will wonder himself where it all came from.

---

[1] 156.2–4: *Alvum si voles deicere superiorem, sumito brassicae quae levissima erit P. IIII inde facito manipulos aequales tres conligatoque. Postea ollam statuito cum aqua. Ubi occipiet fervere, paulisper demittito unum manipulum, fervere desistet. Postea ubi occipiet fervere, paulisper demittito ad modum dum quinque numeres, eximito. Item facito alterum manipulum, item tertium. Postea conicito, contundito, item eximito in linteum, exurgeto sucum quasi heminam in pocillum fictile. Eo indito salis micam quasi ervu et cumini fricti tantum quod oleat. Postea ponito pocillum in sereno noctu. Qui poturus erit, lavet calida, bibat aquam mulsam, cubet incenatus. Postea mane bibat sucum deambuletque horas IIII, agat, negoti siquid habebit. Ubi libido veniet, nausia adprehendet, decumbat purgetque sese. Tantum bilis pituitaeque eiciet, uti ipse miretur, unde tantum siet. Postea ubi deorsum versus ibit, heminam aut paulo plus bibat. Si amplius ibit, sumito farinae minutae concas duas, infriet in aquam, paulum bibat, constituet.* Tr. W. D. Hooper and H. B. Ash, with modification and emphasis mine. J. Hörle (1929) has asserted that the arrangement of *De Agricultura* is due, not to Cato himself, but to a later compiler.

Time and Ancient Medicine: How Sundials and Water Clocks Changed Medical Science. Kassandra J. Miller, Oxford University Press. © Kassandra J. Miller 2023. DOI: 10.1093/oso/9780198885177.003.0003

This prescription is riddled with temporal markers (italicized above) telling the reader when and for how long to perform each step in the process. Cato articulates these points and durations using a wide variety of temporal registers. On the one hand, we see several modes of "time-indication"—that is, marking time by means of a personal, social, celestial, or terrestrial phenomenon. First of all, Cato often tells us to perform an action before, after, or during another action. We are told to engage in activities "after macerating," "before taking a dose," and while "count [ing] to five." Cato also indicates temporal points and durations in terms of the general position of the sun, specifying that we should "let the cup stand in the air through a calm night" and drink our prepared juice "early the next morning." Finally, he sometimes asks us to pay attention to cues within our own or our patients' bodies. A person should lie down and purge "when the desire comes on him and he is seized with nausea," and similarly, one should determine how long to soak a head of cabbage based on one's own perception of "a short time."

Interestingly, though, alongside these general time-indications, Cato also makes reference to a systematic and continuous form of "time-reckoning" when he suggests that the patient walk around "for four *hours*." This raises a question: how did individual healers decide when to privilege hourly time-reckoning over various modes of time-indication? For instance, did Cato use hours in this passage because of the *length of time* involved—i.e., a period that is shorter than a whole day but perhaps inconvenient to measure with common methods of time-indication? Or was his decision dictated by the conventions of the *genre* in which he was writing? Alternatively, his choice may have had something to do with the nature of the *activity* itself. Perhaps it had become a cultural norm to mark the times involved in walking or attending to business using numbered hours. Or perhaps Cato's decision was motivated by the *location* in which the activity was taking place. A man "attending to any business he has" might find himself moving through public spaces, like marketplaces, where one could expect monumental public clocks to be available.

The concept of the "chronotope," a term coined by M. Bachtin, can be particularly helpful for talking about this last intersection, between timekeeping conventions and the spaces in which they are applied. The word "chronotope" is a portmanteau formed from the Greek words for "time" (*khronos*) and "place" (*topos*),[2] and it describes the intimate relationship between a particular location, such as a marketplace or sanctuary, and the temporal norms that govern activities within that space. At a present-day Olympic swimming event, for example, it would be customary to specify athletes' racing times down to the millisecond. At a restaurant, in contrast, to measure the punctuality of your server or your dinner date with that level of precision would be in rather poor taste. Similarly, in the

[2] Bachtin 2014.

ancient world, regardless of the availability of hourly timekeeping tools, there remained spaces and social contexts in which such temporal exactitude was deemed irrelevant. Since chronotopes reflect the values and practices of particular cultures at particular times, historians must be careful not to assume that Greek and Roman chronotopes remained static over the centuries, or that they can be mapped neatly onto our own. In modern, post-industrial societies, for example, hourly timekeeping is integral to many wage-calculations, but in the Greco-Roman world, we have only a couple of examples of a clock being used to determine a worker's pay.[3] In what follows, we will see that, especially over the course of the Hellenistic and Roman periods, several medically relevant chronotopes emerged—including those of the bath-house, the gymnasium, and the dining room—that privileged hourly timekeeping.

But to return, for the moment, to our Cato passage: it is likely that Cato's decisions about what temporal language to employ in which context—whether these decisions were conscious or unconscious—were the result of a combination of the various factors we have considered (e.g., length of time, genre, activity, location or "chronotope"), and probably others as well. The elements involved in such decision-making will be investigated further in the succeeding chapters. Meanwhile, the present chapter will tackle yet another question raised by our Cato passage: namely, when and how did hourly timekeeping come to be used by healers to structure aspects of their medical discourse? This chapter offers a chronological narrative of how medical practices evolved in relation to the timekeeping technologies (sundials, water clocks, hours) that we encountered in the previous chapter.[4] It begins with relevant Pharaonic Egyptian and Assyro-Babylonian precedents and proceeds to outline some of the general innovations and trends in the medical timekeeping of Hippocratic, Hellenistic, and Imperial-period physicians. In constructing this narrative, epigraphic, artistic, and other material evidence for hourly timekeeping is placed in dialogue with selections from medical literature.

The picture that emerges is as follows. We will see that the Hippocratics, building upon Assyro-Babylonian and Egyptian models, introduced to Classical Greek medicine a keen interest in the temporal trajectories of illnesses and in the opportune moment (in Greek, the *kairos*) for administering therapy. In the Classical period, however, whether due to lack of exposure or lack of interest, hourly timekeeping is rarely attested in medical contexts. In the Hellenistic period, physicians continue to develop their interest in medical timekeeping and begin to play with the possibilities offered by clocks, especially for disease prognosis. By the

---

[3] One example comes from the New Testament (Matt. 20: 9), the other from a comic play by Eubolus (Ath. *Deip.* 567d). The eponymous protagonist of this play is the sex worker "Klepsydra," so named because she used a water clock to time her clients' calls.

[4] On how other technologies shaped ancient Greek and Roman medicine and led to the development of the concept of the "organism," see Webster 2023.

Roman Imperial period, however, we see that sundials, water clocks, and hours are regularly integrated into medical theory and practice. In this period, the ubiquity of clocks in public spaces, paired with the popularity of astrology across social classes, led to the development of more and more medically relevant chronotopes that relied on hourly timekeeping. All of these factors spurred intense debate among physicians about whether, when, and how they should interact with clock technologies. We will see that some physicians developed a keen interest in hourly precision, while others pushed back against this trend. The present chapter will take a bird's-eye view of these developments in order to set the stage for the focused case studies that we will consider in later chapters.

Egyptian and Assyro-Babylonian Precedents

Long before the emergence of Classical Greek medicine, Assyro-Babylonian and Pharaonic Egyptian physicians were thinking deeply about time, albeit in different ways. Medical papyri from New Kingdom Egypt (c.1543–1078 BCE)—including, for example, the Ebers, Hearst, London, Berlin, Kahun, and Chester Beatty Papyri—demonstrate that physicians of that time were concerned with the durations of therapeutic treatments and, in certain cases, their timing within the day. Case 10 of the Kahun gynecological papyrus, for example, advises that the physician treat a case of painful urination in the following manner:[5]

> You should treat this: beans, pine nuts, and *mut* part of nutgrass, grind fine, with a *hin* of diluted beer and boil. Drink *for four mornings*. She should *spend a day and a night* fasting; *in the morning*, drink 1 *hin* of the same and *spend the day* fasting *until breakfast time comes*.

In this papyrus, as in our Cato passage above, points in time are typically indicated by the rough position of the sun (e.g., "in the morning") or coordinated with an activity (e.g., eating breakfast). Durations are given in days, nights, or the (unquantified) interval of time between two events (e.g., waking up and eating breakfast). But what may have motivated *this* author, writing in a very different period and social context from Cato, to adopt this kind of temporal language? And to what extent do his choices reflect empirical observation, the application of theory, or symbolic considerations?

While full consideration of these questions is outside the scope of the present study and awaits further attention from Egyptologists, it is interesting for us to note that, across the New Kingdom medical papyri, certain numbers appear with far greater frequency than any others with regard to therapeutic durations. In each

---

[5] 32–4. Tr. Strouhal et al. 2014.

of these papyri, patients are regularly asked to eat, drink, or apply a substance either "for one day" or "for four days" (or, as in the Kahun passage above, "for four mornings"). While other durations occasionally appear,[6] it is clear that they are the exceptions, not the rule. Such consistency of practice suggests that these durations may have been dictated by a larger theoretical or ideological framework. E. Strouhal, B. Vachala, and H. Vymazalová suggest that a kind of numerology may have been in play: "In the ancient Egyptian conception, the number 4 expressed unity and completeness. Its symbolism was apparently influenced by four cardinal directions, which, however, did not always have to be directly manifested in the symbolic use."[7] It may have been the case, then, that intervals of one or four days were favored in Pharaonic Egyptian therapeutics because they symbolically represented the condition of wholeness or integrity to which physicians hoped to restore their patients.

The idea of administering therapeutics for a set interval of time seems to have retained its popularity in Egypt for so long that two Greek authors attest to its continued importance in the Classical and Roman Imperial periods. Aristotle, in his *Politics*, says that, "in Egypt, doctors have the right to alter their prescription only after four days; and if one of them alters it earlier, he does so at his own risk."[8] Three centuries later, Diodorus Siculus asserts that "in order to prevent diseases, [Egyptians] treat their bodies by means of enemas and vomiting, sometimes every day and sometimes at intervals of three or four days."[9] Diodorus goes on to say that, in administering these treatments, Egyptian physicians are acting in accordance with "a written law which was composed in ancient times by many famous physicians" and could be consulted in "the sacred book."[10] These testimonies derive, of course, from outsiders to the cultural sphere of Egyptian medicine, but they suggest that the idea of the set therapeutic interval had great longevity in Egyptian tradition and was available to influence contemporary Greek and Roman physicians. It is possible, for instance, that the physicians known as the Methodists, to whom we will return in Chapter 6, developed their own system of therapeutic intervals (the *diatritos*), at least in part, as a response to such models.

But to return to the pre-Classical Mediterranean: while the authors of the New Kingdom medical papyri paid attention to timing within therapeutic contexts, they left no record of interest in the temporal trajectories of diseases themselves

---

[6] E.g., "Drink it for three days" at Kahun 16.54.
[7] 2021, 170. Their perspective is informed by Wilkinson 1999, 133–5.
[8] *Pol.* 3.15, 1286a12–14. Tr. Jouanna 2012, 12.
[9] 1.82.1: Τὰς δὲ νόσους προκαταλαμβανόμενοι θεραπεύουσι τὰ σώματα κλυσμοῖς καὶ νηστείαις καὶ ἐμέτοις, ἐνίοτε μὲν καθ' ἑκάστην ἡμέραν, ἐνίοτε δὲ τρεῖς ἢ τέτταρας ἡμέρας διαλείποντες. Tr. Jouanna 2012, 11.
[10] 1.82.3: οἱ γὰρ ἰατροὶ τὰς μὲν τροφὰς ἐκ τοῦ κοινοῦ λαμβάνουσι, τὰς δὲ θεραπείας προσάγουσι κατὰ νόμον ἔγγραφον, ὑπὸ πολλῶν καὶ δεδοξασμένων ἰατρῶν ἀρχαίων συγγεγραμμένον. κἂν τοῖς ἐκ τῆς ἱερᾶς βίβλου νόμοις ἀναγινωσκομένοις ἀκολουθήσαντες ἀδυνατήσωσι σῶσαι τὸν κάμνοντα.... Tr. Jouanna 2012, 12.

(e.g., the times at which illnesses manifest, advance to new stages, and resolve). This, however, was a great preoccupation of Assyro-Babylonian physicians whose medical practices were deeply influenced by their wider cultural commitment to seeking out and interpreting omens. Fundamental to these practices was the idea that certain events in the heavens and in the world around us encode information about the course of future events. In order to decode that information accurately, one must pay attention not only to the kind of ominous phenomena that occur, but also where they take place and *when*. Timing and timekeeping were thus integral components of this omenological mindset, which inflected most Assyro-Babylonian social practices, from war and politics to trade, travel, and of course, medicine.

The interaction between Babylonian medicine and omen-interpretation is displayed in The Diagnostic and Prognostic Series (DPS), a handbook redacted by the eleventh-century BCE Borsippan Esagil-kīn-apli.[11] This handbook, running to 40 cuneiform tablets, synthesized a vast quantity of medical knowledge and reorganized it according to a variety of principles. The first subseries (DPS 1–2), for example, focuses on the interpretation of omens that the physician might encounter en route to his patient, while the second subseries (DPS 3–14) addresses physical ailments *a capite ad calcem,* from head to toe. The beginning of the third subseries, however, Tablets 15–17, is organized around diseases' temporal patterns. As J. Scurlock describes:[12]

> DPS 15–16 contained entries mentioning the number of days the patient had been sick, whereas DPS 17 was concerned with phases of illness and times of day. This organization facilitated the addition of later material as when, in DPS 16, a section organized by the number of five-day weeks a patient had been sick ('one to two five-day weeks, one to four five-day weeks') was apparently inserted, set off by paragraph lines, into the appropriate slot (between 'more than five days' and 'more than six days').

Tablet 17, with its focus on medically significant times of day, is of particular interest for our present study. Let us consider two entries from that Tablet, each of which reflects ideas that also become important in Greek and Roman medical timekeeping. The first passage reads as follows:[13]

> If *at the beginning* of his illness, he has sweat (and) *bubu 'tu*-blisters and that sweat does not reach from his shins to his ankles and the soles of his feet, if that illness has reached the *critical stage* (lit. "becomes ill")[14] by *the second or third day*, he will get well.

---

[11] Scurlock (2014) provides an English translation of and critical commentary on the DPS, to which the following discussion is indebted. On the development of Babylonian medicine over time, see Gellar 2010.
[12] Scurlock 2014, 9.   [13] DPS 17.1–3. Tr. Scurlock 2014.
[14] I am indebted to J. Steele (pers. comm.) for his explanation of the cuneiform.

This entry encourages physicians to pay attention to the points in an illness' temporal trajectory at which certain symptoms arise, worsen, and/or abate. Physicians would have used this information to predict the outcomes of illnesses or, in other words, to prognosticate. This entry tells us, for example, that the arrival of sweats and blisters *at the beginning* of an illness can indicate that a patient will recover. These symptoms must also appear in conjunction with another phenomenon: by the second or third day since its onset, the illness must reach a "critical stage" in Scurlock's translation or, more literally, the illness must "become more ill" (LÍL-ma DIN) so as to cause him to recover. As we will see, particularly in Chapter 5, many Hippocratic authors of the fifth and fourth centuries BCE became interested in the idea that the temporal trajectory of an illness is punctuated by decisive turning points (*kriseis* in Greek; *krisis* in the singular) at which the patient's condition either improves or worsens dramatically and which can be used to decipher the illness's ultimate outcome (namely, recovery or death). DPS Tablet 17.1–3 and similar passages prefigure this notion by suggesting that the timing of a patient's turn for the worse can encode information about that patient's chances of recovery.[15]

Another entry in Tablet 17 displays the high level of attention that Babylonian physicians sometimes paid to timing within the day. It reads:[16]

> If he is sick *in the morning* and *in the late afternoon* his illness leaves him and then suddenly it returns on him [and he has] a reduced form of his illness the second day *until noon*, the third day *until late afternoon*, the fourth day *until evening*, the fifth day *until its normal term* [i.e., sundown], the sixth day *until the middle of the [first] watch [of the night]*, the seventh day *until the middle watch [of the night]*, the eighth day *until the dawn watch* [and] the ninth day *until it is light*, if he gets up on the tenth day, he will get well.

This entry describes an illness that seems to come and go, reappearing each day in a form that is a little bit weaker but lasts a little bit longer. The author of this entry carefully notes the timing of symptoms within the day, but it is interesting that, in charting the temporal course of this illness, he does not employ the most mathematically precise system of time-reckoning available—namely, the system of fixed-length units (*bēru*, UŠ, and NINDA) which were discussed in the previous chapter and which equate, respectively, to two of our hours, four of our minutes, and four of our seconds. This physician did not deem these units—which were so essential within the context of astronomy, for example—to be relevant for medical practice. Instead, in structuring his narrative of the illness, the author of Tablet 17.95–9, identifies nine non-quantitative moments or spans within the (twenty-

---

[15] E.g., Tablet 17.4–7 and 8–9.   [16] 95–9. Tr. Scurlock 2014.

four-hour) day: three points that mark the limits of the sun's daily journey (sunrise, noon, and sunset), three watches of the day (morning, afternoon, and evening), and three watches of the night (the first, middle, and dawn watches). In order for a physician to locate himself or a symptom within this kind of temporal series, he would not need to rely on much specialist knowledge or equipment. On a clear day, he could simply look up at the position of the sun in the sky, and on a cloudy day or during the night, he could consult a water clock or perhaps listen for an announcement of the watch.

This entry reflects the idea that greater temporal precision—whether facilitated by a physical device like a water clock or simply by a lexicon for talking about the sun's movement—can lead to diagnoses and prognoses that are more precise and accurate. However, it also demonstrates that the most quantitatively precise timekeeping system available is not always considered the most practical or relevant for the task at hand. Just as the author of Tablet 17.95–9 rejected the rigorous astronomical system of fixed-length units in favor of the more variable but more widely used system of watches, we will see later Greek physicians grapple with the relative advantages of employing the equinoctial hours of the astronomers, the seasonal hours of certain public spaces, or no hours at all.

Some Assyro-Babylonian medical practices, however, did incorporate the quantitative timekeeping of the astronomers. These practices pertained to the tradition of "astral medicine" and required the physician to apply to his craft the tools of mathematical astronomy. While direct links between Babylonian astronomy and medicine are attested as early as the first half of the second millennium BCE, the practice of astral medicine was revolutionized by the concept of the twelve-sign Zodiac, which was introduced in the Achaemenid era (a period roughly contemporaneous with the early Hippocratics, whom we will discuss in the following section). As M. Rumor describes:[17]

> As was the case with contemporary Hippocratic writers, in an attempt to increase certainty Babylonian scholars were seeking a way to fit disease and therapy within predictable structures. The zodiac looked as though it could provide that structure. Within it, any natural cycle, such as the rhythm of the calendar, could not only be observed but also calculated, and from there, the leap to *calculating* and thus predicting *human* natural cycles, such as patterns of disease, would not have required too much imagination.

Assyro-Babylonian astral medicine operated on the assumption that signs of the zodiac could cast a benefic or malefic influence on the courses of diseases, on individual parts of the human body, and on specific *materia medica*, such as the

---

[17] Rumor 2021, 50.

stones used in amulets, the plants used for ointments, and the wood burned in fumigations. Practitioners were particularly concerned to know the position within the zodiac of certain celestial bodies (especially the sun or moon) at an illness's onset, since these positions were thought to both influence and encode the disease's trajectory and ultimate resolution. In order to calculate these zodiacal positions and deduce their effects, astro-medical physicians relied on hemerologies, menologies, so-called *Kalendertexte*, and other tables that used arithmetical progressions to model the movements of celestial bodies on daily, monthly, yearly, and ultimately, hourly bases. In what follows, we will see how some of the general ideas and even specific tools of Assyro-Babylonian astral medicine were adopted and adapted by Greco-Roman physicians eager to make their own diagnoses and prognoses ever more precise and accurate.

Hippocratic Evidence

At about the same time that Assyro-Babylonian physicians were beginning to restructure their astral medicine around the idea of the twelve-sign zodiac, physicians living in the eastern Greek island group known as the Dodecanese were wrestling with some of the same fundamental problems that worried their Assyro-Babylonian and Egyptian colleagues: what were the best times to administer various therapeutics? What signs and signals (in the patient's body or in the wider world) should they look for in order to most accurately identify diseases and anticipate their phases? And more broadly, how could the behavior of natural phenomena (like celestial bodies, plants, or animals) help them to explain, predict, and respond to disease phenomena? Like their Assyro-Babylonian counterparts, many of the Greek physicians who have come to be known collectively as the Hippocratics saw potential answers to these questions in the temporal trajectories of diseases. They suspected that by knowing something about the timing and duration of various symptoms, and by developing systems for organizing and interpreting these time-data, they could refine their therapeutic practices and classifications of disease types and do a better job of distinguishing acute illnesses from chronic, and treatable cases from lost causes. However, while the contributors to the Hippocratic Corpus, on the whole, seemed to agree that time was a medically significant variable, that was about as far as the consensus went. They disagreed, often dramatically, about which kinds of time were relevant in which contexts, and which systems should be used to structure and interpret temporal information.

It is noteworthy that this Hippocratic interest in medical timing coincided, more or less, with the introduction of hourly timekeeping technologies like sundials and water clocks to the Greek-speaking world. Yet, despite this correspondence, numbered hours appear very rarely in the Corpus. Instead, the vast majority of temporal markers within the Hippocratic texts are forms of "time-indication" rather than "time-reckoning," which rely, like the examples from our Cato passage, on a variety of personal, social, celestial, and terrestrial phenomena. As a representative case study, let's consider a single treatise: *On Regimen* 3, dated to c.400 BCE.[18] The primary goal of *On Regimen* is to recommend daily regimens (or *diaitai*, whence our word "diet") that can help different types of people to maintain their health throughout the year. In designing these regimens, the author focuses on four main activities—eating, drinking, exercising, and bathing—and takes five variables into consideration—namely, the patient's age, body type, and geographical location; the dominant winds; and the *hōrai* of the year (which must here mean the "seasons"):[19]

> And it is necessary, as it appears, to discern the power of the various exercises, both natural and artificial, to know which of them tends to increase flesh and which to lessen it; and not only this, but also to proportion exercise to bulk of food, to the constitution of the patient, to the age of the individual, to the season of the year, to the changes of the winds, to the situation of the region in which the patient resides, and to the constitution of the year.

Throughout this text, *hōra* in the sense of "season" appears as a central concept, but the term is never used to indicate a *seasonal* hour (or an equinoctial one, for that matter). Instead, when the Hippocratic author wants to describe a moment or span within the day, he does at least one of two things: he establishes more imprecise windows of time and/or uses different forms of personal or "body" time to determine the right moment within a temporal window for a patient to perform an action. The following passage includes a series of representative examples (with the temporal language marked in *bold*):[20]

> Exercises should be many and of all kinds: running on the double track increased gradually; wrestling *after being oiled*, begun with light exercises and gradually

---

[18] For the dates of the Hippocratic works, see Jouanna 1999, 373–416.
[19] Hippoc. *Vict.* 3.2.21–35 L: Δεῖ δέ, ὡς ἔοικε, τῶν πόνων διαγινώσκειν τὴν δύναμιν καὶ τῶν κατὰ φύσιν καὶ τῶν διὰ βίης γινομένων, καὶ τίνες αὐτῶν ἐς αὔξησιν παρασκευάζουσι σάρκας καὶ τίνες ἐς ἔλλειψιν, καὶ οὐ μόνον ταῦτα, ἀλλὰ καὶ τὰς ξυμμετρίας τῶν πόνων πρὸς τὸ πλῆθος τῶν σιτίων καὶ τὴν φύσιν τοῦ ἀνθρώπου καὶ τὰς ἡλικίας τῶν σωμάτων, καὶ πρὸς τὰς ὥρας τοῦ ἐνιαυτοῦ καὶ πρὸς τὰς μεταβολὰς τῶν πνευμάτων, καὶ πρὸς τὰς θέσεις τῶν χωρίων ἐν οἷσι διαιτέονται, πρός τε τὴν κατάστασιν τοῦ ἐνιαυτοῦ. Tr. Jones (1931) with modification.
[20] Hippoc. *Vict.* 3.68.30–9 L: τοῖσι δὲ πόνοισι πουλλοῖσιν ἅπασι, τοῖσί τε δρόμοισι καμπτοῖσιν ἐξ ὀλίγου προσάγοντα, καὶ τῇ πάλῃ ἐν ἐλαίῳ, μακρῇ, ἀπὸ κούφων προσαναγκάζοντα· τοῖσί τε περιπάτοισιν ἀπὸ τῶν γυμνασίων ὀξέσιν, ἀπὸ δὲ τοῦ δείπνου βραδέσιν ἐν ἀλέῃ, ὀρθρίοισί τε πολλοῖσιν ἐξ ὀλίγου ἀρχόμενον.... καὶ σκληροκοιτίῃσι καὶ νυκτοβατίῃσι καὶ νυκτοδρομίῃσι χρέεσθαι ξυμφέρει.... Ὁκόταν δὲ ἐθέλῃ λούσασθαι, ἢν μὲν ἐκπονήσῃ ἐν παλαίστρῃ, ψυχρῷ λουέσθω· ἢν δὲ ἄλλῳ τινὶ πόνῳ χρήσηται, τὸ θερμὸν ξυμφορώτερον.... Διδόναι δὲ καὶ τῷ ψύχει ἑωυτὸν θαρσέων, πλὴν ἀπὸ τῶν σιτίων καὶ γυμνασίων, ἀλλ' ἔν τε τοῖσιν ὀρθρίοισι περιπάτοισιν, ὁκόταν ἄρξηται τὸ σῶμα διαθερμαίνεσθαι.... ὑπερβολὴν γὰρ οὐκ ἔχει, ἢν μὴ οἱ κόποι ἐγγίνωνται. Tr. Jones (1931) with modification.

made long; sharp walks *after exercises,* short walks in the sun *after dinner;* many walks *in the early morning....* It is beneficial to sleep on a hard bed and to take *night* walks and *night* runs.... *Whenever one desires to bathe,* let it be cold *after exercise* in the *palaistra; after any other exercise,* a hot bath is more beneficial.... One may expose oneself confidently to cold, except *after food and exercise,* but exposure is wise in *early-morning* walks, *whenever the body has begun to warm up....* There is no risk of excess unless *fatigue-pains follow.*

In this passage, we see how the author indicates general temporal windows by reference to the position or absence of the sun, using words like *orthrioisi* ("in the morning") or the prefix *nykto-* ("at night"). He also indicates times within the day by means of the activities and experiences of the patient's body. These can be broken down into three general categories:

1. First, the *signs* (or *tekmēria*) that are perceptible on or within the patient's body. In this passage, for example, we see that during a morning walk, the right moment for a person to remove an outer layer is indicated by his body temperature. Likewise, the moment at which a workout transitions from "suitable" to "excessive" is signaled by the patient's personal perception of fatigue.
2. Second, the *sequence* of the patient's physical activities. We see here, for instance, that dinner should always *precede* one's short walk, and that one should only take a cold bath *after* having exercised in the *palaistra*.
3. Third, the patient's personal *desires*, as prompted either by habit or by the present state of the patient's body (e.g., whether he or she is tired, hungry, cold, etc.). This indicator appears explicitly when the author invites one to bathe "whenever one desires," and is further implied by the fact that the author leaves out so many details. He sketches out broad guidelines but seems to trust the patient to choose for her- or (more likely) himself precisely how and when to exercise.

Although the author of *On Regimen* 3 does not avail himself of "clock time" per se, two aspects of his work set the stage for later physicians like Galen, who were committed to presenting themselves as neo-Hippocratics, to feel that there were Hippocratic precedents for incorporating hourly timekeeping into their discussions of hygienic regimen. First of all, the Hippocratic author demonstrates a keen interest in identifying the *kairos*, or "opportune moment," at which to administer various therapies. This concept, which has a long history in Greek medicine and rhetoric, will be explored more fully in Chapter 7. Most important for our present discussion is the fact that, with Hippocratic writings like *On Regimen*, we see the idea of "time within the day" assume a new importance in defining *kairoi*. The "time within the day" is no longer simply a neutral backdrop but has become something like an active gameboard, in which different windows of time, like different boardgame squares, could offer specific risks or rewards to the discerning physician. Tools, such as sundials and water clocks, that could help physicians to track and describe these opportune moments with even greater precision could therefore be considered in keeping with "Hippocratic" tradition, even if the Hippocratic authors themselves made little use of them. In Chapter 5, for example, we will see how Galen uses hourly timekeeping to refine the systems that Hippocratic authors developed to hunt for medical *kairoi* by charting the temporal rhythms of diseases. This act of reading hourly timekeeping back into the Hippocratic texts may also have been facilitated by the fact that, by the Roman Imperial period, the word *hōra*—which the Hippocratics often identify as an important medical variable—could denote an "hour" as easily as a "season."

But when *do* numbered hours appear in the Hippocratic Corpus? As mentioned in the previous chapter, it has been asserted that the Hippocratic Corpus is the source of some of the earliest examples of the word *hōra* meaning "hour," a claim that, if true, might imply that certain physicians were early and perhaps influential adopters of clock technology.[21] The Hippocratic texts in question are *On Internal Affections* and *Epidemics* 4, each of which seems to contain a single reference to numbered hours. These references, however, are each problematic in their own ways. In the case of *On Internal Affections*, the reference to numbered hours has not been challenged, but the date of the text itself remains an open question. In the case of *Epidemics* 4, the dating window is more secure, but the numbered-hour reference itself has been questioned by at least one prominent editor.

The author of *On Internal Affections*, in a series of recommendations on treating liver disease, offers the following advice to fellow physicians: "And if he [i.e., the patient] vomits up some bile or phlegm, it is necessary to do [the procedure] again in four hours, for it will help."[22] This nosological treatise cannot be dated with precision, particularly because few ancient authorities make mention of it. As F. Adams observed in his 1886 publication, *The Genuine Works of Hippocrates*, "Erotian has omitted [this text] in his list of the works of Hippocrates; Palladius does not mention it; and Galen notices it in a confused manner under a variety of titles."[23] Instead, scholars of the Hippocratic Corpus have had to rely on shared elements of style, content, and medical worldview to group this text with other Hippocratic works, and thereby to make an educated guess as to its authorship and date. E. Craik, for example, suggests that *On Internal Affections* belongs in a grouping with two other nosological works, *Affections* and *Diseases* 2, and that, within this set, "it may be possible to hear echoes of

---

[21] See Langholf 1990.
[22] Hippoc. *Int.* 27.7.238 L: ἢν δὲ μὴ ἔμετος ἔχῃ, ἐπιπιὼν μελικρήτου χλιαροῦ κύλικα δικότυλον, οὕτως ἐμείτω. καὶ ἢν τι ἀπεμέσῃ χολῆς ἢ φλέγματος αὖτις ταὐτὰ χρὴ ποιέειν ἐπὶ τέσσερας ὥρας· ὠφελήσει γάρ.
[23] Adams 1886, 71.

Euryphon," a slightly older contemporary of Hippocrates who was active on Cos's neighboring island of Cnidos.[24] Were *On Internal Affections* to have been authored directly by Euryphon, this would situate the text in the late Classical period.[25] However, our evidence for this is largely speculative, and the very fact that other references to numbered hours in this period are scarce, or perhaps even nonexistent, may itself militate against interpreting *On Internal Affections* as a Classical text.

The dating of our second numbered hour reference within the Hippocratic Corpus is more secure, but the reading itself has been called into question. The passage appears in *Epidemics* 4, a collection of patient case histories that has been grouped, in both ancient and modern scholarship, with *Epidemics* 2 and 6. Together, this set is now typically dated to the period between 427/6 and 373/2 BCE. The manuscript tradition (as well as many subsequent editions) preserves the phrase "in the third hour" (*tritēn hōrēn*).[26] E. Littré presents the whole line in the following manner: "His [i.e., the patient's] neck became hard; he felt the same amount of pain later in the third hour."[27] W. Smith, however, who has edited this text for the Loeb Classical Library, emends "third" (*tritēn*) to "the same" (*tēn autēn*), and reads the line in the following way: "His neck became hard perhaps about the same time. Afterward his pain recurred."[28] This emendation seems sensible because, while no other numbered hour is mentioned in *Epidemics* 4 (or anywhere else within the *Epidemics*, for that matter), the phrase "at the same time" (*tēn autēn hōrēn*) appears on three other occasions within this text specifically[29] and seven times within other books of the *Epidemics*.[30] How to weigh the sensibility of this emendation against the evidence from the manuscript tradition is a fraught question that cannot be resolved definitively given the current state of our evidence. Thus, the claim that our earliest Greek references to numbered hours can be found in two late Classical Hippocratic texts ought to be taken with a healthy grain of salt. What should be emphasized instead is that the Hippocratics introduced to Greek medicine a new fascination with the temporal patterns of disease and the right time (*kairos*) for therapeutic treatment—though at this early stage, references to hours were rare, and there was little consensus about which temporal units, technologies, and theoretical frameworks could be most usefully applied.

---

[24] Craik 2018, 36. Soranus (*Vit. Hipp.* 5) is our oldest source for the dating of Euryphon.

[25] The following studies are devoted to *Int.*: Wittern 1978; Roselli 1990; Pérez Cañizares 2002, 2005.

[26] Hippoc. *Epid.* 4.12, 5.150 L. *Vaticanus graecus* 276 (= V), which dates to the twelfth century CE, is the earliest to include *Epidemics* 4. Other medieval manuscripts to do so include *Parisinus graecus* 2140 (= I), *Parisinus graecus* 2142 (= H), and *Vaticanus graecus* 277 (= R), which are later versions of the tenth-century *Marcianus graecus* 269, which only picks up with *Epidemics* 5. For further discussion of this manuscript tradition, see W. D. Smith 1994, 11–12.

[27] τράχηλος σκληρὸς ἐπεγένετο· τρίτην ὥρην ἴσως ὠδυνήθη ὕστερον.

[28] τράχηλος σκληρὸς ἀπεγένετο τὴν αὐτὴν ὥρην ἴσως. ὠδυνήθη ὕστερον.

[29] *Epid.* 4.1.9.4, 4.1.13.2, and 4.1.13.6 L.

[30] *Epid.* 1.2.9.55, 1.2.9.57; 5.1.13.6; 6.3.1.6; 7.1.25.20, 7.1.25.32, 7.1.92.5, 7.1.120.14 L.

Material evidence from the healing sanctuaries that became popular in the fifth and fourth centuries BCE suggest a similarly complicated relationship with the new technologies of hourly timekeeping. For example, one of the earliest-attested monumental sundials and a monumental water clock were discovered at the Amphiareion of Oropus, a sanctuary in honor of the hero Amphiareios that offered oracular and healing services. But what were the functions of these clocks? Were they primarily used for practical purposes, for instance, to time rituals or healing therapies or to coordinate public events, like performances in the sanctuary's theater? An Imperial-period inscription from the Asklepieion at Epidauros suggests that clocks could indeed be used within healing sanctuaries to time ritual activities—specifying, in this case, the exact hour of the day at which a liturgy was to be performed.[31] But it is unclear how widespread this practice was or how early it began.[32]

To complicate matters further, many clocks found in sanctuaries seem to have been dedicated as votive offerings, which could have served practical purposes but may have functioned primarily as symbols—of solar motion, for example, or of broad concepts like "time," "eternity," or "the revolutions of the cosmos."[33] As we will see in Chapter 4, clocks (and particularly sundials) accrued a rich range of symbolism over the Hellenistic and Roman periods, and thus could appear in contexts where their semiotic functions were more important than their time-telling ones. Thus, we should not assume that the presence of a clock within a particular space implies that hourly timekeeping was an essential feature of that space's chronotope, nor that a time-telling device found in a medically coded space (such as a healing sanctuary or a bath-house) was necessarily involved in the healing practices that took place there.

Hellenistic Evidence

Like the evidence for the medical use of clocks in the fifth and fourth centuries BCE, evidence for their use in the third, second, and first centuries BCE is scant and often problematic. However, two textual references suggest that, at least by the third century BCE and within the multicultural city of Alexandria, Egypt, some

---

[31] IG IV²,1 134, which opens with the heading "In the Third Hour" (ὥρᾳ ν τρίτηι). On this inscription, see Wagman 1995; Alonge 2011.

[32] Similar questions can be raised about the beautifully wrought water clock associated with the healing sanctuary of Borvo-Apollo, which was calibrated, similarly to the Egyptian "flower-pot" style of water clock, to read seasonal hours throughout the year. Did it play a practical role within the temple's medical program? Or should we view it as a symbolic or prestige item? On this clock, see Stutzinger 2001.

[33] Hannah suggests that the sundials erected in sanctuaries to Apollo, such as those at Claros and Delphi, reflect "an actualization, through the cultic furniture, of the identification of Apollo with Helios the Sun-God" (2020, 336).

physicians had begun to play with the possibilities offered by clocks and were incorporating timekeeping into their practices more regularly. The first of these references appears in a fragment of the New Comic playwright Machon, who was active in Alexandria during the third century BCE.[34] This fragment describes how the dithyrambic poet and well-known gourmand Philoxenos died of dyspepsia after eating the better part of an enormous octopus—two full cubits in length! After Philoxenos had taken to his bed, Machon describes how a doctor came to see him and prognosticated that he would die, specifically within the seventh hour. While Machon himself was certainly not a doctor, nor does this text claim to be a medical treatise, it offers us a valuable example of a medical stereotype: that of the physician who is unable to suggest an effective treatment, but who can predict with great precision the very hour of a patient's death. Audience members would not have read this character as "funny" if they had failed to recognize the physician's behavior as a trope and were not generally familiar with the practice of hourly timekeeping. Therefore, this fragment suggests that, at least in Alexandria at this time, the idea of doctors using clocks for prognostication was somewhat well-established.

We also have a single testimonium, albeit from an Imperial-period source, that a Hellenistic Alexandrian physician used a clock for patient *diagnosis* rather than the kind of prognosis depicted in the Machon fragment. The second-century CE medical writer Marcellinus tells us that Herophilus designed a special kind of "water clock" that would allow him to measure the frequency of a patient's pulse against what he had established as the "norm" for that patient's age group. The relevant passage reads:[35]

> Herophilus declared that a patient had fever whenever the pulse became more frequent, larger, and more vehement with much internal heat. If therefore [the pulse] should lessen its vehemence and size, the fever is in remission. He says that the frequency of the pulses is first established when fevers begin and remains so until their final resolution. The story goes that Herophilus was so encouraged in using the frequency of the pulse as a secure sign that he constructed a water clock holding an expressed measurement for the natural pulses of each age-group and

---

[34] Machon Fr. 9, line 74 Gow: Φιλόξεν' ἀποθανῇ γὰρ ὥρας ἑβδόμης. These fragments were preserved in Ath. *Deip.* (341a8–b9). For bibliography, see Gow 1965, 35–56.

[35] Marcellin. *De puls.* 255–67 Schöne: Ὁ δὲ Ἡρόφιλος πυρέσσειν ἀπεφήνατο τὸν ἄνθρωπον, ὁπόταν πυκνότερος καὶ μείζων καὶ σφοδρότερος ὁ σφυγμὸς γένηται μετὰ πολλῆς θερμασίας ἔνδον, εἰ μὲν οὖν προαπαλλάξειε τὴν σφοδρότητα καὶ τὸ μέγεθος, ἔνδοσιν τοῦ πυρετοῦ λαμβάνοντος· τὴν δὲ πυκνότητα τῶν σφυγμῶν ἀρχομένων τε τῶν πυρετῶν πρώτην συνίστασθαι καὶ συμπαραμένειν μέχρι τῆς τελείας αὐτῶν λύσεως λέγει. οὕτω δὲ τῇ πυκνοσφυξίᾳ τὸν Ἡρόφιλον θαρρεῖν λόγος ὡς βεβαίῳ σημείῳ χρώμενον, ὥστε κλεψύδραν κατασκευάσαι χωρητικὴν ἀριθμοῦ ῥητοῦ τῶν κατὰ φύσιν σφυγμῶν ἑκάστης ἡλικίας εἰσιόντα τε πρὸς τὸν ἄρρωστον καὶ τιθέντα τὴν κλεψύδραν ἅπτεσθαι τοῦ πυρέσσοντος· ὅσῳ δ' ἂν πλείονες παρέλθοιεν κινήσεις τῷ σφυγμῷ παρὰ τὸ κατὰ φύσιν εἰς τὴν ἐκπλήρωσιν τῆς κλεκινήσεις τῷ σφυγμῷ παρὰ τὸ κατὰ φύσιν εἰς τὴν ἐκπλήρωσιν τῆς κλεψύδρας, τοσούτῳ καὶ τὸν σφυγμὸν πυκνότερον ἀποφαίνειν, τουτέστι πυρέσσειν ἢ μᾶλλον ἢ ἧττον. Tr. Berrey (2017, 202) with modifications.

that on entering in to the patient and setting down the water clock he felt the patient [for his pulse]. By as much as the greater movements of pulses overshot the <magnitude> natural for the filling up of the water clock, by so much did he reveal the pulse to be more frequent, that is, either more or less feverish.

Exactly how this water clock functioned has been a matter of some debate. H. von Staden, for example, acknowledges "several feasible constructions (e.g., a set of four perforated 'sinking bowls' of different sizes, one for each age group, or a single traditional overflowing or draining container with four receptacles of different sizes, or a draining or receiving container with a 'dipstick' marked for different ages)."[36] More recently, M. Berrey has argued specifically for this latter reconstruction. He suggests that Herophilus took as the model for his *klepsydra* an Egyptian water clock of the "flower-pot" style which, as the reader may recall, tracked numerical seasonal hours by means of a series of hour-scales on its interior (one for each month, so as to approximate the changing length of a seasonal hour over the course of the year). Berrey proposes that Herophilus adapted this model so that his *klepsydra*'s internal scale was calibrated not to hourly time, but rather to a sequence of "normative times" corresponding to the amount of water, appropriate for each age group, that Herophilus expected to have flowed into the clock by the time he had counted a specific number of "heartbeats."[37]

While Berrey's reconstruction strikes me as plausible, I would like to probe more deeply what it means to say, as Berrey does, that Herophilus "*quantif[ies]* the pulse."[38] It is not clear from Marcellinus's testimony precisely how Herophilus articulated the relationship between the heartbeats he felt under his fingertips and the amount of water that had flowed into the vessel. Was this relationship expressed with a single number, as an absolute quantity, or with a pair of numbers, as a ratio? Or perhaps, rather than expressing this relationship numerically, Herophilus simply spoke of a quantity of water being "either more or less" than a gradient line on the vessel? In other words, one would like to know more about the roles that numbers played both in Herophilus's "data-gathering" process and in the way that he expressed his results. Furthermore, what level of quantitative or qualitative precision did Herophilus consider necessary for this task? Unfortunately, these questions cannot be answered with certainty on the basis of our current evidence. This "story" (*logos* in Greek) was reported by a medical writer who was active nearly 500 years after Herophilus's own day, and it is the only textual evidence for this kind of medical *klepsydra* to have come down to us.

---

[36] Von Staden 1989, 283.
[37] On how interpretations of the pulse differed in antiquity, see Berrey 2017, 192. On the ways in which ancient Greek understandings of the pulse differed from classical Chinese understandings, see Kuriyama 1999.
[38] Berrey 2017, 206, emphasis added.

Even Galen, who, as we shall see, is very interested in the medical applications of clocks, never mentions using one to time a patient's pulse. Nor has the archaeological record yet yielded an example of a *klepsydra* that matches Berrey's reconstruction.

Ultimately, the anecdote about Herophilus's *klepsydra* tells us little about how such physicians incorporated hourly timekeeping into their medical practice. As Berrey points out, Herophilus's "water clock" does not seem to have kept continuous time throughout the day or night, as measured with numbered hours. Instead, it operated more like an egg-timer, which measures short intervals of normative time. In this regard, then, Herophilus's *klepsydra* bears a stronger resemblance to the "water clocks" used to time speeches in Athenian and, later, Roman courtrooms than, for example, to the monumental water clocks that stood outside of these law courts and told citizens when it was time for a session to begin or adjourn. However, the Herophilus anecdote may tell us something about the wider temporal landscape through which Alexandrian physicians moved—namely, one in which they were regularly exposed to Egyptian water clocks that reckoned hourly time.

On the basis of these textual references and given what we have learned about Egyptian and Babylonian medical timekeeping practices, it is not unreasonable to suppose that the Mousaion of Alexandria may have served as one of the interdisciplinary, cross-cultural crucibles in which Greek-speaking physicians first forged their own hourly timekeeping habits. Physicians at work in Alexandria would have had exposure not only to Egyptian-style water clocks, but also to Egyptian-style sundials and to the "equal hours" with which Babylonian astronomers organized their tables. However, the absence of hourly timekeeping evidence from other Hellenistic communities may simply be the product of differential excavation and preservation. Thus, we can do no more than speculate as to the uniqueness of Alexandria's role in horological transmission and exchange. Thankfully, though, our evidence for hourly timekeeping in the Roman period is more abundant and informative. Before we turn, in the succeeding chapters, to the textual evidence for medical timekeeping during this period, the last section of this chapter will offer a brief overview of how and where Imperial-period physicians and patients might have encountered clocks.

The Roman Period

By the Roman period, physicians would have had access to the full range of hourly timekeeping technologies introduced in the previous chapter.[39] This meant that,

---

[39] According to Webster (2023, 270), the Roman period "evince[s] a culture of innovation surrounding the material implements of medicine." Medical timekeeping tools ought, I submit, to be included alongside specialized surgical tools and the like as part of this culture.

broadly speaking, their interactions with clocks could come in three different flavors. On the one hand, a physician could carry with him his own portable sundial, such as the one discovered in the "tomb of the physician" near Este. In order to own and use such a dial, the physician would need to be able to afford it, be well-versed in its operation (or possessed of a slave or assistant who could perform this task), and be able to hang the dial nearby in a position where it could catch the sun's rays appropriately. This sort of tool would not have been useful at night or in cloudy weather, which raises further questions as to how the physicians who refer, in their writings, to numbered hours of the night were able to determine the time with such precision.

Instead of or in addition to a portable sundial, a physician attending a wealthy client, either in their urban home or in their rural villa, might have been able to avail himself of his patient's domestic sundial. Such a dial would likely have been fixed in a garden or courtyard where it could receive maximum light, making it inconvenient to consult from the patient's sickroom. Thus, a physician relying on this timekeeping method would probably have had to send one of his assistants or a member of his patient's household staff outside to read the dial and report back the time. The success of this enterprise would, in turn, depend on the ability of these emissaries to correctly read the sundial and on the amount of available light. Wealthy clients may also have been able to boast their own water clocks, as does the character of Trimalchio in Petronius's *Satyricon*.[40]

That reporting the time of day was the duty of at least certain domestic slaves is attested in the Augustan-era poetry of Juvenal and Martial. Juvenal, for instance, in a satire on the woes of old age, paints a vivid picture of an elderly man who is hard of hearing: "There is need of shouting in order that he might perceive with his ears whom the slave says has come and how many hours he announces."[41] This passage suggests that, at least among the wealthier slave-owning classes of Rome, an enslaver might expect to hear the time of day reported by an enslaved member of his or her household. Martial's epigram to Caecilianus supports this idea, even using the same words for "slave" (*puer*) and "announce" (*nuntiat*) that we see in Juvenal: "Your slave has not yet announced to you five hours." Martial, however, goes on to describe another way in which an elite Roman man might have learned the time of day: "yet you, Caecilianus, have already come to dine with me; even though hoarse [voices] have just announced the time to adjourn the bail courts at the fourth hour."[42] These lines seem to imply that there were also individuals in public service, perhaps public slaves, who were given the task of first observing

---

[40] Petron. *Sat*. 26. Other testimonia to the private use of water clocks among the wealthy include Dig. 33.7.12 section 23; Pliny, *Ep*. 3.1.8; Sid. Apoll. *Epist*. 2.9.6 and 2.13.4.
[41] *Sat*. 10.215–16: *clamore opus est, ut sentiat auris / quem dicat venisse puer, quot nuntiet horas*.
[42] *Epig*. 8.67.1–4: *horas quinque puer nondum tibi nuntiat, et tu / iam conviva mihi, Caeciliane, venis, / cum modo distulerint raucae vadimonia quartae / et Floralicias lasset harena feras*.

and then shouting out to listening bystanders the times of day that were important for the coordination of judicial activities.

A terracotta statuette (Figure D), discovered in the necropolis of Myrina and provisionally dated to the Hellenistic period, may also support the notion that slaves were frequently responsible for reading and announcing the time of day. This statuette depicts a male slave, clad solely in a loincloth, leaning against a pillar that supports a representation of a conical sundial. As we will see in Chapter 4, sundials appear frequently in the funerary iconography of Roman elites, and this statuette may have been intended to depict an enslaved person grieving at the hour of his master's passing. However, the artist's specific decision to represent a grieving *slave* alongside the sundial—rather than, for example, a grieving member of the deceased's immediate family—may reveal an underlying assumption that reading sundials was considered, at least in certain contexts, to be the job of an enslaved person. Thus, "hourly timekeeping" should be added to the list of specialized skills (like writing, accounting, etc.) that could be acquired by people enslaved within wealthy households. This material also raises broader questions, largely outside the scope of the present volume,[43] about how enslaved persons experienced day-to-day time and engaged with timekeeping devices in ways that differed from their enslavers.[44]

Martial's testimony suggests that, if a doctor was working in a sufficiently urban environment—particularly, in one of the hubs of the Roman Empire, such as Pergamon, Athens, Alexandria, or Rome—he need not have relied either on himself or on his patients to supply a timekeeping device. Instead, he could have personally consulted, or listened for the time to be announced from one of the monumental sundials or water clocks that adorned many a marketplace, healing sanctuary, bath-house, or gymnasium—each of these a prime location for a doctor to attract or treat a patient. The convenience of this solution would have depended largely on where the physician was performing his consultations and the proximity of that place to a monumental clock. It would also have depended on how, exactly, these monumental clocks communicated the time. If passersby were expected to read the hour from the clock's "face," as it were (i.e., by searching for the place where the gnomon's shadow or the water clock's pointer intersected with the net of hour and date lines), then the ease of doing so would depend very much on the size and legibility of the clock's markings and on the proficiency of the viewer, or his or her attendants, at reading clocks.[45] If, on the other hand, the "striking" of the hour was also announced aurally—either by a

---

[43] Though I touch upon it briefly in Chapter 7.

[44] For a discussion of how slaves in the American South had different experiences of "clock time" than their enslavers, see M. M. Smith 1996.

[45] Traces of red paint on some extant sundials suggest that their nets of hour and date lines were sometimes painted to enhance visibility. One early example is MANN: 2541 (A. Jones 2016, fig. III-13), a conical dial found in the Stabian Baths at Pompeii and dated to c.100 BCE.

Figure D Statuette of grieving slave with sundial, Hellenistic (?), terracotta, Necropolis, Myrina, Greece, H. 18.7 cm, National Archaeological Museum, Athens: 5007/D.95. Image © Institute for the Study of the Ancient World, photographer Orestis Kourakis.

mechanism, such as a bell or whistle, or by a person, such as a public slave—then those interested in learning the time would have required less expertise with the technology and, assuming that their hearing was not impaired and that ambient noise levels were sufficiently low, they could have discovered the hour from farther away.

To learn with greater precision when, how, and why individual healers made use of clocks, we must turn to our literary sources. Occasional references to numbered hours pop up in the medical writings of several Roman-period authors, including Soranus, Celsus, and Rufus of Ephesus, but our only evidence for the *systematic* use of this technology in medical contexts comes from the corpus of Galen. Fortunately, Galen offers us some insight not only into the contexts in which *he* felt hourly timekeeping to be important, but also into the contexts in which *other doctors* (often frustratingly unnamed) considered this tool to be relevant. Thus, the rest of the present book will perform close readings of Galenic texts in an effort to reconstruct the temporal landscape of the medical world in which he moved. It will also contextualize themes within Galen's works by putting them into conversation with larger social and cultural trends under the Roman Empire.

To prepare ourselves for our encounters with Galen, however, it is salutary to consider two texts from the Roman period written by individuals who were not themselves medical experts but, for various reasons, were interested in portraying the experiences of a certain (or a certain kind of) medical patient. We will come to see that these texts—one an autobiographical account by Aelius Aristides of his personal struggles with illness and treatment, the other a caricature by the poet Juvenal of a superstitious woman—provide interesting complements and foils to the perspectives on medical timekeeping that emerge from Galen's writings. By putting Galen's texts into conversation with texts like these, we will be better able to appreciate both the ways in which Galen and his rival physicians were men of their times, who internalized and reflected certain elements of their sociocultural environments, and also the ways in which they innovated, pushing the boundaries of those environments and imagining new uses for sundials, water clocks, and the unit of the hour within medical contexts.

The physician Galen and the patient Aelius Aristides had much in common. Aristides, who was born in 117 and died in 181 CE, was only twelve years' Galen's senior and came from a similarly privileged socioeconomic background.[46] Born the son of a wealthy landowner in northwest Asia Minor (not far from Galen's hometown of Pergamon), Aristides too traveled to places like Smyrna and Alexandria to complete an expensive education. But whereas Galen's studies focused on medicine and philosophy, Aristides was especially interested in

---

[46] There is a rich bibliography on Aelius Aristides. Some important works include Behr 1969; Harris and Holmes 2008; Israelowich 2012; Downie 2013.

rhetoric, and when he arrived in Rome in 143 CE, he intended to seek his fortune as an orator. Ultimately, he succeeded in earning great renown for his oratorical skills, winning the favor of the very same emperor, Marcus Aurelius, who first elevated Galen to the status of imperial physician. However, between his arrival in Rome and his peak success there as an orator in the mid-170s CE, Aristides's career trajectory was repeatedly interrupted by bouts of severe illness, which inspired him to turn with great devotion to the Greco-Roman god of healing, Asclepius/ Aesculapius. In the 140s CE, Aristides even made a pilgrimage to the great sanctuary of Asclepius in Galen's home city, where Aristides, like other suppliants, slept (or "incubated") within the temple precinct in the hope of receiving visitations and guidance from the god himself within his dreams. Aristides's *Sacred Tales* (or *Hieroi Logoi*) is an autobiographical account of his journey from sickness into health, structured as an almost-daily diary of his symptoms, the dreams he received from Asclepius, the medical instructions he received from other physicians, and how he responded to various treatments. While the narrative style of the *Sacred Tales* borrows, in true Second Sophistic fashion, from a variety of Classical genres (including, among others, epic poetry and the Hippocratic case histories), there is nothing else quite like it among our extant material. Hence, the *Sacred Tales* offers us an invaluable glimpse into the behavior and perspectives of an elite male patient under the Roman Empire—including his approaches, at least within this specific work, to timekeeping and time-indication within the day.

Numbered hours appear at several points within the *Sacred Tales*, suggesting that Aristides was accustomed to consulting clocks within his environments. However, he does not use them systematically to pinpoint, for example, the duration of a symptom or the time at which a treatment was administered. For such purposes he, like the Hippocratic authors examined earlier in this chapter, tends to use more general temporal terms, such as "morning," "evening," or "midday." He notes, for instance, "After this, at night, stomach trouble, which reached such a pitch that it scarcely stopped a little before midday," and says, "I bathed at evening, and at dawn I had pains in my abdomen."[47] In certain contexts—as, for example, when recounting his experience of a fever—Aristides betrays particular concern for temporal markers, though even in such cases he avoids using hours to structure his narrative:[48]

---

[47] *Hier. log.* 1.8: μετὰ δὲ τοῦτο λελοῦσθαί γε καὶ τῆς γαστρὸς φαύλως ἔχειν, καὶ φάναι πρὸς τὸν Ζώσιμον, ἔδει γὰρ ᾐσιτηκέναι; καὶ τὸν εἰπεῖν, ἔδει· ᾐσίτησα αὖθις. καὶ τῆς ἐπιούσης ᾔμουν πάλιν εἰς ἑσπέραν. Tr. Behr (1969) with modification.

[48] *Hier. log.* 3.16–20. συσπειραθεὶς δὲ ὡς εἶχον εἰσῇξα εἰς τὸ δωμάτιον οὗ κάθευδον, καὶ κατακλίνομαι ὅντινα δὴ τρόπον, καὶ πῦρ ἐπιγίνεται πολὺ καὶ σφοδρόν.... καὶ δὴ μετὰ τοῦτο ἡλίου τε ἦν δύσις, ἢ καὶ ἔτι πορρωτέρω, καὶ ἐπιγίνεται σπασμὸς ἐπὶ τῷ πυρετῷ.... ἀλλ' ἦν ἐπὶ πλεῖστον ἡ κίνησις καὶ ἡ τῶν ὀδυνῶν κατάτασις ἄρρητος, οὔτε σιωπᾶν ἐῶσα καὶ πρὸς τὴν φωνήν ἔτι μεῖζον ἀπαντῶσα. Ταῦτα δ' ἐγίγνετο καὶ εἰς μέσας νύκτας καὶ πρόσω μηδὲ μικρὸν ἀνιέντα· ἔπειτα τῆς μὲν ἀκμῆς ἀφείλέ τι, οὐ μὴν παντάπασί γε ἐπαύσατο.... καὶ πρὶν ἡμέραν εἶναι, ἔθει τις ἰατρὸν καλῶν. ἥκεν εἴτε δὴ τῆς ἐπιούσης εἴτε καὶ ἡμέρᾳ ὕστερον. καὶ μεσημβρίας μοι δοκεῖν πάλιν προσβολή τις προσῄει.... νύξ τ' ἐπιγίνεται, οἷα δὴ ἐν κλυδωνίῳ καὶ σάλῳ, καὶ κατέδαρθον σχεδὸν ὅσον εἰς ὄναρ. Tr. Behr (1969) with modification.

> Doubled up, as I was, I rushed into my bedroom, and lay down in some fashion, and a great and strong fever arose.... And after this, when the sun set, or even still later, a convulsion was added to the fever.... Mostly there were convulsions and the unspeakable state of pain, which did not let me be silent, and ever more ended in screams. These things continued up to midnight or later and abated not a bit. Then it lost some of its force; however, it did not wholly stop.... And before it was daytime, someone ran to summon the doctor. He came, either on the following day or even a day later. And at midday, I think, I again had an attack.... Night arose, as in a pitching, rolling sea, and I slept enough to dream.

We will see (particularly in Chapters 5 and 6) that Galen likewise viewed febrile illness as a circumstance in which careful timekeeping was particularly important. Unlike Aristides, however, Galen enlists hourly timekeeping in his fever narratives.

When Aristides does use numbered hours to indicate time in the *Sacred Tales*, he does so within very specific contexts: namely, when discussing activities that take place in bath-houses and in law courts. We will focus on the former. Aristides describes one bath-house visit in the following terms: "I decided upon the sixth hour [for my visit], as then it is safest to move about. When the hour came, I accused Bassus of procrastinating. 'You see,' I said, 'how the shadow is passing by?' indicating the shadow of the columns."[49] While neither a sundial nor a water clock is specifically mentioned here, this passage tells us not only that the precise timing of his bath was important to Aristides, but also that he used the position of a shadow (in this case, a column's rather than a gnomon's) to indicate the degree of his tardiness in bathing.[50] Even greater temporal precision can be seen in a later passage, in which Aristides describes a dream he received about the timing of his bath:[51]

> Two doctors came and at the doorway discussed other things, and, I believe, cold baths. One inquired, and the other answered. "What does Hippocrates say?" he said. "What else than to run ten stades to the sea, and then jump in?"... After this, the doctors themselves, in fact, came in, and I marveled at the precision of

---

[49] *Hier. log.* 1.21: οὕτω δὴ συνθέσθαι εἰς ὥραν ἕκτην ὡς τηνικαῦτα ἀσφαλέστατον ὂν κινεῖσθαι· ἐπεὶ δὲ ἀφίκετο ἡ ὥρα, αἰτιᾶσθαι Βάσσον ὡς διατρίβοντα. ὁρᾷς, ἔφην, ὡς καὶ δὴ παρέρχεται ἡ σκιά; δεικνὺς τὴν τῶν κιόνων. Tr. Behr (1969).

[50] A similar concern about bathing outside of the appropriate time (in this case, about bathing too early) is evidenced at *Hier. log.* 1.34: "On the seventh, I dreamed that I saw, in some dressing room of a bath, the orator Charidemus, from Phoenicia, shining and just bathed, and I said to him, with my greeting, that he bathed too early, and at the same time I also undressed myself."

[51] *Hier. log.* 5.49–51: ἡκέτην ἰατρὼ δύο καὶ διελεγέσθην ἐν τῷ προθύρῳ ἄλλα τέ μοι δοκεῖν καὶ περὶ ψυχροῦ λουτροῦ ἠρώτα μὲν ὁ ἕτερος, ὁ δ' ἀπεκρίνετο, τί λέγει, ἔφη, Ἱπποκράτης; τί δ' ἄλλο γε ἢ δραμόντα δέκα σταδίους ἐπὶ θάλατταν οὕτως ῥῖψαι;...μετὰ δὲ τοῦτο ἐπελθεῖν ὡς ἀληθῶς αὐτοὺς τοὺς ἰατρούς, θαυμάσαι τε δὴ τοῦ ἐνυπνίου τὴν ἀκρίβειαν καὶ πρὸς αὐτοὺς εἰπεῖν, ἄρτι γε ὑμᾶς ἐδόκουν ὁρᾶν καὶ ἄρτι ἥκετε.... ἐρομένου δὲ τοῦ ἑτέρου τῶν ἰατρῶν περὶ τῆς ὥρας τοῦ λουτροῦ, φάναι ὅτι πέμπτην δέοι κινηθῆναι, συμβήσεσθαι γὰρ εἰς τὴν ἕκτην· πάλιν γὰρ αὖ τὸ πρῳαίτερον τοῦ δέοντος ἢ λοῦσθαι ἢ ἐσθίειν ἐργῶδες εἶναι· οὐδὲ γὰρ εἰς αὐτὸ τοῦτο φέρειν ὃ δοκεῖ αὐτοῦ χρήσιμον εἶναι, τὸ πέψαι ῥᾷον, ἀγρυπνίας γὰρ ἀπ' αὐτοῦ γίγνεσθαι. Tr. Behr (1969) with modification.

the dream, and said to them, "Just now I dreamed that I saw you and just now you have come...." When one of the doctors asked about the time of the bath, I said that it was necessary to get going at the fifth hour of the day, for it would take place at the sixth. I added that it is troublesome either to bathe or to eat earlier than is suitable, for it does not contribute to that feature which seems to be useful in it, to have an easier digestion, for insomnia comes from it.

We will see, particularly in Chapter 7, that Galen too believed it was important to consider the timing of one's bath relative to the timing of one's meals. This idea, articulated already in the Hippocratic Corpus, seems to have been in wide circulation by the second century CE and may have informed many individuals' decisions about when to visit the bath-house. In any case, these passages from the *Sacred Tales* suggest that the chronotope of the bath-house was particularly attuned to clock time.

Epigraphic and archaeological evidence from Greek and Roman bath-houses and gymnasia support this understanding. Many sundials and water clocks of varying sizes have been found in such locations, and inscriptions tell us that they could be used, for example, to regulate visiting hours,[52] to keep young athletes punctual,[53] and to help responsible parties know when to supply oil.[54] The proliferation of clocks in bath-houses and gymnasia may have been due to the fact that the activities performed in such spaces required a high degree of temporal coordination: athletes needed to meet up with their trainers and masseurs, and bathers wanted to convene with their friends. It is also possible, though, that the chronotopes of bath-houses and gymnasia, which were sites of healing and personal care, actually developed in dialogue with theories about medical time-keeping. In other words, as healers came to emphasize the importance of bathing, exercising, and getting massaged at the right time (*kairos*) within the day, sundials and water clocks may have been viewed increasingly as essential bath-house and gymnasium apparatus.

Two inscriptions, one from Akades on Crete and one from Aljustrel in what is now Portugal, also suggest another motivation for integrating clocks into the bath-house experience which may have affected the timing of treatments for different patients.[55] That motivation was to segregate female bathers from their male counterparts. The *Lex Metalli Vipascensis*, for instance, tells us that the renter of the bath-house or his associate ought to "place [the bath] at the disposal of women from first light to the seventh hour of the day and at the disposal of men from the

---

[52] *SEG* 26: 1044. Cf. *CIL*² 05181, 19–23.
[53] One humorous graffito from an Ephesian gymnasium's toilet warns the reader, "On time or death (τὴν ὥραν ἢ τὸν θάνατον)!" (*IEph* 561.1). On gymnasium punctuality, see also *SEG* 27: 261.
[54] For the presence of clocks in gymnasia: *ID* 1417; *MDAI(A)* 32 (1907) 257–8. On the provision of oil: *IG* V¹ 1390; *IG* X² 2 323; *IG* X² 2 3261 [= *SEG* 28: 680]; *IPriene* 112.
[55] *CIL*² 05181 and *SEG* 26: 1044.

eighth hour until the second hour of the night."[56] It is likely that the bath-house in question was too small to accommodate separate rooms for male and female bathers, a problem which was solved by segregating the sexes temporally rather than spatially. This inscription demonstrates how hourly timekeeping could be used to reinforce gender divisions and power dynamics through the creation of different daily schedules for men and women—ones that reserved for male bathers the choicest and most stimulating windows of access.[57] But how representative might the opening hours of this bath-house have been? To date, I am aware of only one other inscription that specifies separate bathing times for men and women, this one from Cretan Akades.[58] Unfortunately, the maker of this inscription left the exact bathing hours blank, leaving us no way to compare them with those of the Aljustrel bath-house. Thus, we must be careful about generalizing from the Aljustrel example and assuming, for instance, that all small bath-houses segregated their opening hours, and by extension, that physicians would have recommended different bathing hours for their male and female patients, hours that perhaps reflected the relevant access windows available to each patient at his or her preferred bath-house.

Let us return, however, to Aelius Aristides. It appears that even though Aristides, like other members of the second-century urban elite, had abundant access to sundials and water clocks, his decisions about when to incorporate hourly timekeeping into his narrative of healing were largely informed by chronotopic norms. When describing events that took place in bath-houses and council chambers—places where, as we have seen, clocks were readily accessible and where it had become customary to talk about time in terms of numbered hours—Aristides not only uses such hours, but even betrays an especial concern

---

[56] *CIL*² 05181, 20–1: *praestare debeto a prima luce in horam septim[am diei mulieribus et ab hora octava] / in horam secundam noctis viris....*

[57] Note that, according to this inscription, women could only visit the bath during the morning, when most men would still have been engaged in labor or business. By the time that many men would have been at liberty and keen to unwind with a social bath, we might imagine that most women were expected to be back at home attending to their family's dinner preparations. It is probably significant that this inscription reserves the darker hours of the late evening and early night exclusively for male bathers. Perhaps this injunction was meant to "protect" men's wives and daughters from the "high-risk" activities of walking to and from the bathhouse in the dark and, once there, bathing naked by lamplight in view of other men. A similar motivation likely underlay Augustus's decision, reported by Suetonius (*Aug.* 44.2) to prevent women from attending the theater prior to the fifth hour of the day, so that they could be shielded (from the violence? from the male nudity?) of certain wrestling matches: "He excluded however, the whole female sex from seeing the wrestlers: so that in the games which he exhibited upon his accession to the office of high-priest, he deferred producing a pair of combatants, which the people called for, until the next morning. And he intimated by proclamation, his pleasure that no woman should appear in the theatre before the fifth hour." Tr. Thomson 1889 with modification.

[58] *SEG* 26: 1044, dated to the Imperial period. The relevant portion reads: "that the women bathe from the ___ hour until the ___ hour, and that the men [bathe] from the ___ hour" (6–8). This appears to be a template that an individual bath-house operator could fill in according to his needs. That templates like this were available may suggest that, at least in this part of Crete, there was some demand for signs indicating bath-house opening hours according to sex.

for punctuality.[59] But in other contexts, such as describing the sequence and duration of his symptoms or the timing of his treatments, he is satisfied with looser temporal language that does not imply the use of a clock. During our upcoming adventures in the Galenic corpus, we will see that Galen and his rival physicians share several of Aristides's assumptions, while also suggesting new ways in which clocks might be incorporated into medical thought and practice.

Juvenal's portrayal, in Satire 6, of a superstitious woman offers us a rather different perspective on how patients seeking treatment might engage with hourly timekeeping. Both this text and the woman it describes contrast in significant ways with Aelius Aristides and his *Sacred Tales*. While the voice we hear in the *Sacred Tales* is that of the author's own male, elite persona, the patient we encounter in Satire 6 is a caricature, drawn by an elite male author, of a woman of uncertain socioeconomic status who is clearly literate but, in the author's view, somewhat irrational. Juvenal paints her portrait for us as follows:[60]

> But beware of ever encountering one whom you see clutching a well-worn calendar in her hands as if it were a ball of clammy amber... [I]f there is a sore place in the corner of her eye, she will not call for a salve until she has consulted her horoscope: and if she be ill in bed, deems no hour so suitable for taking food as that prescribed to her by Petosiris.

Here we see a patient who, rather than consult a physician or visit a temple of Asclepius, prefers to seek medical help by astrological means. In the Hellenistic and Roman periods, two forms of astrology made their way from Mesopotamia and Egypt to the Greek- and Latin-speaking worlds, where they became wildly popular. These are horoscopic astrology, which predicts certain aspects of an individual's life experience based on the alignment of significant asterisms at the time of his or her birth, and catarchic astrology, which is used to determine the most auspicious and inauspicious moments within the day and week for performing various activities.[61]

Underlying catarchic astrology was the idea of the "chronocrator" or "time-ruler." Within this framework, each of the seven days of the astrological week (the precursor, by way of Norse mythology, to our own seven-day week) was "ruled" by a deity that corresponded to the Sun, Moon, or one of the five planetary bodies

---

[59] For an example of his use of hours in reference to the council chamber, see *Hier. log.* 5.30-2.

[60] *Sat.* 6.572–81: *illius occursus etiam vitare memento, / in cuius manibus ceu pinguia sucina tritas / cernis ephemeridas.../... si prurit frictus ocelli / angulus, inspecta genesi collyria poscit; / aegra licet iaceat, capiendo nulla videtur / aptior hora cibo nisi quam dederit Petosiris.* Tr. Ramsey (1928).

[61] On Greco-Roman astrology, see Barton 1994; A. Jones 1999; Heilen 2020; and esp. Greenbaum 2020c. On its Babylonian precedents and their evolution into the Hellenistic period, see Rochberg 2020a. For our purposes here, it may also be noteworthy that, during the Hellenistic period, "Berossus, a priest of Bēl, is said to have founded a school on the isle of Cos, where Babylonian tradition was translated for a Greek audience" (Hübner 2020, 332).

recognized in antiquity (Mercury, Venus, Mars, Jupiter, and Saturn). Likewise, each seasonal hour within a given day had its own "ruler." The ruler of the first hour was, by definition, the chronocrator of that day, and the rulership of each successive hour cycled through the remaining astrological deities in a predetermined sequence (usually progressing in descending order of the celestial bodies' positions in the cosmos: Saturn, Jupiter, Mars, Sun, Venus, Mercury, Moon).[62] Every seven hours, a new cycle would begin, starting with the chronocrator of that day. Some of these celestial bodies were understood to exert a benefic influence during their periods of chronocratorship, making the days and hours that they governed particularly auspicious times. Other celestial bodies, meanwhile, exerted a malefic influence, making the days and hours of their chronocratorships particularly inauspicious. I suggest that, in the Imperial period, the idea of the astrologically auspicious (or inauspicious) moment had come to influence the concept of a *kairos* ("opportune moment"). Regardless of whether one views time within the day through an astrological or a "kairotic" lens, time ceases to be a neutral backdrop against which actions happen to take place and becomes, again, more like an active gameboard, within which each individual player must think carefully about how to optimize the timing of his or her activities.

Sundials and water clocks, which enabled individuals to position themselves upon this temporal gameboard more precisely, were thus essential tools not only for the physician interested in specifying medical *kairoi* with numerical exactitude, but also for healers and patients with an astrological view of medicine. The woman that Juvenal describes in Satire 6 relies on both horoscopic and catarchic astrology to determine the right moment to apply a topical treatment or to eat a meal of prescribed foods. To aid in her calculations, she is said to consult "a well-worn calendar," a series of tables that would have permitted her to keep track of the chronocrator of each day and hour over the course of the seven-day week. The epigraphic record preserves several stone fragments of such calendars,[63] and it is likely that calendars written on more perishable material, like papyrus scrolls, were in broad circulation.[64] Juvenal's superstitious woman, then, allows us to understand a particular kind of patient experience, one where the patient's astrological mindset makes him or her more keenly aware of clock time, used specifically for prescriptive and proscriptive purposes: the clock, as a reflection of the current state of astrological influence, indicates when it is or is not appropriate to perform an action. Such patients seem to have been less interested in using clocks for *descriptive* purposes—for example, to gather data on when, and for how long, their symptoms manifested. In what follows, we will see that physicians like

---

[62] Heilen 2020, 243.
[63] E.g., *CIL* 9.4769 and 5808. On this latter inscription, see Heilen 2020, 244–5.
[64] For both literary and archaeological evidence for such astrological calendars, see Heilen 2020, 244 n. 19.

Galen used hourly timekeeping for prescriptive, proscriptive, *and* descriptive purposes, seeing clocks as tools not only for translating theory into practice, but also for generating new theories on the basis of time-stamped observations.

With certain exceptions, which we will discuss in Chapter 5, Galen himself eschews astrological explanations for medical phenomena. Yet many healers among his rough contemporaries incorporated elements of astrological thinking into their medical practice. A fragment attributed to Thessalus, for example, shows how this first-century CE physician applied the doctrine of chronocratorship to the task of harvesting medicinal herbs:[65]

> If, however, you do not have on hand plants from such *klimata*, and you will be compelled to take them from other places, see that you take each of these in accordance with one of the days of the week, whichever one is of that star to which the plant also belongs, but also in accordance with the hour of that day which belongs to the same star, as will be set out by us here. Pulling this up from the earth, utter the associated prayer and after taking it, hurl into the hole in which the plant was found a measure of grain or barley. The days of the planets are these:
>
> Day 1 belonging to Helios and hour 1: its plants are *kichorion* and knot-grass
>
> Day 2, hour 1: of Selene. Peony and *kunobate*.
>
> Day 3, hour 1: of Ares. *Peudekanon* and plaintain.
>
> Day 4, hour 1: of Hermes. Sage and cinquefoil.
>
> Day 5, hour 1: of Zeus. *Sancharonion* and agrimony.
>
> Day 6, hour 1: of Aphrodite. Panacaea and holy vervain.
>
> Day 7, hour 1: of Kronos. Houseleek and asphodel.
>
> In addition to these things, also let Selene be full or when she is going through the zodiac of the planet of the day.

Texts like this remind us that, while Galen himself rarely appeals to astrological principles, he worked in an environment in which many healers did. Doctors like Galen, who offered purely physical explanations and solutions for medical

---

[65] *De virt. herb.* 2, prol. 1.1–20. Friedrich 1968: Εἰ μέντοι οὐκ εὐπορεῖς βοτανῶν ἀπὸ τῶν τοιούτων κλιμάτων, διὰ δὲ χρείαν εἰς ἀνάγκην ἔρχῃ λαμβάνειν ἀφ' ἑτέρων τόπων, σκόπει λαμβάνειν ἑκάστην τούτων κατὰ μίαν τῶν τῆς ἑβδομάδος ἡμερῶν, ἥτις ἐστὶ τοῦ ἀστέρος ἐκείνου, οὗ ἔστι καὶ ἡ βοτάνη, ἀλλὰ δὴ καὶ κατὰ τὴν ὥραν τῆς ἡμέρας ἐκείνης, ἥτις ἐστὶ τοῦ αὐτοῦ ἀστέρος, καθὼς ἐκτεθήσονται καὶ παρ' ἡμῖν ἐνταυθοῖ. ἀνασπῶν δὲ ταύτην ἐκ τῆς γῆς ἐπίλεγε καὶ τὴν ἐγκειμένην εὐχὴν καὶ μετὰ τὸ λαβεῖν ῥίπτε εἰς τὸν λάκκον ἐν ᾗ εὐρίσκεται ἡ βοτάνη κόκκον σίτου ἢ κριθῆς. Εἰσὶ δὲ τῶν πλανητῶν αἱ ἡμέραι αὗται·/ ἡμέρα α' ἤτοι κυριακὴ Ἡλίου καὶ ὥρα α', βοτάναι δὲ αὐτῆς κιχώριον καὶ πολύγονον / ἡμέρα β', ὥρα α', Σελήνης, ἀγλαοφάντον καὶ κυνοβάτη / ἡμέρα γ', ὥρα α', Ἄρεως, πευδέκανον καὶ ἀρνόγλωσσον / ἡμέρα δ', ὥρα α', Ἑρμοῦ, φλόμος καὶ πεντάφυλλον / ἡμέρα ε', ὥρα α', Διός, σαγχαρώνιον καὶ εὐπατόριον / ἡμέρα ϛ', ὥρα α', Ἀφροδίτης, πανάκεια καὶ περιστερεών / ἡμέρα ζ', ὥρα α', Κρόνου, ἀείζωον καὶ ἀσφόδελος / πρὸς δὲ τούτοις ἔστω καὶ ἡ Σελήνη πανσέληνος ἢ ὅτε διέρχεται τὸ ζῴδιον τοῦ πλάνητος τῆς ἡμέρας.

problems, were regularly exposed to, and had to compete with, healers whose methods accounted for divine and/or celestial influence. The popularity of astrology in the Roman period—as well as the extensive archaeological and textual evidence for amulets, prayers, esoteric rituals, etc. that were aimed at healing—suggests that many patients were attracted to metaphysical interpretations of and responses to disease. In order for physicians like Galen to draw prospective clients to themselves and away from metaphysically informed healers, they needed to persuade those clients that their approaches to medicine were even more effective. In the case studies we are about to investigate, we will see how Galen adapts a core tool of astrological medicine, the clock, as well as one of its core concepts, that of the "auspicious" moment, to support his own non-metaphysical theories of disease, health, and scientific methodology, as well as to present himself, above rival healers of all persuasions, as the ideal doctor.

3
Clocks as Symbols 1
In Galen's Thought

In the present chapter and those that follow, we will investigate in greater detail how one prominent physician of the Roman Imperial period, Galen of Pergamon, interacted with these timekeeping tools and concepts. Chapters 3 and 4 will focus specifically on Galen's discussions of clocks (both sundials and water clocks), while Chapters 5, 6, and 7 will examine his use of numbered hours. Through close readings of a series of Galenic treatises, I hope to illuminate the complex ways in which Galen's thoughts about clocks and hours intersected with his other ideological commitments and with wider trends that formed the social and intellectual architecture of his world. I assert that Galen's writings offer us important windows not only onto his own timekeeping views and practices, but also onto those of a variety of other people: namely, physicians, philosophers, and laymen among Galen's predecessors and contemporaries. But first: who was Galen himself?

Galen's Chronotopic Biography

Galen's *curriculum vitae* has been rehearsed numerous times in recent scholarly literature.[1] Therefore, this brief overview will focus particularly on the various chronotopes through which Galen would have moved during his life. This will help us to begin thinking about how the contexts in which Galen encountered clocks and hours might have affected the ways that he integrated them into his medical thought and practice.

Galen was born in 129 CE on the northwestern coast of modern-day Turkey, in the bustling city of Pergamon. This city was home both to a renowned sanctuary honoring the healing god Asclepius and, by Galen's time, to a number of sundials and water clocks in a range of sizes.[2] Thus, from his earliest days, Galen was navigating environments in which clocks were prevalent and healing was a prominent industry. It is even possible that his father was in the business of making

---

[1] On Galen's biography, see, e.g., W. D. Smith 1979, 62–72; Pearcy 1985; Barnes 1991, 54–5; Gourevitch 1999, 118–20; Schlange-Schöningen 2003; Hankinson 2008b; Mattern 2013; Nutton 2013, 222–35; 2020.
[2] Berlin Sundial Collaboration, Ancient Sundials, Dialface IDs 70, 71, 154, 155, 303, 594, 595, 698 (Graßhoff et al. 2015).

clocks. Galen describes his father as someone with expertise in *arkhitektonia* ("architecture" or, more broadly, "engineering"), and Vitruvius tells us that *gnomonice*, the art of clock-construction, was a sub-category of that discipline.[3] Thus, Galen may have benefitted from even more exposure to the processes of clock-construction and the ins and outs of hourly timekeeping than a typical man of his day.

During his adolescence and early adulthood, Galen studied broadly and traveled extensively, thereby exposing himself to a variety of individuals' and cultures' differing orientations to clock time. Galen's father, generally identified as a professional *arkhitekton*, was determined that his son should receive a well-rounded early education, particularly in logic, arithmetic, geometry, and of course, *arkhitektonia*.[4] But when Galen was a teenager, a series of dreams persuaded his father that Galen should shift his focus and pursue advanced training in the arts of medicine and philosophy.[5] This Galen proceeded to do, initially in Pergamon itself under the tutelage of figures such as Satyrus, Stratonicus, Aeficianus, and Aeschrion.[6] After his father's death in or around 149 CE, Galen continued his studies in a variety of geographical locations, including Smyrna, where he studied under Pelops and Albinus,[7] and the Egyptian city of Alexandria, where he engaged with thinkers like Heraclianus, Numisianus, and Julian the Methodist.[8] In total, Galen received about a decade of advanced education from many different tutors who hailed from a variety of cultural backgrounds and adhered to diverse schools of medical and philosophical thought. As we will see, this broad exposure ultimately helped Galen to create his own synthetic methodologies for practicing medicine, doing good science, and engaging with hourly timekeeping.

In 157 CE, at age 28, Galen returned to Pergamon where he officially began his medical practice as a physician to the city's gladiators. This post, which he held for approximately four years, would have required him to report for duty at a specific hour of the day and would have given him intimate experience with the chronotopes of performance venues. Then, around 162 CE, Galen made his first voyage to Rome, a city which not only had monumental sundials and water clocks in its *fora*, bath-houses, and sanctuaries, but also—as we will see in Chapter 4—featured a towering obelisk that acted as a *gnomon* and transformed part of the Campus Martius into a giant timekeeping device.

In Rome, Galen quickly became both famous and wealthy—certainly wealthy enough to afford a portable sundial, though he makes no explicit mention of

---

[3] Vitr. 1.3.1.
[4] Gal. *Aff. pecc. dig.* 5.40–1 K. On the possible identification of Galen's father with the "Aelius Nikon" attested in Pergamene inscriptions, see n. 42.
[5] *Ord. lib. prop.* 19.59 K; *MM* 10.609 K; *Praen.* 14.608 K.
[6] Satyrus and Ephicianus were Rationalists, Stratonicus a Pneumatist, and Aeschrion an Empiricist. On these designations, see below, Galen's Scientific Method. On Galen's education, see Nutton 2020: 10–22.
[7] Pelops wrote commentaries on Hippocratic texts, while Albinus was a philosopher in the Platonist tradition (Johnston and Horsley 2011, p. xvii).
[8] Numisianus was Pelop's teacher and the father of the anatomist Heraclianus. See Nutton 2020: 19.

owning one. Upon his arrival in the city, Galen succeeded in attaching himself to the renowned Peripatetic philosopher Eudemus and impressing the equally famous Boethus with his public dissections. With the help of these elite social connections, Galen began to make such a name for himself that, despite his return to Pergamon in 166 CE, Galen was swiftly summoned back to Rome in 168 by Marcus Aurelius, the emperor himself. Ultimately, Galen was appointed personal physician to several members of the imperial family. Because of this privileged position, Galen would often have worked in wealthy patients' well-appointed homes—the kinds most likely to boast domestic sundials and water clocks—and would have been attuned to the rhythms of elite daily life, which typically involved bathing and eating at specific hours.

During his time with the imperial family, Galen was also exposed to the regimented chronotopes of military camps, where precise coordination was vital. Galen accompanied the imperial army on a campaign in northern Italy, although the soldier's life clearly did not suit him. Once the army's mission was aborted in 169, he hurried back to Rome, where he spent most of the rest of his life, produced much of his medical writing, and died, perhaps in 216 or 217 CE, in his mid-eighties.

Having made Galen's acquaintance, let us turn now to one of the few texts in which Galen mentions sundials and water clocks explicitly, rather than implicitly by referencing numbered hours.[9] This text is *On the Diagnosis and Treatment of the Affections and Errors of Each Individual Human Soul*, or simply *Affections and Errors*.[10] A close reading of this text will reveal that Galen viewed clocks not only as useful tools but also as rich, complex symbols, with resonances within the intellectual circles of mathematics, *arkhitektonia*, and astronomy, as well as within Roman society more broadly.

Introducing *Affections and Errors*

> Thus reason, after seeking by the analytical method, found a design for the water clock, the test of which is clear even to laymen. For the topmost line, which indicates the twelfth hour of the day, has the greatest height where the water clock measures out the longest day,

---

[9] The only other extant description of water clock construction, outside of Galen's treatise, is Vitr. 9.5–15, in which Vitruvius provides instructions for designing both regular and "anaphoric" water clocks.

[10] Galen refers to this work at *Lib. prop.* 19.45.10–47.11 K (= 169–171 BM), where he lists it under the heading of "Works on Ethical Philosophy." I refer to both Kühn's (= K) and De Boer's (DB) editions. De Boer's text derives primarily from the medieval Florentine manuscript *Laurentianus* 74.3 fol. 149v–171r (twelfth or thirteenth century). On the challenges of the manuscript tradition for this text, see Singer 2013, 232–6. English translations of *Affections and Errors* can be found in Harkins 1963 and Singer 2013, and the Budé series has produced a French translation and commentary (Barras et al. 1995).

and has the shortest height where it measures out the shortest day. In between the two is the line that marks the days at the equinoxes. In between the equinoctial sections on the edge of the water clock you will see the four days after those mentioned.[11]

This passage, which proceeds into a lengthy description of how to design and test a water clock, would be very much at home in the technical writings of the *arkhitekton* Vitruvius or the astronomer and mathematician Ptolemy.[12] Yet, surprisingly, these lines were penned by Galen, in a treatise devoted not to engineering or celestial geometry but rather to epistemology and ethical psychology. In fact, in *Affections and Errors*, Galen elects to walk his reader through the construction of not one clock, but two: first a sundial, and then the inflow water clock described above. Why might Galen have chosen to include passages like these in a philosophical treatise? And what, in his opinion, could clock-construction teach us about how to pursue truth and live our lives?

Over the next two chapters, we will approach these questions from two different angles. The present chapter will examine the local roles that these clock-making paradigms play within *Affections and Errors* itself by asking how they function to support the treatise's central claims. Specifically, we will see how Galen engages with clocks in ways that highlight the positive characteristics of his own method of scientific inquiry in contrast to the (in his view) unfortunate and error-riddled strategies employed by Stoics, Peripatetics, Epicureans, and other sectarian philosophers. The following chapter will contextualize Galen's decision to focus on clock construction in particular by comparing his portrayal of clocks in *Affections and Errors* with the clocks described, depicted, or produced by contemporary authors, artists, and artisans. These investigations will help us to understand what clocks might have signified within Galen's thought and how representative his thought might have been for his time.

The dating of *Affections and Errors* is problematic.[13] Since the text, as we have it, makes reference to *Character Traits*, a psychological treatise that postdates the

---

[11] *Aff. pecc. dig.* 57.10–20 DB = 5.84.16–85.10 K: οὕτω δὲ καὶ κλεψύδρας καταγραφὴν ὁ λόγος εὗρεν ἀναλυτικῇ μεθόδῳ ζητήσας, ἧς πάλιν ἡ βάσανος ἐναργής ἐστι καὶ τοῖς ἰδιώταις. ἡ γὰρ ἀνωτάτω γραμμὴ <ἡ> δωδεκάτην ὥραν σημαίνουσ᾽ ἡμέρας μέγιστον μὲν ὕψος ἔχει, καθ᾽ ὃ μέρος ἡ κλεψύδρα τὴν μεγίστην ἡμέραν ἐκμετρεῖ, βραχύτατον δὲ καθ᾽ ὃ τὴν ἐλαχίστην, ἐν τῷ μέσῳ δ᾽ ἀμφοῖν ἐστιν ἡ τὰς ἰσημερινὰς μετροῦσ᾽ ἡμέρας · τὸ δὲ μεταξὺ τῶν [ἰσημερινῶν] τμημάτων ἐπὶ τοῦ χείλους τοῦ <τῆς> κλεψύδρας δηλοῖ σοι τὰς <μετὰ τὰς> εἰρημένας τέτταρας ἡμέρας. ἀφ᾽ ὧν τμημάτων ὁρμηθεὶς τὸ μὲν ἐφεξῆς τῷ σημαίνοντι τὴν μεγίστην εὑρήσεις δηλοῦν, ἄχρι τίνος μέρους τῆς ὑψηλῆς γραμμῆς ἐπὶ τῇ κλεψύδρᾳ τὸ ὕδωρ ἀναβήσεται <τῇ ἐφεξῆς ἡμέρᾳ> τῆς δωδεκάτης ὥρας συμπληρουμένης....

[12] Both of these authors do, indeed, discuss strategies for constructing sundials and water clocks. See Vitr. 9.1–15; Ptol. *Anal.*

[13] It is also quite lacunose in some places (see Donini 1988 and Singer's response to Donini at Singer 2013, 230). The passages that we will consider in this chapter are undamaged, but in thinking about the argument of the text as a whole, it is important to recall that some material may be missing.

fire of 192 CE,[14] *Affections and Errors* has been tentatively dated to the very end of the second century CE and thus is included among Galen's later works.[15] However, the treatise's opening sentence suggests that the text may, in fact, constitute a later write-up of an earlier oral presentation; how much earlier, we cannot say.[16] We do know that Galen composed this treatise for a friend who had asked him to assess Antonius the Epicurean's theories on managing the affections of the human soul.[17]

Galen's major criticism of Antonius's work is that it fails to distinguish clearly between "affections" (*pathē*), which threaten the *irrational* part of the soul, and "errors" (*hamartēmata*), which endanger the soul's *rational* part.[18] According to Galen's definition,[19] a man who shovels half a cake into his mouth could be said to be in the grip of an "affection" or "irrational impulse" because he is yielding to a gluttonous urge.[20] An "error," on the other hand, derives from the creation of a "false opinion," and Galen seems to suggest that these come in two flavors.[21] An individual can err either by misidentifying which goal is most appropriate for him to pursue or by drawing incorrect conclusions about how to attain the appropriate goal. Thus, our cake enthusiast could be said to err if he either (a) believed that gluttony was a virtue worth pursuing (hence, wrong goal, right strategy), or (b) sought to maintain good health, but concluded that devouring cake was the best way to achieve this aim (right goal, wrong strategy).[22] In the first half of *Affections and Errors,* Galen focuses on the different types of affections that afflict human beings, and encourages his readers to band together in order to combat these impulses. He points out that men tend to be blind to their own faults but hawk-eyed when it comes to the faults of others,[23] and thus he advocates fostering a community in which men can exchange criticism freely for mutual benefit.

In the second half of the treatise, Galen shifts his attention to the mistakes of judgment that can prevent human rationality from functioning correctly, and it is in this context that we find his extended discourse on clock-construction. Galen's

---

[14] Galen kept his personal "library" in the Temple of Peace. When this building burned down in 192 CE, Galen lost many of his own works. See *Comp. med. gen.* 13.362 K; *Lib. prop.* 19.19 and 41 K. For discussion of the fire and Galen's response to it in *On Avoiding Distress*, see Tucci 2008. On the relationship between this text and *Affections and Errors*, see Singer 2013, 38–40.

[15] Singer 2013, 2, 34, 38–40.

[16] *Aff. pecc. dig.* 3.5–7 DB = 5.1.1–4 K: "You have asked me to write up, in the form of a commentary, my response to your question about Antonius the Epicurean's book, *On Guarding One's Affections.* Therefore, I shall indeed do so and thus make this beginning." This would also help to explain why Galen makes no mention of the fire in *Affections and Errors*, nor references his earlier work *On Avoiding Distress*, despite its relevance to the subject at hand.

[17] On *Aff. pecc. dig.* as a work of moral instruction, or *parainesis*, see Curtis 2014, 51.

[18] *Aff. pecc. dig.* 4.2–7 DB = 5.2.11–5.3.4 K.

[19] On the complexities of the term "affections," see Tieleman 1996, p. xxiv n. 42b.

[20] *Aff. pecc. dig.* 6.26 DB = 5.7.10 K: τιν' ἄλογον ὁρμήν.

[21] *Aff. pecc. dig.* 6.25 DB = 5.7.9 K: τὴν ψευδῆ δόξαν.

[22] On the problems inherent to Galen's distinction between affections and errors in this text, see Donini 1988, 66–9.

[23] *Aff. pecc. dig.* 6.5–10 DB = 5.6.4–5.7.8 K.

argument in this half of the treatise is two-pronged. He argues, first, that many of the beliefs which undergird the philosophical schools of his day are actually false beliefs, and thus "errors;"[24] second, that his own scientific method (to be described shortly) offers the only sure way of avoiding such errors entirely. "All the things that different [philosophical] sects do and say to contradict each other over the course of time arise on account of false judgment," Galen says. "It becomes clear that all of these are bad, and that the mistakes which pervade every sect are errors, now that someone else [namely, Galen himself] has found not only the aim but also the way of life that is in conformity with truth."[25] But what, we may ask, *is* that way of life? What methodology does *Galen* promote for distinguishing truth from falsehood and generating accurate knowledge of the world? And what does it have to do with clocks?

Galen's Scientific Method

As attested in the title of his work *That the Best Doctor is also a Philosopher*, Galen felt that the professions of medicine and philosophy were closely interrelated.[26] In fact, over the course of his extensive, interdisciplinary training, Galen became increasingly convinced that practitioners of both disciplines could learn a lot from one another's methodologies. Contemporary Skeptics, Stoics, Epicureans, Peripatetics, and Platonists, in Galen's opinion, ultimately failed in their efforts to persuade because they did not know how to subject their propositions to the appropriate logical and empirical tests[27]—this in spite of the fact that Aristotle himself had invented the categorical syllogism and the Stoics had contributed the field of propositional logic.[28] On the other hand, according to Galen, the most popular medical sects of his day erred by tending to favor *either* logic *or* empiricism, each to the near exclusion of the other.[29] He describes the Rationalists, for example, as relying too heavily on theoretical inference and the Empiricists as

---

[24] Galen's point is that a person cannot know whether a belief is true or false until he has tested it for himself.

[25] *Aff. pecc. dig.* 52.15–53.1 DB = 5.76.11–5.77.4 K: καὶ δῆλον ὅτι πολλὰ παραδείγματα τῶν ἁμαρτημάτων ὧν ἁμαρτάνουσιν ἐν αὑτοῖς γέγραπται. πάντα γὰρ ὅσα καθ' ὅλον τὸν βίον ἐναντίως ἀλλήλοις πράττουσίν [ὅσα] τε καὶ λέγουσιν οἱ ἀπὸ τῶν διαφόρων αἱρέσεων ἀναγόμενοι, κατὰ κρίσιν ψευδῆ γίγνεται καὶ πρόδηλόν γ' ἐστίν, ὡς κακῶς τε γίγνεται πάντα καὶ ἁμαρτήματά ἐστι τὰ κατὰ πᾶσαν αἵρεσιν πλημμελήματα τῆς ἀληθοῦς εὑρηκυίας οὐ μόνον τὸ τέλος, ἀλλὰ καὶ τὸν ἀκόλουθον αὐτῷ βίον.

[26] He was not the only second-century CE doctor to feel this way. See, e.g., Sextus Empiricus and Heraclitus of Rhodiapolis.

[27] Cf. *Aff. pecc. dig.* 5.70.4–7 K; 5.102.8–13 K.

[28] On Galen's relationship to philosophical traditions, see Barnes 1993; Tieleman 2008; Lloyd 2008. On the Hippocratic precedent for doctors denouncing the unfounded hypotheses of philosophers, see *Prisc. med.* 1.620 L. Cf. the attack on Melissus in *Nat. hom.* 1.22–5 L.

[29] Members of these sects would not necessarily have agreed with Galen's portrayals of them. Later Empiricists, e.g., did not see themselves as entirely avoiding the use of logic.

giving too much weight to experience and observable phenomena.[30] Then there were the Methodists, for whom Galen had nothing but contempt.[31] He presents them as rejecting both theory and observation altogether, opting instead for a hasty, one-size-fits-all method that could be mastered within six months.[32] No contemporary philosophical or medical school had, in Galen's opinion, discovered the sweet spot, that perfect marriage of rationality and empiricism which would allow practitioners to make progress in their respective arts.[33] In Galen's view, the only member of his generation to have succeeded was, of course, himself.[34]

Galen's stance was as much a competitive strategy as a conviction. As discussed in the Introduction, he lived in a world without any formal systems of accreditation, which meant that, in order to attract and retain both patients and students, healers had to rely on their reputations and their skill at self-promotion. The situation was much the same among philosophers. Aspiring students were not required to enroll exclusively in one "school" or another but were free to hop from teacher to teacher as they pleased. If a philosopher or physician wanted to outcompete his rivals, it was his responsibility to persuade students that his own system was more effective and offered greater benefits than anything else on the market. Galen hoped to be a successful competitor in both the medical and the philosophical arenas, but if he was to avoid aligning himself with a particular school of thought, he had some work to do. Galen needed to develop his own method—his own brand, as it were—and sell it with such confidence that his audience would feel compelled to buy.

Galen's scientific method relies on a *combination* of logical reasoning and observational experience,[35] as he states in the following passage from *On the Composition of Medications According to Kind*:[36]

---

[30] On the Empiricists, see Nutton 2013, 147–9. For an incisive discussion of how Rationalists and Empiricists distinguished themselves with regard to sign-inference and the role of reason in medical discovery, see Allen 2001, 89–97.

[31] Although Thessalos was not the actual founder of Methodism, Galen likes to represent him as the "refounder" of the school. See Nutton 2013, 189–94.

[32] *Sect.* 1.83 K.

[33] On other ancient medical sects, such as the Pneumatists and Eclectics, see W. D. Smith 1979; Gourevitch 1999; Nutton 2013, 207–11. For references to doctors within our epigraphic sources, see Samama 2003.

[34] Nutton describes Galen's thought-world as being dependent "on a basically Aristotelian epistemology and on a combination of data drawn from the Hippocratic Corpus and Plato and inserted into a world of Aristotelian physics" (2013, 120).

[35] On Galen's scientific epistemology, see Barnes 1991; Hankinson 1991a; 1991b; Lloyd 1996; Tieleman 1996; Lloyd 2006; Hankinson 2008a; Morison 2008b; Tieleman 2008; Hankinson 2009; Chiaradonna 2014; 2019; Hankinson and Havrda 2022. It also bears mentioning that Galen did not invent axiomatic reasoning or its pairing with empiricism, nor did he claim to have done so. Nor was he original in bridging the gap between philosophy and medicine via logical reasoning. Plato himself actually praises Hippocrates for his methodology at *Phaedr.* 270c–d. For discussion, see Tieleman 2008, 52–3.

[36] *Comp. med. gen.* 13.886–8 K: Ὅτι τῶν βοηθημάτων ἔνια μὲν ὁ λόγος εὑρίσκει μόνος, ἔνια δὲ ἡ πεῖρα, καὶ αὕτη τοῦ λόγου μὴ χρῄζουσα, τινὰ δ' ἀμφοῖν ἀλλήλοις συνεργούντων δεῖται, πολλάκις ὑμῖν

That *logos* (reason) alone discovers some remedies, *peira* (experience) without employing *logos* [discovers] others, while some [remedies] require both of them working together, we have shown on many occasions. Furthermore, the ones discovered through both *logos* and *peira* employ a conjectural path to the discovery of what is sought, where what is expected on the basis of *logos* is confirmed by *peira*.... In everything to do with the results of the art [of medicine], which is made more precise by constant practice, one needs only to learn the generalities of the case by way of *logos*, but also to be familiar with the particular cases [encountered through experience].... For this reason...I have not simply been satisfied with general theorems, but I have also written down numerous practical examples, through which I hoped that you would acquire a clearer and more secure understanding.

Here, in the context of medicine, Galen recognizes three ways in which solutions can be found and new knowledge generated: (a) by logical reasoning, without observational experience; (b) by observational experience, without logical reasoning; and (c) by a combination of reason and experience in which the latter helps to verify and emend conclusions drawn by the former. A doctor engaging in strategy (c) first makes logical inferences based on accepted axioms, such as the axiom—assumed within Neo-Hippocratic humoralist circles—that all human bodies and therapeutic substances can be classified according to the four qualities: hot, cold, moist, and dry. The physician then checks his logical conclusions against observational experience for confirmation, rejection, or further refinement.[37] Crucial to Galen's scientific methodology, then, is the ultimate interdependence of reasoning and experience, of theory and practice, and while Galen acknowledges the effectiveness of strategies (a) and (b) in certain contexts, he considers strategy (c) to be the most versatile and effective. Furthermore, to reduce the chance that semantic disagreements will hinder this pursuit of scientific knowledge, Galen also emphasizes the importance of establishing clear, standardized definitions of medical subjects and of differentiating accurately between those subjects' genera and species.[38]

Throughout *Affections and Errors*, Galen critiques the epistemologies of sectarian philosophers, contrasting what he considers to be the purely conjectural and

---

ἐπιδέδεικται, καὶ πρός γε τούτοις ὅτι τὰ διὰ λόγου καὶ πείρας εὑρισκόμενα στοχαστικῇ μὲν ὁδῷ χρῆται πρὸς τὴν τῶν ζητουμένων εὕρεσιν, ἐλπισθέντα δὲ τῷ λόγῳ βεβαιοῦται τῇ πείρᾳ... πάντων δὲ τῶν κατὰ τὰς τέχνας ἔργων ὑπὸ συνεχοῦς ἀσκήσεως ἀκριβουμένων οὐκ ἀρκεῖ τὸ καθόλου μόνον ἐν αὐταῖς καταμαθεῖν, ἀλλὰ κἂν τοῖς κατὰ μέρος ἐνεργῆσαι χρὴ πολλάκις.... διὰ τοῦτ' οὖν... οὐκ ἠρκέσθην μόνον τοῖς καθόλου θεωρήμασιν, ἀλλὰ καὶ γυμνάσματα πολλὰ προσέγραψα, δι' ὧν ἤλπιζον ὑμῖν ἐναργεστέραν τε καὶ βεβαιοτέραν ἔσεσθαι τὴν γνῶσιν αὐτῶν. Tr. Hankinson 2022, 104.

[37] On the importance to Galen of empirical testing, see also *SMT* 11.459–61 K. For discussion, see, e.g., Hankinson 2008a, 102; 2022.

[38] See, e.g., *MMG.* 11.4 K. For an explanation of how Galen defines species-forming differentiae, see Hankinson 1991b, 102. On the role of clear definitions in Galen's epistemology, see Havrda 2022.

insufficiently defensible kinds of "knowledge" they generate with the more "error-free" kinds of knowledge available to the mathematical or "exact" sciences. True to his practice, as described in the previous passage, of supporting theoretical claims with illustrative examples, Galen proceeds to highlight the contrasts between his scientific method and those of sectarian philosophers through a series of *paradeigmata*. The longest of these, which occurs at the central, fulcrum point of the treatise, revolves around the design, construction, and testing of two hourly timekeepers, a sundial and a water clock.[39] He introduces this extended *paradeigma* in the following way:[40]

> It is necessary that the man who wishes to become scientific attend to himself and train in many things successively, as I said before, which are able to bear witness to the ones who discover them. We find such things in the theories of numbers and lines on which astronomy and architecture are based. Come; for the sake of clarity, I will give an example from *arkhitektonia*.

Note how Galen claims the adjective "scientific" (*epistēmonikon*) for the practitioner of his own method, which he is about to describe by analogy to the process of clock-construction. Galen's use of this adjective is an explicit nod to one of his role models, Aristotle, who understood *epistēmē* ("knowledge" or "scientific knowledge") to be both a "demonstrative state (*hexis apodeiktikē*)" of the soul—which is to say, in the words of R. J. Hankinson and M. Havrda, "a developed capacity for providing demonstrations of why things are the way they are"[41]— and as a formal system of logical propositions which are demonstrable by recourse to agreed-upon axioms. Across his oeuvre Galen, like Aristotle, argues that the idea of scientific knowledge is intimately entwined with the idea of demonstration (*apodeixis*), and he often describes the mathematical or "exact" sciences as "apodeictic," i.e., "permitting of demonstration."[42]

Now, why might Galen have chosen to draw his central examples of successful scientific method from the domain of *arkhitektonia*, a Greek term that, again, encompasses both architecture and what we might call general engineering? To answer this question, most scholars have pointed to the fact that Galen's father[43] was himself an architect and—as Galen stresses in the first half of *Affections and*

---

[39] On these clock-making examples, see also Singer 2022, 18–24.
[40] *Aff. pecc. dig.* 54.14–21 DB = 5.80.1–5 K: χρὴ παρακολουθεῖν τὸν ἐπιστημονικὸν γενέσθαι βουλόμενον, ἐπὶ πολλῶν τῶν κατὰ μέρος, ὡς ἔμπροσθεν ἔφην, γυμναζόμενον, ἃ μαρτυρῆσαι δύναται τοῖς εὑρίσκουσιν αὐτά· τοιαῦτα δ' ἐστὶ τὰ κατὰ τὴν ἀριθμητικήν τε καὶ γραμμικὴν θεωρίαν, αἷς οἷον ἐπιβάσεσιν ἀστρονομία καὶ ἀρχιτεκτονία χρῆται.
[41] Hankinson and Havrda 2022: 2. They go on to note that Aristotle also "uses the word to refer to a structured system of propositions pertaining to and fully describing a particular domain."
[42] On the role of *apodeixis* in Galen's thought, see also Balalykin 2020.
[43] Galen does not include his father's name in any of his extant writings, but the *Suda* gives it as Aelius Nikon. *Suidae lexicon* 1 p. 506 Adler, s.v.: "Galen: son of Nikon, the geometer and architect."

Errors—a very influential figure within Galen's own life.[44] Galen may well have sought to pay homage to his father here, and perhaps to lend prestige to the *tekhnē* of architecture by showcasing it as a mathematically grounded discipline.[45] Yet Galen's choice had to do, first and foremost, with what he considered to be the "apodeictic" features of *arkhitektonia* as a science and, I will argue, with the particular semiotic range that clocks and hours had developed under the Roman Empire (a topic we will continue to explore in Chapter 4).[46] In the rest of this chapter, I will demonstrate how Galen uses his clock-construction *paradeigmata* to defend certain propositions about apodeictically inflected methods such as his own, namely that:

(1a) The results of such methods are achieved through rigorous, rational theory and are easily verified by sense perception.

(2a) The practitioners who employ these methods make useful contributions to the intellectual and civic communities within which they are embedded.

(3a) Such practitioners advance scientific "progress," which is presented here as a gradual accumulation of knowledge over time, beginning with the revered ancients and extending, potentially without limit, into an exciting future.

In this treatise, Galen also seeks to present sectarian philosophers as the inverse of scientific thinkers,[47] asserting that:

(1b) Their conclusions are insecure, resulting from poorly informed conjecture.

(2b) They isolate themselves from their civic and intellectual communities and contribute nothing of practical value.

(3b) They fail to immerse themselves in the current of scientific "progress," neither building upon the knowledge of their predecessors nor passing down their own knowledge effectively to the next generation.

---

[44] Galen establishes a sharp contrast between his father's temperance, probity, and generosity and his mother's tendency toward wild excess, spite, and parsimony: "I enjoyed the great good fortune of having a father who was particularly free from anger and extremely righteous, worthy, and benevolent. Meanwhile, my mother was very hot-tempered, to the extent that she sometimes bit her servants.... Seeing the noble actions of my father juxtaposed with the shameful affections of my mother, I came to welcome and love the former, and to avoid and despise the latter" (27.22–28.4 DB = 5.40.13–5.41.5 K). Galen goes on to imply that his father's moral rectitude was grounded in his ability to approach all situations rationally. Furthermore, it was specifically his schooling in "geometry, mathematics, *arkhitektonia*, and astronomy" (28.19–20 DB = 5.42.5–6 K) that allowed his father to hone this skill.

[45] In so doing, Galen expands the mathematics curriculum that Plato outlines at *Rep.* 521e–525b. For discussion of the social status of architects in Greco-Roman antiquity, see, e.g., Clarke 1963; Müller 1989; Thomas 2007. A collection of the epigraphic evidence can be found at Donderer 1996.

[46] Clocks are also mentioned in connection with *apodeixis* elsewhere in the Galenic corpus: *Inst.* 12.4.1–4 Kalbfleisch; *Lib. prop.* 19.40.10–18 K; *Cur. rat. ven. sect.* 11.256.10–18 K.

[47] Such a strategy was common in philosophical protreptic. See Curtis 2014, 49.

In what follows, we will perform close readings of Galen's clock-making examples and contextualize them within these larger arguments. In the Conclusion, we will turn to the *paradeigma* that Galen presents as the antithesis of these examples—namely, the philosophers' debate about the nature of the circumcosmic void—and examine the interplay between them.

Defending Propositions 1a and 1b: Testability

> When a city is being built, let [the following problem] be set before those who will inhabit it: they want to know, not by conjecture but with precision, on an everyday basis, how much time has passed, and how much is left before sunset. This problem, according to the analytic method, must be referred back to the first criterion if anyone intends to solve it in accordance with the method we learned in the study of gnomonics [i.e., the art of producing and using sundials].[48]

Galen launches into his clock-making paradigm by presenting us with a *problēma*, a Greek word whose semantic range includes everyday "tasks" or "business," as well as formal "problems" in mathematics and logic. This opening allows Galen to speak, with one side of his mouth, to the practical concerns of laypeople operating within civic communities and, with the other, to the technical concerns of specialists within intellectual communities. The problem at hand is how to construct an accurate clock that will help residents of a given city to orient themselves in time and coordinate their activities.

Galen's emphasis on community here is quite interesting, and we will probe it more fully in the next section. For now, let's follow Galen as he explains how, using the kind of scientific method that he endorses, an *arkhitekton* can indeed make such a clock and confirm its accuracy. Galen points out that the design and construction of an operative sundial must begin from accepted axioms, such as the principles of celestial trigonometry that underlie gnomonics, and must proceed according to the rules of logic. Once the design has been made, in order to double-check that the clock's hour-lines have been drawn appropriately, the clock should be tested in two different ways: both logically (against the original axioms) and empirically (against multiple types of observable phenomena). If all of this is done

---

[48] *Aff. pecc. dig.* 54.21–4 DB = 5.80.7–10 K: πόλεως κτιζομένης προκείσθω τοῖς οἰκήσουσιν αὐτὴν ἐπίστασθαι βούλεσθαι, μὴ στοχαστικῶς ἀλλ' ἀκριβῶς, ἐφ' ἑκάστης ἡμέρας, ὁπόσον τε παρελήλυθεν ἤδη τοῦ χρόνου τοῦ κατ' αὐτήν, ὁπόσον θ' ὑπόλοιπόν ἐστιν ἄχρι δύσεως ἡλίου. On the scope and definition of ancient gnomonics, see Evans 2005.

correctly, the clockmaker, like any practitioner of Galen's scientific method, can achieve a persistently high degree of precision and accuracy.[49]

To demonstrate his point, Galen recounts, in great detail, the order of operations involved in constructing a sundial. According to him, these are as follows: (1) learn the geometric method that will help you to translate the motion of the sun into a system of lines;[50] (2) determine the shape of the sundial body you wish to use (conical, spherical, etc.); (3) figure out how to transfer your diagram of lines onto the shape of the sundial body; (4) use logical reasoning to check your plans; (5) construct the actual sundial and find a level surface on which to mount it; and (6) perform empirical tests to confirm, with your senses, that your sundial is functional.[51] This process advances from logical, mathematical modeling to the creation of a testable, usable object.[52] The method itself he describes as both "universal and common," the product of consensus within an intellectual community.[53]

What seems to excite Galen most about the clock-construction process is its testability. He walks the reader through not one, not two, but three different methods for assessing a clock's accuracy, each of which involves checking the sundial's reading against another celestial or terrestrial time-indicator. The first method is to check one's initial and final hour-lines against the actual rising and setting of the sun.[54] Although this is a task that could be performed by the clockmaker alone, Galen insists that other parties participate in the process: "Then, having inscribed many shapes in succession, we must give them to men for empirical testing, to see whether the aim has been achieved."[55] By subjecting his work to the assessment of educated laypeople, the clockmaker can minimize the preservation and propagation of errors.[56] Galen is less explicit about the social benefits that will accrue to the clockmaker in doing this, but Galen likely found them compelling, as well; in submitting his work for critical review, the clockmaker displays his commitment to truth and thereby enhances his reputation within the community.

The second test that Galen proposes for a completed sundial is one of self-consistency: have all the lines been drawn on the sundial in correct proportions to

---

[49] On the self-confirmation available in gnomonics, see also *Aff. pecc. dig.* 47.12–21 DB = 5.68.12–5.69.6 K.

[50] This presumably involves acquiring (or creating) an *analemma*, a two-dimensional geometric diagram of the celestial movements.

[51] See *Aff. pecc. dig.* 55.3–13 DB = 5.81.1–11 K.

[52] On the marriage of calculation and empirical testing in Galen, see Tieleman 2008, 54–5.

[53] *Aff. pecc. dig.* 55.3 DB = 5.81.1 K: καθόλου τε καὶ κοινῆς.

[54] The start of the first hour should coincide with sunrise and the end of the last hour with sunset.

[55] *Aff. pecc. dig.* 55.17–19 DB = 5.81.15–5.82.2 K: εἶτα ποιήσαντας ἐφεξῆς καταγραφὴν ἐπὶ πολλῶν σχημάτων καὶ δοῦναι τοῖς ἀνθρώποις ἔργῳ πειραθῆναι, <εἰ> γεγονὸς ἤδη τὸ προβληθέν.

[56] Galen makes a similar injunction to team up with good men at *Aff. pecc. dig.* 54.3–6 DB = 5.79.4–7 K. For comparison, on the contribution of public critique to the maintenance of seventeenth-century European clocks, see Sauter 2007, 698.

one another? This can easily be assessed with the aid of a compass and a trained eye. Finally, Galen advocates checking the equality of the sundial's hours against an external measure, specifically, the rate at which water fills a perforated vessel. No specialty equipment is needed, just some "clear water" and a bowl with a small hole in it.[57] The process itself is also quite basic and, Galen implies, could easily be performed by anyone.[58] When the sundial tells you that the first hour of the day has begun, set the perforated bowl in the water and allow it to fill. At the end of the first hour, mark the spot on the vessel to which the water level has risen, then dump the water out, and repeat the process for the second hour, checking to make sure that the water reaches the same level. Repeat again for each hour through the twelfth. Then, Galen insists, "Unless you are completely ignorant, you will be convinced that the sundial has been inscribed well, because what is before your eyes has indicated it."[59] Galen reiterates his instructions for every single hour, in a manner that, while tedious, is also rhetorically serviceable. He stresses the ease and accessibility of the testing process and the clear visibility of its results; it is plain to the observer whether or not the water reaches the same level at the end of each hour.[60] Furthermore, in specifying each step, Galen demonstrates the thoroughness that a scientist must embrace if he is to achieve certainty.[61] Haste has no place on Galen's road to truth.[62]

According to Galen, this kind of self-confirmation is not limited to gnomonics and *arkhitektonia*, but is characteristic of all arts (elsewhere he lists geometry, arithmetic, logic, and astronomy) that are built upon a dialectic between reason and experience and founded on testable axioms.[63] As Galen likes to put it, such sciences "bear witness to those who truly discover them."[64] Galen invites the

---

[57] That is, free of particulates. Presumably, one also needs a larger, unpierced vessel that can contain both the pure water and the perforated vessel.

[58] The full passage appears at 55.17–56.17 DB = 5.81.15–5.83.13 K.

[59] *Aff. pecc. dig.* 56.15–17 DB = 5.83.11–13 K: εἰ μὴ παντελῶς ἀγνὼς εἶ, πεισθήσῃ καλῶς καταγεγράφθαι τὸ ὡρολόγιον, εἴ γε τὸ προκείμενον ἐπεδείξατο.

[60] Galen does not offer any suggestions for how to correct errors that do occur. Presumably, one would have to adjust one's trigonometrical calculations and then build another sundial. It is interesting, though, that Galen glosses over the mathematical underpinnings of clock construction and chooses instead to focus on the ways in which designs can be empirically tested.

[61] Galen also stresses the importance of having a clear order of operations at *MM* 10.31 K: "In order to investigate something according to the method...one must do so with some direction and order, so that something comes first in the investigation, and something second and third and fourth, and so on through the remaining sequence until you arrive at the original proposition."

[62] Cf. *Aff. pecc. dig.* 60.8–11 DB = 5.89.7–9 K: "For those [who are rash in the conceit of their own wisdom] the path to the truth is neither long nor steep, as Hesiod characterized the path to excellence. Instead, it is a shortcut or, rather, it does not actually exist."

[63] For Galen, there are two kinds of self-evident truths: the empirical and the rational.

[64] *Aff. pecc. dig.* 47.11–14 DB = 5.68.9–14 K: μαρτυρούσαις δὲ τοῖς ἀληθῶς εὑρηκόσιν. The alliance between the geometer and the physician is also particularly apparent at *Aff. pecc. dig.* 62.13–18 DB = 5.92.16–5.93.4 K: "While [the philosophers] are eagerly engaging in rivalry on these matters, rehashing the same discourses, often a geometrician will come among them, or some physicians or other men who are fond of discourse and educated in the disciplines but who don't practice a trade anymore on account of wealth and are not enslaved to some philosophical sect."

reader to consider the following example from geometry, which involves drawing one geometric figure around or inside of another:[65]

> Take, for example, an instance where we are enjoined to draw a circle around a given square or, in the same manner, to draw a square around or inside of a given circle.... If anyone is able immediately to circumscribe each of these figures by means of the method which he has learned, then he will give evidence, by doing this very thing, that he has found the object of his search.

All one has to do in order to confirm the validity of an inscribed polygon is to look at it and determine "whether the polygon is clearly seen as inscribed or circumscribed, just as the circle is seen as circumscribed or inscribed with respect to the polygon."[66] Likewise, if the result is mistaken, that fact will be readily apparent. By aligning his scientific method with mathematically based disciplines in this way, Galen claims for it several characteristics inherent in mathematical proofs: they are clear, visibly self-evident, eminently testable, and as a result, trustworthy.

In contrast, Galen associates the methods of rival thinkers with the opposite characteristics: murkiness, confusion, and the damning corollary of these: ignorance. "What blindness in recognizing one's own errors," Galen demands, "is greater than that of these men?" In his opinion, the only thing such men can truly be said to "see" or "know" is the fact that they are "especially dull with respect to understanding and remembering those things discovered by means of arithmetic, geometry, *arkhitektonia*, and astronomy."[67] Elsewhere, Galen also accuses sectarian philosophers of being charlatans and producers of nonsense, unknowing and wise solely in their own conceit.[68] Such practitioners make decisions hastily, even recklessly, and thus pass over, instead of catching, those important factors that allow the diligent practitioners of Galen's scientific method to assess the validity of their results and thereby win the day.[69]

---

[65] *Aff. pecc. dig.* 46.24-7 DB = 5.67.6-11 K: οἷον ὅταν περιγράψαι τῷ δοθέντι τετραγώνῳ κύκλον ἐπιταχθῶμεν, κατὰ ταὐτὰ δὲ καὶ τῷ δοθέντι κύκλῳ περιγράψαι τετράγωνον.... ἐὰν γάρ τις ἕκαστον τῶν τοιούτων δι' ἧς ἔμαθεν μεθόδου περιγράψαι παραχρῆμα δυνηθῇ, πρὸς αὐτοῦ τοῦ πράγματος μαρτυρήσεται τὸ ζητούμενον εὑρηκώς.

[66] *Aff. pecc. dig.* 47.2-5 DB = 5.68.2-4 K: καὶ γὰρ ἐγγραφόμενόν τε καὶ περιγραφόμενον ἐναργῶς ὁρᾶται, ὥσπερ γε καὶ ὁ κύκλος ἐγγραφόμενός τε καὶ περιγραφόμενος τῷ τοιούτῳ πολυγώνῳ. On the importance of testing propositions for oneself, rather than relying solely on someone else's assertions, see *Aff. pecc. dig.* 53.23-6 DB = 5.78.10-14 K and 64.4-25 DB = 96.3-97.4 K.

[67] *Aff. pecc. dig.* 61.11-15 DB = 5.91.7-11 K: τίς οὖν μείζων τυφλότης εἰς τὴν τῶν ἰδίων ἁμαρτημάτων γνῶσίν ἐστι τῆς τοιαύτης τῶν τοιούτων, ὅταν ἄνθρωποι τῶν ἐπιτυγχανόντων ἑαυτοὺς ὁρῶντες ἀφυεστέρους εἴς τε τὸ νοῆσαι καὶ μνημονεῦσαι καὶ τὰ δι' ἀριθμητικῆς τε καὶ γεωμετρίας ἀρχιτεκτονίας τε καὶ ἀστρονομίας εὑρισκόμενα. Cf. *Aff. pecc. dig.* 61.2-5 DB = 5.90.11-15 K, where Galen mocks the supposedly keen vision of philosophers' souls.

[68] *Aff. pecc. dig.* 48.8 DB = 5.70.2 K: διδασκάλοις ἀλαζόσι; 59.24 DB = 5.88.7 K: φλυαροῦσιν; 66.27 DB = 5.101.2 K: ἀνόητοι; 60.8-9 DB = 5.89.6 K: δοξόσοφοι.

[69] See *Aff. pecc. dig.* 62.29-65.15 DB = 5.94.1-5.98.8 K.

Defending Propositions 2a and 2b: Useful Contributions to Intellectual and Civic Communities

Let us return now to the observation, discussed briefly above, that Galen situates his clock-making paradigm within a communal context. Galen claims, you will recall, that clocks are fundamental to the creation of a civic community, using the following words to frame his forthcoming *problēma*: "When a city is being built, let [the following problem] be set before those who will inhabit it...."[70] Galen suggests that these clocks must be constructed *while* the city is still being built (he uses the present participle *ktizomenēs*), even *before* the future citizens move in (he uses the future participle *oikēsousin*). Urban civilization, as Galen defines it here, seems to require precise timekeeping; a community cannot function without its clocks.[71]

Later in the clock-making paradigm, Galen places one of the units that clocks could measure—the seasonal hour—alongside other weights, measures, and units considered essential for day-to-day civic functioning. Galen's goals, in this instance, are twofold. The first is to show how, by means of an apodeictically inflected method such as his own, one can perform the very useful task of subdividing units—a day, a coin, a personal estate—into twelve equal parts. But Galen does not stop there. He goes on to assert that his method can be used to sub-divide units into *any number* of equal parts:[72]

It was proposed that the duration of the whole day be divided into twelve equal parts.[73] They chose this number because it was the most useful. For it has a half, a third, a fourth, a sixth, and a twelfth, which no other number has before nor after it until you reach 24. This number they rejected as too large, and having judged

---

[70] *Aff. pecc. dig.* 54.21–4 DB = 5.80.7–10 K: πόλεως κτιζομένης προκείσθω τοῖς οἰκήσουσιν....

[71] For discussion of the use of temporal markers within poems about the Augustan-era *urbs*, see Wolkenhauer 2011, 112–14. Cf. Hor. *Serm.* 2.6 and discussion in Reckford 1997.

[72] *Aff. pecc. dig.* 56.17–57.9 DB = 5.83.13–5.84.15 K: προκείμενον <δ'> ἦν εἰς δώδεκα μοίρας ἴσας νεμηθῆναι τὸν χρόνον τῆς ὅλης ἡμέρας. ἀλλὰ τοῦτον μὲν ὡς χρησιμώτατον προείλοντο· καὶ γὰρ ἥμισυ ἔχει καὶ τρίμοιρον καὶ τέταρτον ἕκτον τε καὶ δωδέκατον, ἅπερ οὐδεὶς ἄλλος ἔχει πρὸ αὐτοῦ καθάπερ οὐδὲ μετ' αὐτὸν ἄχρι τοῦ εἰκοστοῦ τετάρτου. τοῦτον μὲν ὡς μακρὸν παρῃτήσαντο, σύμμετρον δ' εἶναι κρίναντες τὸν δωδέκατον <εἰς> τοσαῦτα μέρη τὸν χρόνον τῆς ὅλης ἡμέρας ἔτεμον · ὅτι δὲ χρήσιμός ἐστιν ἡ τοιαύτη τομή, τῇ πείρᾳ βασανίσαντες αὐτῇ <γ'> ἄλλοι τε πολλοὶ καὶ Ῥωμαῖοι χρῶνται τὴν οὐσίαν ἅπασαν, ὅταν διατίθωνται, διαιροῦντες [δ'] εἰς δώδεκα μέρη καὶ τῶν ἐν τῷ βίῳ σταθμῶν τε καὶ μέτρων εἰς δώδεκα μοίρας τέμνοντες τὰ πλεῖστα. σὺ δ' εἰ κελεύεις, καταγράψομαί σοι κατὰ τὴν μέθοδον ὡρολόγιον, εἴτ' εἰς <δώδεκα> τὴν ὅλην ἡμέραν διαιρεῖν ἐθελήσεις, εἴτ' εἰς ἄλλον τινὰ τῶν ἐφεξῆς ἀριθμῶν. εὑρήσεις γὰρ ἐπ' ἐκείνου πάλιν ὁ προὔβαλον γεγονὸς ἔκ τε τοῦ συμφωνεῖν τὴν διὰ τῶν τετρημένων ἀγγείων ἐκμέτρησιν κἀκ τοῦ πάντ' ἀλλήλοις ὁμολογεῖν <τὰ> καταγεγραμμένα, πρός τ' <ἐκ> τοῦ τὰς τελευταίας ἐν αὐτοῖς γραμμὰς ὁρίζειν τὰ πέρατα τῆς ἡμέρας. On Galen's attitude toward "utility" and his favoring of applied over theoretical logic, see Barnes 1993, 51.

[73] Just as Galen did not name an individual person or group as the clockmaker(s) in his city-founding example, neither does Galen mention an individual inventor of the twelve-hour system. Thus, we cannot say whether Galen would have agreed with Zeno and Diodorus Siculus—the only other authors known to have discussed the etiology of the hour—who attributed its invention to the Heliadae of Rhodes (Zeno Fr. 2 = Diod. Sic. 5.57).

twelve to be appropriate, they cut the time of the whole day into that many pieces. And because such a subdivision is so useful, the Romans and many others, having tried it out, use it when they manage their estates, dividing the whole into twelve portions. And they cut the majority of their everyday rules and measures into twelve portions, as well. If you were to ask me, I could inscribe any clock for you in accordance with my method, regardless of whether you wished to divide the whole day into twelve or into some other of the successive numbers. For you will find that the task has been accomplished, both from its consistency with the measures made with the perforated vessel and from the agreement of all the lines with one another, as well as from the fact that the outermost and innermost lines mark out the limits of the day.

Galen's method is so useful and versatile that it can determine whether any number (not just the convenient number twelve) has been accurately sub-divided in the first place and, in so doing, can sustain important civic operations.[74]

The examples presented so far focus on how practitioners of Galen's method, and the products they produce, can contribute substantially to their civic communities. However, Galen also seeks to contrast the ways in which sectarian philosophers and followers of his method relate to their wider communities of *knowledge*. In both kinds of communities, civic and intellectual, Galen asserts that practitioners of his method participate actively and meaningfully, while misguided philosophers are doomed to inutility and isolation. One reason for this isolation, Galen repeatedly asserts, is that sectarian philosophers do not participate in processes of consensus-building when they establish their axioms, draw their conclusions, or define their central terms.

One of Galen's favorite soapbox issues is how intellectually unsound, and even dangerous, it can be for thinkers to debate an issue before agreeing on how to define first principles and critical terms. V. Boudon-Millot has suggested that Galen's concerns with community consensus and terminological precision stem especially from his frustrations navigating the polyphony of plant and mineral names that he encountered across the multicultural, multilingual Roman Empire.[75] Galen's efforts to catalogue the wide range of known *materia medica* were frequently stymied by inconsistencies in measurement systems and naming practices, which often left him unclear as to whether, for example, he was

---

[74] It is interesting to observe how Galen, in this passage, links hours to other metrics, including measures of distance, weight, and coinage. Many writers under the Empire, such as Columella, Pliny the Elder, Balbus, and Volusius Maecianus, display an interest in metrology and include these latter measures within its ambit. Yet these Latin authors make no mention of calendrical and horological systems, which suggests that they did not view temporal concerns as aspects of metrology. We might wonder, then, whether Galen's portrayal in this passage is *descriptive*—i.e., a fairly accurate reflection of contemporary understanding—or *prescriptive*, a controversial claim regarding the status of temporal measurement. On weights and measures under Marcus Aurelius, see Cuomo 2007.

[75] Boudon-Millot 2008.

looking at one plant with five names or five similar plants with one name apiece.[76] As a result, Galen became committed to the goal of developing a common scientific lexicon and to the practice of carefully defining his own terms before constructing an argument.[77] Indeed, in the case of *Affections and Errors*, it is Antonius's slippery usage of important words like "error," "passion," and "guard" that seems to have provoked Galen to put pen to papyrus in the first place, in an effort to clean up the Epicurean's semantic mess.[78]

Elsewhere in *Affections and Errors*, before he embarks on his clock-construction paradigm, Galen singles out two intellectual communities in particular whose members can be led astray by imprecision in their studies and definitions. These are philosophers, of course, and medical professionals:[79]

> The very thing that Hippocrates said about those who practice medicine seems to apply to those who practice philosophy. Hippocrates said that strong similarities cause errors and confusion even for good doctors.[80] Inasmuch as not only average doctors but even the very best physicians get tripped up by the similarities of symptoms, it is not unlikely that errors and confusions happen even for good philosophers in matters pertaining to philosophy.

Galen implies here that, if only everyone would carefully train themselves in his scientific method, then doctors and philosophers alike could find themselves speaking the same precise and informed language—one that makes clear, accurate delineations between confoundingly similar things.[81] Because apodeictically inflected methods like Galen's are founded upon truths that are verifiable— empirically, rationally, or both—they have the unique ability to create interdisciplinary communities of individuals who agree about the nature of the world (in so far as their beliefs are based on the same self-evident axioms) and about how to refer to the world's relevant entities and divisions.[82]

---

[76] See esp. *Alim. fac.* and *SMT*.
[77] See *Aff. pecc. dig.* 41.6–11 DB = 5.58.3–5.59.1 K; 46.7–12 DB = 5.66.12–17 K.
[78] *Aff. pecc. dig.* 3.8–4.7 DB = 5.1.5–5.3.5 K.
[79] *Aff. pecc. dig.* 43.25–44.6 DB = 5.62.13–5.63.4 K: ὅπερ οὖν Ἱπποκράτης ἐπὶ τῶν κατὰ τὴν ἰατρικὴν πραγματείαν εἶπε, τοῦτο φαίνεται καὶ κατὰ φιλοσοφίαν ὑπάρχειν. Ἱπποκράτης δ' εἶπε τὰς ὁμοιότητας πλάνας καὶ ἀπορίας ἐργάζεσθαι καὶ τοῖς ἀγαθοῖς ἰατροῖς. ὥστ' οὐ μόνον τῶν ἐπιτυχόντων ἰατρῶν ἐν ταῖς ὁμοιότησι σφαλλομένων, ἀλλὰ καὶ τῶν ἀρίστων οὐκ ἀπεικός ἐστι καὶ τοῖς ἀγαθοῖς φιλοσόφοις ἐν τοῖς κατὰ φιλοσοφίαν ἀπορίας τε καὶ πλάνας γίγνεσθαι.
[80] *Epid.* 6.8.26 L.
[81] For more on Galen's struggles with inconsistent medical terminology, see Wilkins 2007, 76–8. On his attitudes to language in general, see Hankinson 1994b; Morison 2008a.
[82] Dewald, in her discussion of Herodotean and Thucydidean historical narrative, has demonstrated how each of these authors establishes a dialogic relationship between his own critical intelligence and that of his reader (Dewald 2006, 177–82; cf. Dewald 2007). Galen seems to write within a similar framework, in that he not only models critical assessment for his reader but also expects his reader to exercise the same faculty.

School philosophers, in contrast, are unable to point to recognized axioms that support their theories. Instead, these mistaken thinkers hide behind claims of "privileged knowledge" that strike Galen—and other rival philosophers—as weak and evasive. Thus, while sectarian philosophers bicker amongst themselves,[83] the shared worldview of Galen's ideal scientists allows them to understand and build upon one another's contributions to the shared pool of knowledge. Terms such as "common" (*koinos*) and "to agree" (*homologeō*) are buzzwords for Galen, both in this text and elsewhere in his oeuvre.[84] Towards the end of *Affections and Errors*, for instance, Galen praises an architect of his acquaintance who "uses for his demonstrations principles that are clear and with which everyone indisputably agrees,"[85] and Galen promises to describe the complex, finicky process of constructing a water clock in a manner that is "clear even for laymen."[86] Meanwhile, philosophers whose definitions of the truth are incompatible with the communally ratified definition find themselves at odds with each other and with the unified community.[87]

Galen highlights the isolation of these philosophers by regularly employing a syntax that places "everyone" or "all of us" in contrast with "the philosophers alone." For instance, when two squabbling philosophers ask an *arkhitekton* to help them determine whether wood is heavier than water, Galen smugly reports that the *arkhitekton* "explained quickly and clearly, so that all who were present understood, *except for the philosophers alone*."[88] Galen uses

---

[83] For quarreling philosophers, see *Aff. pecc. dig* 52.3–5 DB = 5.76.1–3 K; 60.13–17 DB = 5.89.12–15 K; 62.13 DB = 5.92.16 K; 62.23–4 DB = 5.93.10 K. Cf. *MM* 10.469 K; *Lib. prop.* 19.40.10–18 K; and elsewhere.

[84] According to a *TLG* lemma search, there are 1,767 instances of κοινός in Galen's corpus, and 754 instances of ὁμολογέω.

[85] *Aff. pecc. dig.* 63.21-3 DB = 5.98.15–5.99.1 K: τοιαύταις γὰρ ἀρχαῖς εἰς τὰς ἀποδείξεις οἶδα χρώμενον αὐτὸν ἐναργέσι τε καὶ ἀναντιλέκτως ὑπὸ πάντων ὁμολογουμέναις. Cf. (among many others) *MM* 10.30, 42, and 50 K. Many scholars have pointed out that Galen's notion of "everyone" is not quite as universal as he would have it seem. At *MM* 10.42 K, e.g., he admits that, despite his desire for axioms to be universally accepted, "the discoveries and inquiries and demonstrations of the essence of the objects will depend not on what is believed by most men but on scientific assumptions." Cf. *Aff. pecc. dig.* 60.8–10 DB = 5.89.6–7 K. For a discussion of this contradiction, see Barnes 1991, 78–9. For further discussion of agreement in Galen's work, see Morison 2008b, 71.

[86] *Aff. pecc. dig.* 57.11-12 DB = 5.85.3 K: ἐναργής ἐστι καὶ τοῖς ἰδιώταις. The water clock's design looks like this: as we saw in this chapter's opening passage, the uppermost line on the water clock will indicate the twelfth hour of the day, the bottommost line the first hour of the day (with the ten intermediate lines indicating the ten intermediate hours). These hour-lines will curve to account for the varying lengths of seasonal hours throughout the year (see Introduction). Thus, the twelfth-hour line will be highest at the summer solstice and lowest at the winter solstice. To check the accuracy of a water clock, Galen recommends comparing its reading to that of an accurate sundial (*Aff. pecc. dig.* 57.11–58.16 DB = 5.84.16–5.86.11 K).

[87] On philosophers' unwillingness to convene for healthy debate, see *Aff. pecc. dig.* 61.20–62.10 DB = 5.92.1–13 K.

[88] *Aff. pecc. dig.* 66.20-4 DB = 5.100.11–12 K: ὁ μὲν διῆλθεν ἐν τάχει τε καὶ σαφῶς, ὡς ἅπαντας τοὺς παρόντας νοῆσαι πλὴν μόνων τῶν φιλοσόφων. Cf. *Aff. pecc. dig.* 59.30 DB = 5.88.13 K; 60.23 DB = 5.90.6–7 K; 61.7 DB = 91.3 K; 62.23 DB = 93.9 K.

this technique to great effect at the end of the treatise, where he sums up his argument thus:[89]

> So, who is more likely to know the truth? One who submits his judgments to all those philosophers who are not shams, and submits his judgments also to those who practice all the arithmetical, logistical arts that depend on reason: geometers, astronomers, *arkhitektones*, law-makers, orators, grammarians, musicians? Or one who is judged and crowned solely by himself, yet who, if he were to subject himself to assessment by other judges, would receive no vote?

Here Galen reiterates the message, central to *Affections and Errors* as a whole, that membership in an intellectual community is important because it enables the productive exchange of ideas. In this case, individuals with greater expertise in rational analysis can help those with less experience by providing them with constructive criticism.[90] This will help the less experienced to develop their skills until they can become experts in turn, eligible to receive the benefits that attend community members in high standing. Sectarian philosophers deliberately close their ears to the cogent arguments of peers and experts alike, and thus deprive themselves of the opportunities for moral, intellectual, and social betterment that come with being part of a larger group.[91] Instead, their self-imposed isolation limits them to what they can learn on their own.

Defending Propositions 3a and 3b: Progress and Perpetuity

When Galen shifts the focus of his *paradeigma* from building a sundial to building a water clock, it becomes clear that Galen does not restrict his definition of civic or intellectual "community" to a single time and place—for example, to the philosophers, physicians, and so forth who frequented Imperial Rome in the second century CE. At this point in the treatise, Galen instead presents and defends a view of "community" as diachronic and of "scientific knowledge" as something that

---

[89] *Aff. pecc. dig.* 68.11-7 DB = 5.103.8-14 K: τίνα τοίνυν εἰκός ἐστι μᾶλλον ἐγνωκέναι τὴν ἀλήθειαν, [ἄλλον δὴ] τὸν τοὺς φιλοσόφους ἅπαντας ὑπομένοντα κριτὰς πλὴν αὐτῶν τῶν φαινομένων, ὑπομένοντα δὲ καὶ τοὺς ἀπὸ τῶν λογικῶν τεχνῶν ἁπασῶν ἀριθμητικοὺς λογιστικοὺς γεωμέτρας ἀστρονόμους ἀρχιτέκτονας νομικοὺς ῥήτορας γραμματικοὺς μουσικούς, ἢ τὸν ὑφ᾽ ἑαυτοῦ κρινόμενόν τε καὶ στεφανούμενον, <ὃς> ὑπὸ τῶν ἄλλων κριτῶν, <εἰ> ὑπομείνειε δοκιμάζεσθαι, μηδεμιᾶς ψήφου μεταλαμβάνεται.
[90] Cf. *Aff. pecc. dig.* 8.1-4 DB = 5.9.6-9 K and 43.3-11 DB = 5.61.5-13 K.
[91] Galen says that such people are able "to be shameless" (ἀναισχυντεῖν, *Aff. pecc. dig.* 59.24-5 DB = 5.88.7 K). "Shame" (αἰσχύνη) was fundamental to the Greek notion of a functioning community; it was the fear of feeling shame before fellow citizens that kept individuals from committing selfish acts which could cause harm to others and rend the social fabric.

accumulates over generations.[92] In the following passage, Galen chastises rival philosophers who fail to appreciate this fact:[93]

> Then did you not set your heart on discovering what such a method is [i.e., for building water clocks]? Do you not perceive the conceit of wisdom in yourself, you who, being uneducated in these arguments, could find out nothing in a year or rather, one should say, in your whole life? For the theory of lines [i.e., plane geometry] was not discovered in one lifetime, but came about little by little. First, elementary observations were sought, and once these had been found, the men who came after added to them the most marvelous theory, which is called "analytical," and trained themselves in it, along with those who were especially willing. And indeed, they can produce no handiwork of the sort which I have discussed up to this point regarding sundials and water clocks.

An individual, says Galen, starting from scratch and working alone, cannot hope to discover, even over the course of an entire lifetime, all of the mathematical theories and engineering techniques necessary to produce the kinds of clocks Galen has just described.[94] It is only by actively participating in a diachronic community of scientific thinkers that a clockmaker can hope to achieve results which equal or surpass those of his predecessors. For, with past masters as teachers, contemporary scientists can build upon their teachings and produce even greater marvels. In this way, the theory of lines was built upon the analytical method,[95] upon the analytical method was built the sundial, and upon sundial theory (i.e., gnomonics) was built the water clock. This claim can help us understand why Galen, in *Affections and Errors*, describes the construction not simply of one clock, but of two: by discussing first the sundial and then the even more complex water clock,[96] Galen can better illustrate for his readers the upward

---

[92] On Galen's belief in the accumulation of scientific knowledge over time, see Barras et al. 1995, p. vii.

[93] *Aff. pecc. dig.* 58.17–59.8 DB = 5.86.12–5.87.5 K: εἶτα τὴν τοιαύτην μέθοδον εὑρίσκειν οὐκ ἐπεθύμησας, ὦ οὗτος, τί ποτ' ἐστίν; οὐκ ᾔσθου τῆς ἐν αὐτῷ δοξοσοφίας, ἀμαθὴς ὅστις ὢν τούτων τῶν προβλημάτων οὐδὲν ἂν εὕροις ἐνιαυτῷ, βέλτιον δὲ εἰπεῖν, ὅλῳ τῷ βίῳ; οὐδὲ γὰρ εὗρεν αὐτὰ βίος ἀνδρὸς ἑνός, ἀλλὰ κατὰ σμικρὸν προῆλθεν ἡ γραμμικὴ θεωρία πρῶτον μὲν τῶν στοιχειωδῶν ἐν αὐτῇ θεωρημάτων ζητηθέντων αὐτῇ, ὁπότε δ' εὑρέθη ταῦτα, προσθέντων [αὐτῶν] αὐτοῖς τῶν ἐφεξῆς γενομένων ἀνδρῶν τὴν θαυμασιωτάτην θεωρίαν, ἣν ἀναλυτικὴν ἔφην ὀνομάζεσθαι, καὶ γυμνασάντων ἑαυτούς τε καὶ τοὺς βουληθέντας ἐπὶ πλεῖστον ἐν αὐτῇ· καίτοι μηδὲν ἔχουσι δεῖξαι χειρούργημα τοιοῦτον, οἷα μέχρι δεῦρο διῆλθον ἐπί τε τῶν ἡλιακῶν ὡρολογίων καὶ <τῶν> κλεψυδρῶν.

[94] Galen is, of course, taking another jab at sectarian doctors and philosophers here. This entire passage may also be targeting Methodists, whose school of medical thought offered a speedy six-month training.

[95] It should be noted that Galen does not always use the term "analysis" to refer to the logical process of reducing a problem to a set of initial principles. Here, as elsewhere in his corpus, Galen seems to refer to the geometrical meaning of "analysis," i.e., an apodeictic method of problem-solving. On the meaning of analysis within this text, see Chiaradonna 2014, 67.

[96] This procedure is rather trickier than the one for designing a sundial since, as we have seen, Greco-Roman societies favored seasonal hours over equinoctial, and these are difficult for a water clock

trajectory of scientific progress, here presented as steady, cumulative, and unbounded.[97] "Our ancestors,"[98] Galen assures us, made brilliant discoveries, without which present society would be much the poorer. But those ancestors could not even begin to imagine the marvels that their theories have facilitated in the present day—and may yet facilitate in days to come.

Galen insists that his contemporaries build upon the intellectual achievements of "ancestors" from the Classical and Hellenistic periods, such as (here) Hesiod, Euclid, and Hippocrates[99] and (elsewhere) both Plato and Aristotle.[100] Galen asserts that only a man who "is prudent by nature and who received the education that has been esteemed by the Greeks from the beginning" can hope to master the method that Galen advocates.[101] With claims such as this, Galen conveys the impression that practitioners of his method are the direct inheritors of Classical Greek tradition—a weighty assertion in the time of the so-called "Second Sophistic." Galen's rivals, he implies, have cut themselves off from this heritage.

Galen's notion of an intellectual community extends not only into the past but also into the future. He clearly envisions the readers of *Affections and Errors* as younger men, whose bad moral and intellectual habits have not yet ossified. Men who have long since devoted themselves to a particular school of thought, however, are past help:[102]

> Why do you think that the ignorance and pretense to wisdom of such people can easily be cured? If a man has a tumor that has grown hard over the span of three or four years, that hardness cannot be relieved. Similarly, what old man's soul can be relieved that has maintained the hardness of ignorance and feigned wisdom for thirty or forty years?

---

to measure. The makers of water clocks had to approximate the shift in daylight length over the course of a year by marking their vessels with twelve different sets of hour-lines, one for each month. Recall the water clock discovered in the temple of Borvo-Apollo and discussed in Stutzinger 2001.

[97] On Galen's understanding of scientific progress, see De Lacy 1979; Hankinson 1994a; Asper 2013, 421–5. On the concept of progress in ancient science more generally (a controversial topic to this day), see esp. Finley 1965; Edelstein 1967b; Jouanna 1999, 232–8; Keyser 2013, 48–52; Asper 2013.

[98] *Aff. pecc. dig.* 46.2 DB = 5.66.7 K: τοῖς πρὸ ἡμῶν.

[99] *Aff. pecc. dig.* 42.3–4 DB = 5.59.13 K (Euclid); 43.25 and 44.2 DB = 5.62.13 and 15 K (Hippocrates); 60.10 DB = 5.89.8 K (Hesiod).

[100] Galen believed that both Plato and Aristotle were indebted to Hippocrates. For more on Galen's engagement with Plato, see, e.g., Rocca 2006. On Galen's identification with his Greek roots, see Nutton 2013, 227. On his valorization of Greek *paideia*, see von Staden 1997, and on the growing link between Greek *paideia*, athletic training, and the urban elite in this period more generally, see van Nijf 2001; Mattern 2008, 52.

[101] *Aff. pecc. dig.* 51.24–52.2 DB = 5.75.13–15 K: <οὐ> χείρους γὰρ αὐτῶν οἶδα πάντας ἰδιώτας <τούς> γε καὶ συνετοὺς φύσει καὶ πεπαιδευμένους τὴν παρ' Ἕλλησιν ἐξ ἀρχῆς εὐδοκιμοῦσαν παιδείαν.

[102] *Aff. pecc. dig.* 51.16–20 DB = 5.75.5–9 K: τί οὖν, ὦ πρὸς θεῶν; ἡ τῶν τοιούτων ἀμάθειά τε καὶ δοξοσοφία δοκεῖ σοι δύνασθαι θεραπευθῆναι ῥᾳδίως; ἢ ὅστις μὲν ἐσκίρρωται ἐτῶν ἤδη τριῶν ἢ τεττάρων, ἄλυτον ἔχει τὸν σκίρρον, ἡ δὲ τῶν τοιούτων γερόντων ψυχὴ τὸν τῆς ἀμαθείας τε καὶ δοξοσοφίας σκίρρον ἐτῶν τριάκοντα καὶ τετταράκοντα ἔχουσα δύναται λυθῆναι.

Galen exemplifies this point by pitting young men (*meirakia*) whom he has trained in his scientific method against a group of men who have "grown old in philosophy."[103] He invites these old men to try to stump the youths using sophisms that they themselves were unable to puzzle out. Yet, to the old men's chagrin, the boys turn out to have no trouble spotting the faults in the sophistic arguments. By identifying his method so closely with the next generation, Galen represents it as the way of the future. Those poor old sectarian philosophers, on the other hand, can look forward to nothing but extinction.

The vision that Galen presents here of scientific progress as a steady, communal effort at knowledge aggregation and systematization is not commonly expressed in ancient medical writings. As M. Asper has pointed out, treatises such as Galen's own *On the Method of Healing* more typically adopt a narrative of "progress as story of return." This kind of progress narrative constructs "authority through a privileged access to the past" and "stages the reunion of two great individuals that meet each other across centuries."[104] With respect to Galen specifically, Asper is referring to his tendency, which we will explore more fully in Chapter 5, of presenting himself as the one true successor to the great Hippocrates. From this kind of progress narrative, one receives the impression that "progress has become the business not of communities, but of stars. Progress ends, paradoxically, with the beginning."[105] In other words, Galen's claims regarding his own status as a physician and that of Hippocrates often imply that Galen, through his perceptive readings and refinements of Hippocratic texts, has "solved" medicine; the system that he has constructed need not—and, indeed, cannot—be improved upon.

In *Affections and Errors*, however, Galen's progress narrative more closely resembles what Asper terms "progress as a story of growth." In such accounts, the narrator's contributions "always appear as part of a group-culture that comes with a certain collective authority and social standing. The subject is an important part of progress, but nonetheless just one among many."[106] Asper presents this kind of progress narrative as particularly characteristic of writings on Greek mathematics, such as those penned by Archimedes, Heron of Alexandria, and Proclus.[107] I suggest that Galen's strategy of presenting progress in *Affections and Errors* as a cumulative, communal, and unending process might have been intended to support two larger claims: (a) that his own method for distinguishing between truth and falsehood is as rigorous and effective as mathematical proofs in *more Euclideano*, and (b) that, by extension, the art that Galen practices, i.e., medicine, deserves equal standing with the mathematical sciences. By grounding his scientific method in systems of logic and classification inspired by Aristotelian *apodeixis*, developments in Stoic logic, and the structures of

---

[103] *Aff. pecc. dig.* 60.15 DB = 5.75.4 K: γεγηρακότων ἐν φιλοσοφίᾳ.   [104] Asper 2013, 421–4.
[105] Asper 2013, 424–5.   [106] Asper 2013, 417.   [107] Asper 2013, 414–17.

Hellenistic mathematical proofs,[108] Galen seeks to affirm medicine's status as a rigorous "scientific" discipline—that is, as a *tekhnē* organized around a coherent set of logical principles and standards of proof—and, at the same time, to validate and create a role for the kind of practical, experiential knowledge that results from actually *doing* medicine. I concur with Hankinson's assessment that this idea, that "medicine, indeed practical medicine, should be put on all fours with the theoretical, axiomatized sciences like arithmetic and geometry," can be considered "Galen's great claim to originality as a theoretician."[109] Galen's decision, within *Affections and Errors*, to present "progress as a story of growth" helps him to bolster this claim.

Conclusion

Toward the end of his discussion of clock design, and again at the end of *Affections and Errors* as a whole, Galen brings together and sharpens the various contrasts we have identified between sectarian philosophers and practitioners of Galen's scientific method (represented here, again, by clock-making *arkhitektones*). In the first instance, Galen explicitly pits these two groups against each other to highlight the confusion and isolation that attends the former as they, unlike the clock-makers, are unable to verify their results:[110]

> [T]he very prospect of using it [i.e., a demonstrative method] for the greatest tasks makes it a good thing to practice, on account of its unique ability, as I said, to provide evidence of its own success—an ability that does not exist among those who make discoveries via philosophy. And because of this, it is possible for those who talk nonsense foolishly in philosophy to act shamelessly. For, whereas the one who has badly marked a sundial or water clock becomes plainly aware of his error by means of the fact itself, it is not the case that there is similarly clear

---

[108] On Galen's relationship with Aristotelian philosophy, see Singer 2013. On Galen and mathematics, see the introduction to Kieffer 1964; Lloyd 2006. On Aristotelian conditional syllogisms and their relationship to mathematical proofs, see Hintikka 1980; McKirahan 1992; Goldin 1996; R. Smith 2009. On Euclidean production proofs and their analogs in Plato, see Patterson 2007; Wolfsdorf 2008.

[109] Hankinson 1991b, 120.

[110] *Aff. pecc. dig.* 59.19–60.7 DB = 5.88.2–5.89.5 K: αὐτό γε τὸ μέλλειν αὐτῇ χρῆσθαι πρὸς τὰ μέγιστα καλῶς <ἂν> εἶχεν ἀσκηθῆναι κατ' αὐτὴν ἐξαίρετον ἔχουσαν, ὡς ἔφην, τὸ μαρτυρεῖσθαι πρὸς αὐτῶν τῶν εὑρημένων [εὐφρανθέν], ὅπερ οὐκ ἔστιν ἐν τοῖς κατὰ φιλοσοφίαν εὑρισκομένοις. καὶ διὰ τοῦτο τοῖς προπετῶς ἐν αὐτῇ φλυαροῦσιν ἔξεστιν ἀναισχυντεῖν · οὐ γάρ, ὥσπερ ὁ κακῶς καταγράψας ὡρολόγιον ἢ κλεψύδραν ὑπ' αὐτοῦ τοῦ πράγματος ἐλέγχεται προφανῶς, οὕτω καὶ τοῖς ἐν φιλοσοφίᾳ θεωρήμασιν ἐναργής ἐστιν ὁ ἔλεγχος, ἀλλ' ἔξεστι λέγειν, ὥσπερ ἂν ἐθέλῃ τις, ὅταν γε ἅπαξ ἀναισχύντως ἄνευ μεθόδου λογικῆς ὑπ' αὐτῶν τῶν πραγμάτων διδάσκεσθαι <δια>τείνωνται. ἀλλ' εἰ μὲν ἐκείνοις μόνοις διελέχθη τὰ πράγματα λαβόντα φωνήν, εὐλόγως ἀλαζονεύονται, σιγώντων δ' αὐτῶν ἀεὶ καὶ μήθ' ἡμῖν μήτ' ἐκείνοις διαλεγομένων εὔδηλον ὅτι μόνος ὁ ἐν ἡμῖν λόγος εὑρήσει τὴν τῶν πραγμάτων φύσιν.... <ἢ> οὐκ ἂν ἔτι πιστὸς ἐπὶ τῶν ἀδήλων εἴη.

refutation for theories in philosophy. Rather, it is possible for people to speak as they like, once they have maintained shamelessly and without logical method that they were taught by the very matters in question. Certainly, if the "facts," taking up a voice, have indeed spoken to these men [i.e., philosophers] alone, then they are bragging with good reason. But since the matters for discussion are always silent, and speak neither to us nor to these philosophers, it is clear that our reason alone will discover their nature. Therefore, let the one who can first discover what that nature is provide evidence clearly to himself about the matters at hand.... Otherwise, he will no longer be trustworthy in matters pertaining to unclear things.

Philosophers, Galen remarks, claim that the secrets of the cosmos have been whispered to them alone. Unless one accepts this claim, however, it is often impossible to assess the validity of a philosopher's propositions. Unlike clockmakers, whose errors can be spotted easily by anyone, there is no agreed-upon rubric against which to measure most philosophers' commitments to the truth. And without a means of adjudication, philosophical inquiry can break down into chaos, with each philosopher saying whatever he likes to the point that all sense disappears. With his trustworthiness demeaned, such a philosopher disregards—and thus no longer merits—respect.[111] In short, because this kind of philosopher does not pursue testable knowledge via the appropriate balance of logic and empiricism, he fails to earn a place within the larger community.

Toward the end of *Affections and Errors*, Galen develops the contrast between clockmakers and philosophers even further by recounting an imaginary debate between a Stoic, an Epicurean, and a Peripatetic that highlights the speculative nature of their philosophical contributions. Galen himself acts as the adjudicator, while the philosophers argue over an obscure topic: the nature of the void thought by many to encompass the cosmos. The first-person perspective is, of course, Galen's:[112]

---

[111] At *Aff. pecc. dig.* 68.8–9 DB = 5.103.5–6 K, Galen describes these philosophers as "held in low esteem by all, both by laymen and by other philosophers." In several places, Galen laughs (the Greek verb is γελάω) at the philosopher's foolishness (66.15 = 5.100.3 K; 67.6 DB = 101.8–9 K).

[112] *Aff. pecc. dig.* 67.11–68.1 DB = 5.102.1–13 K: "διαφέρει γε μήν", ἔφην, "τῶνδε τῶν φιλοσόφων διττὴν διαφορὰν ἑκάτερος τούτων" (ἐδείκνυον δὲ τόν τε Στωϊκὸν καὶ τὸν Ἐπικούρειον), "ὁ μὲν γὰρ Στωϊκὸς οὐκ ἔνδον εἶναί τι κενὸν <λέγων>, ἔξωθεν δὲ τοῦ κόσμου ὑπάρχειν αὐτό. ταῦτα δ' ἄμφω συγχωρῶν ὁ Ἐπικούρειος ἐν ἄλλῳ τινὶ διαφέρεται πρὸς αὐτούς· οὐ γὰρ <ἕνα> ὁμολογεῖ <τὸν> κόσμον εἶναι τόνδε, καθάπερ ὁ Στωϊκὸς οἴεται, κατά γε τοῦτο τοῖς Περιπατητικοῖς ὁμοδοξῶν, ἀλλ', ὥσπερ γε καὶ τὸ κενὸν ἄπειρον τῷ μεγέθει φησὶν ὑπάρχειν, οὕτω καὶ τοὺς ἐν αὐτῷ κόσμους ἀπείρους τῷ πλήθει. ἐγὼ δ' ἤκουσα μέν, ἃ λέγουσιν οἱ τρεῖς συναγορεύειν βουλόμενοι τοῖς ἰδίοις ὀνείροις, ἀκριβῶς δ' οἶδα μηδένα λόγον ἀποδεικτικὸν ἔχοντας αὐτούς, ἀλλ' ἐνδεχομένους τε καὶ εἰκότας, ἐνίοτε δὲ μηδὲ τοιούτους. γνώσεσθε δ' ὅτι μὴ ψεύδομαι παρακαλέσαντες. εἰπεῖν τιν' ἀπόδειξιν ἕκαστον αὐτῶν εἰς [τ]αὐτὸ τοῦτο <τὸ> προκείμενον." The void debate is also referenced at *Aff. pecc. dig.* 62.13–26 DB = 5.92.16–5.93.12 K; 65.16–21 DB = 5.98.9–14 K.

"Each of these two philosophers," I said, "differs in two ways from the position of the Peripatetics." I pointed to the Stoic and the Epicurean. "The Stoic says that there is no void inside the world, but that there is a void outside of it. The Epicurean agrees with both of these statements but differs from the rest in another respect. He does not concede that this is the only world, as the Stoic thinks, who shares this belief with the Peripatetics. Rather, just as he says that the void is unbounded in size, so he also says that the worlds within it are unbounded in their number. I listened to the things which the three said in their desire to advocate for their own dreams, but I think it is very clear that they possess no demonstrative argument, only ones based on possibility and likelihood—and sometimes they don't even have these. You will know that I am not lying if you ask each of them to provide some demonstration for the issue at hand."

Here, Galen's philosophers are literally fighting over nothing—an empty space that is far beyond the realm of sense perception and verification and has little to no bearing on human life. Their arguments, too, are represented as insubstantial, mere "dreams" born of wishing. Galen portrays the philosophers as completely incapable of employing any demonstrative reasoning. Instead, they are reduced to caricatures within Galen's own demonstration, in which he attempts to show, with sound reasoning and clear examples, that these philosophers are unworthy of trust. An *arkhitekton*, Galen argues, one such as might construct a clock, would never even think of making claims about the void "until, after having actually gone out into the void part of the cosmos, he had tested the matter by means of experience and had clearly observed whether each of the bodies in it remain in the same place or change position."[113] Without real evidence to support one stance over another, the Stoic, Epicurean, and Peripatetic are doomed to quarrel interminably. The work of the *arkhitekton*, on the other hand, can easily be ratified by the community, thus freeing him to advance to other useful projects. For Galen, then, *arkhitektonia* in general, and clock-making in particular, provides an excellent foil for the discussion of the void. While the latter emphasizes nothingness, insubstantiality, and uselessness, clock-making is a rational and empirically testable process, grounded securely in mathematics, that produces tangible objects for use within civic and intellectual communities.

To review, we have seen that, in *Affections and Errors*, Galen aims to persuade his readers that his own scientific method, with its combination of rationalism and empiricism, is more successful at differentiating between truth and falsehood than any method offered by the philosophical schools of his day. To drive home his point, Galen repeatedly contrasts the positive characteristics of his own method

---

[113] *Aff. pecc. dig.* 65.18–21 DB = 5.98.11–14 K: ὁ μὲν γὰρ ἀρχιτέκτων οὗτος οὐκ ἂν ἀπεφήνατο, πρὶν εἰς αὐτὸ τὸ κενὸν ἐξελθὼν τοῦ κόσμου τῇ πείρᾳ βασανίσαι τὸ πρᾶγμα καὶ θεάσασθαι σαφῶς, εἴτε μένει κατὰ τὸν αὐτὸν τόπον ἕκαστον τῶν ἐν αὐτῷ σωμάτων εἴτε καὶ μεθίσταται.

with the negative characteristics of alternative dogmas. He associates his method with testability and clarity, utility within civic and intellectual communities, and scientific progress. The dogmas of contemporary philosophical schools, on the other hand, Galen associates with the reverse: unsupportable conjecture and confusion, uselessness and isolation, stagnation and exclusion from the stream of progress. We have also observed that Galen likes to draw examples of successful scientific methodology from the realms of both theoretical and applied mathematics in order to strengthen the connections between his own methodology (and his medical *tekhnē* as a whole) and the methods employed in the highly respected mathematical arts.

The dichotomy that Galen sets up in *Affections and Errors*, between sectarian philosophers and practitioners of his own scientific method, raises further questions. In order for clocks to be maximally effective as foils to the philosophers' pointless debate about the void, Galen's audience must have shared this view of clocks as useful, familiar tools and impressive examples of human ingenuity—but did they? How representative *was* Galen's portrayal of clocks and timekeeping in the second century? We will pursue these questions in the following chapter by comparing Galen's treatment of clock-making in *Affections and Errors* to the symbolic and rhetorical valences of clocks in contemporary art and literature.

4
Clocks as Symbols 2
Among the Roman Elite

We have now seen *how* Galen uses his horological examples in *Affections and Errors*—namely, as a means of sharpening the contrast between those who practiced Galen's scientific method and adherents of contemporary philosophical schools. But we have not yet sufficiently examined why Galen elected to use horological examples in particular, instead of focusing on any other process grounded in *apodeixis* and a combination of rationalism and empiricism. If Galen's scientific method was as universally applicable as he claimed, then the number of potential examples at his disposal must have been myriad. The question at the heart of this chapter is: what can Galen's decision to use clock-making examples within a treatise on ethical psychology and scientific methodology tell us about the semiology of clocks within Galen's broader sociocultural environment?

This chapter seeks to demonstrate that, by Galen's time, clocks—and particularly sundials—had become well-known, overly determined symbols among the Roman elite, symbols that were both morally and politically connotative and closely associated with Greek *paideia* and scholarship. I argue that, in *Affections and Errors*, Galen appropriates and adapts these symbols such that, for him, the process of clock construction becomes a model for the process of correct and morally upstanding decision-making. In order to get a sense of the semiotic fields of sundials and water clocks under the Empire, we will consider the development of these symbols over time, in both literary production and material culture. This will enable us to better appreciate the range of lenses through which members of Galen's elite readership may have interpreted his accounts of sundial and water clock construction. This investigation will explicate why sundials and water clocks were potent symbols, well-suited to his argument in *Affections and Errors*, and how Galen may have been manipulating these symbols to serve his own rhetorical and intellectual purposes.

A Very Public Gnomon: Timekeeping and the Imperial Program

In Rome's Campus Martius, a two-kilometer-square assembly ground with powerful martial and political connotations, there stood a giant obelisk, transported from

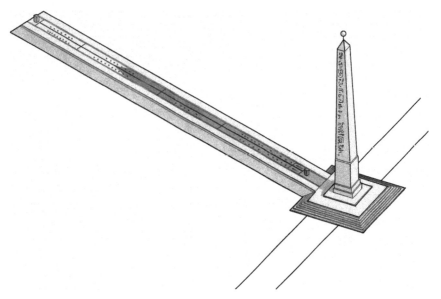

Figure E A schematic representation of the Montecitorio obelisk acting as a meridian dial. Image © Kenneth Jackson.

Egypt and erected by Rome's first emperor, Augustus, in 10 BCE. As Galen, or any other visitor to Rome, walked in view of this structure, he would have noticed that the obelisk cast an enormous shadow over the surrounding stone pavement, a shadow that, at the sixth hour of the day, would intersect with a ruled line marked out on the pavement in bronze (Figure E). Pliny, writing in the mid-first century CE, describes the form and function of this line in the following way:[1]

> The [obelisk] that has been erected in the Campus Martius has been applied to a singular purpose by the late Emperor Augustus; that of marking the shadows projected by the sun, and so measuring the length of the days and nights. With this object, a stone pavement was laid, the extreme length of which corresponded exactly with the length of the shadow thrown by the obelisk at the sixth hour on the day of the winter solstice. After this period, the shadow would go on, day by day, gradually decreasing, and then again would as gradually increase, correspondingly with certain lines of brass that were inserted in the stone; a device well deserving to be known, and due to the ingenuity of Facundus Novus, the

---

[1] Plin. *NH* 36.15: *ei, qui est in campo, divus Augustus addidit mirabilem usum ad deprendendas solis umbras dierumque ac noctium ita magnitudines, strato lapide ad longitudinem obelisci, cui par fieret umbra brumae confectae die sexta hora paulatimque per regulas, quae sunt ex aere inclusae, singulis diebus decresceret ac rursus augeresceret, digna cognitu res, ingenio Facundi Novi mathematici. is apici auratam pilam addidit, cuius vertice umbra colligeretur in se ipsam, alias enormiter iaculante apice, ratione, ut ferunt, a capite hominis intellecta.* Tr. Bostock and Riley (1855).

mathematician. Upon the apex of the obelisk he placed a gilded ball in order that the shadow of the summit might be condensed and agglomerated and so prevent the shadow of the apex itself from running to a fine point of enormous extent; the plan being first suggested to him, it is said, by the shadow that is projected by the human head.

Several scholars, from Athanasius Kircher in the seventeenth century to E. Buchner and his supporters in the twentieth, have proposed that this obelisk was the gnomon of a massive sundial, and that the line described by Pliny was part of that sundial's vast net of hour- and date-lines.[2] However, the textual and archaeological evidence since provided by scholars such as E. Rodríguez-Almeida, M. Schütz, and P. Heslin indicate that this line was, in fact, a solar meridian.[3] A solar meridian is a line running north–south that, each day at noon, receives the shadow cast by a gnomon and allows a viewer to track the waxing and waning of daylight throughout the year (since the shadow cast by the gnomon will be shortest at the summer solstice, when the sun reaches its highest point, and longest at the winter solstice, when the sun reaches its lowest point). Thus, even though this prominent monument was not itself a sundial, it represented an important timekeeping tool that enabled a viewer to locate him- or herself within the solar year. By extension, this also enabled the viewer to approximate the length of a seasonal hour at that point within the year (since, we recall, the length of a daytime seasonal hour was determined by dividing the total length of daylight into twelve equal parts, yielding longer hours in the summer and shorter hours in the winter).

What symbolic resonances might this colossal timepiece have sent thrumming through the minds of passersby, including those of Galen's elite readers? Our passage from Pliny connects the Campus Martius obelisk to several semiotic domains, each of which, I argue below, has bearing on Galen's decision to use sundials and water clocks as *paradeigmata* in *Affections and Errors*. The first of these, moving sequentially through Pliny's passage, is the semiotic domain of imperial power—specifically, the power of the Emperor Augustus. As Schütz and Heslin have observed, the obelisk was erected only two years after Augustus was elected Pontifex Maximus and received the attendant responsibility of seeing to the Roman civil calendar. This calendar had been recently reformed by Julius Caesar so as to last 365¼ days,[4] but in 9 BCE, only one year after the obelisk-*cum*-solar-meridian was erected, it was officially recognized that this new civil calendar had slipped and was now noticeably in error. That same year, Augustus commanded that the appropriate corrections be made to the calendar, thereby

---

[2] A. Kircher, *Obeliscus Pamphilius* (1650); Buchner has published extensively on this topic, beginning with Buchner 1976b.
[3] Rodriguez-Almeida 1978; Schütz 1990; Heslin 2007. Some more recent treatments include Lowrie 2009; Hannah 2011; Haselberger 2011; Baiocchi et al. 2016; Frischer et al. 2017.
[4] It had previously been a lunar calendar.

restoring harmony between celestial time and Roman civic time. Thus, the obelisk itself could be read as a symbol of the emperor's commitment to maintaining this harmony between state and cosmos—and, in a sense, collapsing the distinction between the two.[5]

The importance of this obelisk to Augustus's public relations program is underlined by the fact that its position within the Campus Martius placed it in visual alignment with two other monuments that celebrated the emperor's life and achievements: namely, the Ara Pacis and Augustus's Mausoleum. As Heslin has pointed out, "it is clear that this trio of monuments was deliberately arranged in the form of a right triangle, and that the orientation of the individual monuments emphasized this mutual relationship."[6] Heslin has further argued that the meridian, through its close association with a monument of war (the Mausoleum) and a monument of peace (the Ara Pacis), "not only measured the progress of the sun through the year; it also recorded the dedicator's progress from triumvir to princeps, from man of war to man of peace."[7] According to this interpretation, the obelisk became a symbol not only of harmony between Rome and the heavens but also of harmony within the Roman empire itself. Orderly and correct temporal regulation was thereby portrayed as an important ingredient of the *pax Romana* and as a defining feature of the new Roman society inaugurated by Augustus.

Other of Augustus's actions highlight the harmonious relationship with both celestial and terrestrial timekeeping that he personally enjoyed. Suetonius, for example, tells us that Augustus published his own horoscope and issued coins that featured zodiacal imagery derived from it.[8] These moves would not only have reminded viewers of Augustus' calendrical accomplishments, but might also have suggested to them that Augustus had a personal horoscope to be proud of, and that his reign had received the imprimatur of celestial approval. Suetonius also describes how Augustus was in the habit of meticulously time-stamping his personal correspondence, recording not only the day but also the very hour at which his letters were written.[9] This detail provides further support for the idea that Augustus wished to cultivate a reputation for temporal exactitude and perhaps for rigorous self-regimentation, prefiguring other "great men" like Benjamin Franklin, whose carefully structured daily schedules have been widely admired and seen as indicators of a well-regulated life.[10]

---

[5] For this point and the following discussion of the obelisk's relationship to the Ara Pacis, I am particularly indebted to Heslin 2007. For more on this relationship, see also Frischer et al. 2017.

[6] Heslin 2007, 15.   [7] Heslin 2007, 16.

[8] Suet. *Aug.* 94.12. For examples of Augustan-era coins with astrological imagery, see, e.g., BMC 12.3.80 (reproduced in Evans 1999, 297) and for further discussion, see Barton 1994. On the use of astrology in Roman Imperial rhetoric more generally, see Cramer 1954; Barton 1995; Oestmann et al. 2005, 1–94.

[9] Suet. *Aug.* 50. This and other references to the hour in Latin letters, are discussed by Wolkenhauer (2018).

[10] On how this attitude toward scheduling was influenced by Benedictine monasticism, see Zerubavel 1980. On the concept of the "ordered day" in Rome, see Ker 2023.

Certain later emperors who wished to establish a close connection between their own reigns and that of Augustus did so, in part, by emphasizing their investment in cosmic and civic time. The archaeological evidence from the Campus Martius, for example, may indicate that Domitian refurbished Augustus's solar meridian at the end of the first century CE.[11] Furthermore, the base of the Column of Antoninus Pius, a cenotaph for the deceased emperor erected in 161 CE by his successors Marcus Aurelius and Lucius Verus, depicts Augustus's obelisk prominently. On this base, within a larger scene of the emperor's apotheosis, a personification of the Campus Martius reclines with Augustus's obelisk projecting, like a powerful and pyramidal phallus, from his groin.[12]

Many of the aforementioned ideas—such as the link between temporal regulation and the Roman imperial program, as well as the value placed on a well-regimented lifestyle—are also manifest among the Roman elite of this period. For instance, a trope develops within Augustan-era literature of contrasting the leisurely chronotope of the Roman villa and countryside with the briskly regimented chronotope of the Roman *urbs* or city center.[13] In one of Martial's epigrams, for example, the poet walks us through a "typical" day in Rome, which is structured hour by hour:[14]

> The first and the second hour of the day fatigue the greeting clients.
> The third exercises the booming voices of the lawyers.
> Until the [end of the] fifth, Rome pursues various activities.
> The sixth offers rest to the exhausted; the seventh will end this rest.
> The eighth serves for sport, shining with sweat, until the [end of the] ninth;
> the ninth commands us to drop onto the readied couches.
> The hour of my little books is the tenth, Euphemos,
> when your benevolence offers divine dishes,
> and the good emperor relaxes with celestial nectar
> and holds a modestly measured cup in his powerful hand.
> Then let the jocular poetry come to him: because my Thalia dreads
> With her casual step to approach morning Jupiter.

---

[11] For discussion of Domitian's possible motivations, see Heslin 2007, 16–19. Frischer and Fillwalk (2013), however, are skeptical of a Flavian phase.

[12] Rome, Vatican Museums, Cortile delle Corazze Inv. No. 5115. The base inscription is *CIL* 6.1004.

[13] For further discussion of the role that timekeeping plays in this poem, see Wolkenhauer 2011, 112–13; 2020, 229–31.

[14] Mar. *Ep.* 4.8: *Prima salutantes atque altera conterit hora, / exercet raucos tertia causidicos, / in quintam uarios extendit Roma labores, / sexta quies lassis, septima finis erit, / sufficit in nonam nitidis octaua palaestris, / imperat extructos frangere nona toros: / hora libellorum decuma est, Eupheme, meorum, / temperat ambrosias cum tua cura dapes / et bonus aetherio laxatur nectare Caesar / ingentique tenet pocula parca manu. / Tunc admitte iocos: gressu timet ire licenti / ad matutinum nostra Thalia Iouem.* Tr. Wolkenhauer (2020, 230) with modification.

That a connection was perceived between temporal regimentation, on the one hand, and the idea of imperial "Romanness," on the other, is further illustrated by the fact that, as J. Bonnin has demonstrated, many elites with magistracies in Roman colonies (particularly in the West) developed the habit of erecting sundials in public spaces as acts of euergetism and demonstrations of allegiance to Roman temporal norms.[15]

Thus, by Galen's day, the time-telling gnomon and the tools that relied on it (e.g., solar meridians and sundials) had become associated with concepts like cosmic and civic harmony, the authority of the imperial program, and the rhythms of elite urban life. Therefore, by using the process of sundial-construction as a central *paradeigma* in *Affections and Errors*, Galen can make several implicit claims about the status of his own scientific method. He can assert, for instance, that, in its careful balance of rationalism and empiricism, his method likewise cultivates harmony between humans and nature, in that it clarifies the relationship between observed phenomena and the logical theories developed to explain them. He can contend that his method also promotes concord between humans because its universally accepted axioms encourage broad scientific consensus. Furthermore, Galen can claim that his scientific method is capable of structuring individual decision-making in a productive and well-regulated fashion, and that, therefore, it deserves to be the "ruler," or ultimate authority, that governs individuals' actions.

Our earlier passage from Pliny indicates that gnomons in general and sundials in particular acquired other connotations, as well. Pliny tells us not only that the Campus Martius obelisk was a sign of the close connection between well-regulated timescapes and imperial power, but also that the obelisk was a product of "the ingenuity of Facundus Novus, the mathematician," inspired in particular "by the shadow that is projected by the human head." In what follows, we will see how, by Galen's time, sundials (and, to a certain extent, water clocks) could also be used as symbols of celestial geometry, Greek scholarship, and aspects of the human body. Furthermore, we will see how underlying many of these connections is the idea of the clock or hour as a microcosm of a larger temporal concept, such as celestial time, eternity, or the human lifespan. Certain Stoic authors of the first and second centuries CE, such as Seneca and Epictetus, developed this microcosm–macrocosm relationship in their writings, using clocks and hours as metaphors for larger concepts. I suggest it is no coincidence that Stoic thinkers are among the very groups that Galen, in *Affections and Errors*, is keen to take to task, and I illustrate how Galen, in this text, is deploying a set of symbols that already bore significant weight within contemporary intellectual communities.

---

[15] See Bonnin 2015, 247–50.

Clocks and Celestial Geometry: Symbols of Mathematical Ingenuity

The quintessential ancient Greek sundial design—not necessarily the earliest one—consisted of a concave spherical surface sculpted in a block of stone and a shadow-casting metal rod, or gnomon, whose tip was located at the sphere's geometrical center. The surface thus acted as an inverted but otherwise exact copy of the celestial sphere, on which the shadow point marked the sun's current location.... The spherical sundial was not merely an instrument for time-telling but a vivid didactic image of the foundations of geometrical astronomy.[16]

To accurately tell time within the day was not the only—or, in some cases, even the primary—goal of all Greco-Roman clocks. As the quotation from A. Jones illustrates, sundials were, fundamentally, ways of representing the arc of the sky and the sun's daily and annual progress across it. Thus, in addition to being time-telling tools, such clocks could also serve important didactic purposes, helping viewers to visualize solar motion and to understand how it affected their own perceptions of temporal rhythms. In doing so, Greco-Roman clocks also made implicit claims about the structure of the cosmos. Inherent in the design of these sundials and water clocks is a set of assumptions often collectively called the "Two-Sphere Model."[17] This model is founded on the beliefs that (a) we, as observers, are standing on a terrestrial sphere (the Earth) and that (b) the stars and planets reside within a second spherical shell that rotates above us uniformly.[18] This is a fundamentally *geometric* view of the cosmos and requires that anyone hoping to track solar, lunar, planetary, or sidereal motions be well-versed in a form of mathematics known as celestial geometry.[19]

Certain sundial and water clock designs seem to foreground the mathematical virtuosity of their designers—even, on occasion, at the expense of the clock's legibility. Planar and spherical dials, for example, presented designers with mathematical challenges, since they required the designer to translate the arcs of solar motion from a hemisphere (that of the observable sky) onto another geometrical shape. Roofed dials, meanwhile, playfully inverted the very idea of a gnomon. Instead of casting a linear shadow onto an illuminated net of hour and date lines, as most sundials do, the net of hour- and date-lines on a roofed dial is entirely in shadow, and the time is indicated by a pinprick of light shining through a hole in

---

[16] A. Jones 2016, 25.
[17] A. Jones 2016, 24. On Greek astronomy more generally, see Evans 1999.
[18] Though certain celestial bodies (e.g., planets) were understood to have independent uniform motions within the celestial sphere.
[19] On spherical trigonometry in Greek celestial mathematics, see Van Brummelen 2009, 33–93; 2013, 42–58.

Figure F Roofed spherical sundial with Greek inscriptions. First–second century CE, marble, Roman. H. 30 cm, Diam. 73 cm. Inv. MA5074. Image © RMN-Grand Palais/Art Resource, NY, photographer Hervé Lewandowski.

the dial's roof (Figure F). While such clocks are very impressive, they are difficult to read and thus impractical as timekeeping tools. The relative abundance and diversity of such virtuosic pieces suggests that many clocks acted primarily as prestige symbols that celebrated mathematical ingenuity. We see this idea reflected in Cicero's *On the Nature of the Gods*, for example, when he declares, "When we see something moved by machinery, like an orrery or clock or many other such things, we do not doubt that these contrivances are the work of reason."[20]

That this kind of ingenuity was worthy of celebration is further illustrated in book 9 of Vitruvius's *On Architecture*, where he lists the inventors of various types of sundial and water clock:[21]

---

[20] Cic. *De nat. deo.* 2.97: *an, cum machinatione quadam moveri aliquid videmus, ut sphaeram ut horas ut alia permulta, non dubitamus quin illa opera sint rationis.* Tr. Rackham (2005). Cf. *Rep.* 1.14.21–2, *Tusc.* 1.63.

[21] Vitr. 9.8.1–2: *Hemicyclium excavatum ex quadrato ad enclimaque succisum Berosus Chaldaeus dicitur invenisse; scaphen sive hemisphaerium Aristarchus Samius, idem etiam discum in planitia; arachnen Eudoxus astrologus, nonnulli dicunt Apollonium; plinthium sive lacunar, quod etiam in circo Flaminio est positum, Scopinas Syracusius;* πρὸς τὰ ἱστορούμενα *Parmenion,* πρὸς πᾶν κλῖμα

The semicircular form, hollowed out of a square block, and cut under to correspond to the polar altitude, is said to have been invented by Berosus the Chaldean; the Scaphe or Hemisphere by Aristarchus of Samos, as well as the disc on a plane surface; the Arachne, by the astronomer Eudoxus or, as some say, by Apollonius; the Plinthium or Lacunar, like the one placed in the Circus Flaminius, by Scopinas of Syracuse; the *pros ta historoumena*, by Parmenio; the *pros pan klima*, by Theodosius and Andreas; the Pelecinum, by Patrocles; the Cone, by Dionysodorus; the Quiver, by Apollonius.... Methods of making water clocks have been investigated by the same writers, and first of all by Ctesibius the Alexandrian, who also discovered the natural pressure of the air and pneumatic principles.

By carefully listing each clock type and its inventor by name, Vitruvius acknowledges these inventions as noteworthy achievements and their inventors as brilliant men who deserve commemoration.[22] It is also significant that, to the best of our knowledge, most (if not all) of the men in Vitruvius's list were native Greek-speakers, a fact that concords with the general perception, during the Roman period, of mathematics and astronomy as peculiarly "Greek" arts.[23] Thus, while we have seen that sundials and water clocks could, from one perspective, be viewed as symbols of Roman imperialism, they could also, from another, be seen as symbols of a markedly Greek form of learning.

Galen's choice to use sundial and water clock construction as his central paradigm in *Affections and Errors* allows him to capitalize upon these connections and to present his scientific method as not simply an example of geometrical virtuosity but as the *very basis* for this kind of ingenuity. Vitruvius himself actually makes a similar claim for the status of the principles underlying his own *tekhnē* of *arkhitektonia* (or *ars* of *architectonice* in Vitruvius's Latin):[24]

> The analemma [i.e., the mathematical diagram that enabled clockmakers to produce appropriate nets of hour- and date-lines] is a basis for calculation deduced from the course of the sun and found by observation of the shadow as it increases until the winter solstice. By means of this, through architectural principles and the employment of the compasses, we find out the operation of the sun in the universe.

Theodosius et Andrias, Patrocles pelecinum, Dionysodorus conum, Apollonius pharetram.... Item sunt ex aqua conquisitae ab eisdem scriptoribus horologiorum rationes, primumque a Ctesibio Alexandrino, qui et vim spiritus naturalis pneumaticasque res invenit.

[22] The progress narrative that Vitruvius provides here aligns most closely with Asper's narrative of accumulation, the style most popular within Greek mathematics (Asper 2013, 414–18).

[23] A. Jones also notes that "time-telling devices from the western, Latin-speaking parts of the Roman Empire not infrequently bear inscribed labels in Greek" (2016, 23, with an example at fig. III-14).

[24] Vitr. 9.1.1: ἀνάλημμα *est ratio conquisita solis cursu et umbrae crescentis ad brumam observatione inventa, e qua per rationes architectonicas circinique descriptiones est inventus effectus in mundo.*

Here Vitruvius claims that "architectural principles" are at the foundation of sundial and water clock construction, thereby claiming for *architectonice* the prestige and intellectual sophistication associated with those devices. I suggest that, in *Affections and Errors*, Galen is playing a similar game, attempting to coopt for himself and for his scientific method the reputation for ingenuity that attended sundials and water clocks. This strategy also allowed Galen to emphasize his connection to Greek intellectual heritage, an important theme within his writings to which we will return in the next section, as well as in the following chapter.

Clocks and the Educated Elite: Symbols of the *Pepaideumenoi*

The connection between clocks and ingenuity allowed these devices to participate in another semiotic trend that developed within the iconography of the Roman Imperial period. During this time, it became common for elites to represent themselves in their portraits as *pepaideumenoi*, individuals who had successfully completed the course of Greek-inflected education known as *paideia*.[25] Particularly in the second and third centuries CE (i.e., during Galen's lifetime), we see an increase in the number of portrait busts that depict their subjects clothed in a *himation*, a garment which, in earlier periods, was closely associated with "Greekness" in general and with Greek learning in particular. Regarding the thoughtful expression on many of these portraits' faces, B. Borg has observed that, "when depicted in combination with a bare chest and *himation*, its most obvious association will be with Greek *paideia*, and in this context also the beard will add to the overall picture of someone advertising his Greek education. The same is true of papyrus rolls [i.e., *volumines*], so often depicted either carried in one hand or gathered in a bundle or in a box near the patron's feet."[26] Hence, by Galen's time, the ingredients which, when assembled in various combinations, yielded a *pepaideumenos* could include a beard, a *himation* (preferably covering a bare chest), one or more papyrus scrolls, and an intense expression.

As early as the first century BCE, sundials start to be incorporated into this assemblage. It is at this time that we begin to see engraved gemstones which play with the motif of a bearded thinker sitting near a sundial. Lang has developed a typology for these images (Figure G).[27] On "Type A" gems, this thinker "Denker" in Lang's German is seated before a sundial and reading from a papyrus scroll.[28]

---

[25] The bibliography on *paideia* is extensive. For overviews of the concept and its connection to the period known as the Second Sophistic see, e.g., the contributions in Borg 2004b; Johnson 2017.
[26] Borg 2004a, 162.   [27] Lang 2012, 80–109.
[28] E.g., *Abdruck der Impronte Gemmarie dell'Istituto*, 4 (1837), 81; Landesmuseum Mainz. Inv. R 6036; Bibliothèque Nationale of Paris, Cabinet des Médailles, collection of Slg. Duc de Luynes, No. 127; the Römisch-Germanisches Museum of Cologne, Inv. 8827; the Rheinisches Landesmuseum of Bonn, Old inv. G139, Inv. U1816; the Staatliche Münzsammlung of Munich, Inv. A.197.

Figure G Gem with sitting, reading thinker in front of a sundial (Lang Type A). Staatliche Münzsammlung, Munich, Inv. A.197. Reproduced with permission from J. Lang, *Mit Wissen geschmückt? Zur bildlichen Rezeption griechischer Dichter und Denker in der römischen Lebenswelt*, MAR 39 (Wiesbaden 2012), 170, Abb. 165, Kat. Nr. G TypA65, Taf. 21. Image © Köln Digital Archaeology Laboratory, photographer Jörn Lang.

In "Type B," he sits contemplatively with his chin in his hand,[29] while the seated man on "Type C" gems gestures in an argumentative manner toward the sundial in front of him.[30] The man on "Type D" gems maintains this same posture while holding a pointer, as if he were indicating something to a student or disputant (who is not pictured).[31] Finally, on "Type E" gems, we see him seated before a sundial and scribbling away on a papyrus scroll.[32] This trope of thinker and

---

[29] E.g., the Antikensammlung, Staatliche Museen of Berlin, Inv. FG 4523.
[30] E.g., University of Pennsylvania Museum, Inv. 29-128-1149.
[31] E.g., Staatliche Museen of Berlin, Antikensammlung, Altes Museum, Inv. FG 4520, FG 1262; Museo Archeologico Nazionale di Napoli, Inv. 158777; Kunsthistorisches Museum of Vienna, Inv. XIB538; Museo Archeologico Nazionale of Aquileia, Inv. 48593; Archäologisches Institut of Göttingen, Georg-August-Universität, Inv. G443; Museum August Kestner in Hannover, Inv. K1033.
[32] E.g., Geldmuseum of Utrecht, Inv. RRC 2075.

sundial is analogous to the numismatic motif in which a bearded thinker appears seated before a *sphaera*.[33]

By the Imperial period, related motifs, also incorporating a sundial, begin to appear on certain mosaics and sarcophagi of the elite.[34] In some cases, the sundials depicted there may have been intended to help the viewer identify a particular thinker from Greek antiquity. The famous "Anaximander Mosaic" (third century CE) depicts a *himation*-clad thinker holding a multi-faced sundial on his lap, and may represent Anaximander's supposed invention of the gnomon (although this identification is controversial) (Figure C, in Chapter 1).[35] A mosaic from the villa of T. Siminius Stephanus in Pompeii (mid-first century BCE/CE) has been said to represent Plato's Academy, featuring seven sages gathered in discussion around a globe, with a sundial perched on a column behind them.[36] Most significantly, though, on many elite sarcophagi of the Imperial period, the attributes of the bearded intellectual, including the sundial, are transferred to the deceased. Traversari has grouped these scenes into three general types.[37] In Type I scenes, the Muses gather round the deceased,[38] who is presented in the guise of a poet or philosopher, while a sundial stands mounted on a column in the background.[39] Type II scenes build upon the gem motifs, showing the deceased standing or seated upon a curule chair, often wearing a Greek-style *himation* with no undergarment, facing a sundial, and holding a *volumen* (Figure H).[40] Finally, Type III scenes portray a scholarly lesson or other intellectual discussion. The deceased is

---

[33] E.g., SNG von Aulock 733 (Valerian II; Nikaia, Bithynia; depicting Hipparchos) and Gardner, Period IX, nos. 15 and 16 (Trajan Decius and Commodus; Samos; depicting Pythagoras). Interestingly, I have been unable to find even a single coin on which the *sphaera* has been replaced with a sundial.

[34] Though a few examples are known from earlier. The funerary *stēlē* of Theodotos, e.g., now at the Istanbul Archaeological Museum, has been dated to the second century BCE (Gibbs 1976, No. 1051G).

[35] The mosaic has been dated on stylistic grounds to the third century CE and is currently in the Rhineland Museum, Trier. The tradition that Anaximander was the first to set a *gnomon* on a shadow-casting surface is preserved at DL 2.1. However, the identification of the figure in the Trier mosaic is based solely on the fact that he is holding a sundial. Traversari has argued that the figure should rather be identified as Patrocles, the inventor—so Vitruvius tells us—of the very double-winged *pelecinum* dial that rests on the figure's lap (1991, 69).

[36] For discussion of this mosaic, see Gaiser 1980. On temporal themes in Roman mosaics, see Kondoleon 1999.

[37] Traversari 1991, 66–7. This classification has also been adopted by Bonnin 2013, 482–3. The designation of "Type Number" is my own. Extensive bibliography on these sarcophagi can be found at Traversari 1991, 71–3. For further discussion of the "philosopher" as a motif on Imperial-period sarcophagi, see Ewald 1999.

[38] E.g., a sundial dedicated to the "Nine Muses" (Gibbs 1976, No. 1004 = *I.Aeg. Thrace* 436).

[39] Sarcophagi of Type I are preserved in the crypt of the Duomo of Palermo (Wegner 1966, 33–4, no. 68, pls. 66, 138a–b); the Castello di Agliè, Piemonte (Wegner 1966, 9, no. 2, pls. 35, 45b, 47a, 49a); the Villa de Medici at Rome (Wegner 1966, 82, no. 215, pls. 27a, 29, 30, 43a); the Museo Maffeiano of Verona (Wegner 1966, 87, no. 227, pl. 127b); the William Randolph Hearst Estate in San Simeon, CA (Wegner 1966, 83, no. 219, pl. 31a); the Santa Maria del Priorato in Rome (Wegner 1966, 71–2, no. 183, pls. 36a, 44b, 46a, 47b, 49b); the Cortile del Belvedere in the Vatican Museum (Wegner 1966, 56–7, no. 137, pl. 37b); and the cathedral of Murcia in Spain (Wegner 1966, 28–9, no. 57, pls. 104a, 107c).

[40] Other sarcophagi of Type II are preserved at the Museo Profano Lateranense ("Sarcophagus of Plotinus," Wegner 1966, 47, no. 116, pls. 64b, 70, 71; a second sarcophagus, Grabar 1967, 140, fig. 144);

Figure H Muse sarcophagus. Detail of side panel, depicting a seated "philosopher" looking up at a sundial (Traversari Type II). ANSA I 171, Roman, 180–200 CE, marble, Roman, findspot unknown. H. 67 cm, L. 67 cm. Image © KHM-Museumsverband.

seated at the center, often holding a *volumen* and sporting a long beard and *pallium*, a woolen cloak akin to the Greek *himation* and favored by members of the Imperial-period elite. In these scenes, as in those of Type I, a sundial stands watch in the background.[41]

---

the museum of the Villa Torlonia in Rome (Wegner 1966, 53ff., no. 133, pls. 60, 61, 62, 64a, 73a); in the cathedral of Cagliari (G. Koch and Sichtermann 1982, 205); and in the National Museum at Naples (Himmelmann-Wildschütz 1973, 5, pl. 3).

[41] Type III sarcophagi can be found at the National Museum in Rome (two sarcophagi; see Marrou 1938, 64ff., no. 54, fig. 9, and 97ff., no. 98, fig. 13); the museum of the Castello Sforzesco in Milan (Wegner 1966, 25–6, no. 47, pl. 111b); the catacombs of San Callisto in Rome, inside the Tricora Orientale (two fragments; Wegner 1966, 42, nos. 89 and 91, pl. 80a); the catacombs of Praetestato (Wegner 1966, 44, no. 105, pl. 80b); the Museo Terme at Rome (Wegner 1966, 52, no. 129, pl. 119a, 124a); and the cloister of San Paolo (extramural) in Rome (Wegner 1966, 73, no. 187, pls. 107d, 119c).

Whether or not this symbol system should be associated specifically with Greek *philosophy*—rather than with Greek *paideia* more generally—has been a matter of some debate.[42] It is equally uncertain whether the sundial should be understood as the attribute of a particular *kind* of scholar—perhaps a philosopher, astronomer, or astrologer—or whether it should simply be read (like the beard, *himation,* and papyrus roll) as another broad, non-disciplinary symbol of the *pepaideumenos*.[43] What is clear, however, is that within such visual discourse, the sundial could be a symbol of affluence, erudition, and a particular brand of Greek learning. In light of this awareness, it may seem unsurprising that Galen, himself an elite *pepaideumenos* writing for other members of this class, would trade upon the social currency of sundials in *Affections and Errors*.[44] In what follows, I will develop the idea that, in this text, Galen is laying claim to a recognizable symbol of elite "Greek" intellectualism and adapting it to support his central claim that his own method of scientific inquiry is more intellectually valid than the dogmas of contemporary philosophical schools.

Clocks and Philosophers: Symbols of the Human Lifespan

Interestingly, sundials and water clocks also appear as symbols within the writings of two Stoic philosophers from the first and second centuries CE, Seneca (c.4 BCE–65 CE) and Epictetus (c.55–135 CE). An investigation of these passages will illuminate not only the general connection between clocks and scholarship that existed during this period, but also the ways in which two specific thinkers incorporated such symbols into their philosophical discussions. It is also worth recalling that, in *Affections and Errors,* Stoicism itself comes under direct fire. Thus, by exploring how certain Stoics engaged with clocks and hours, we can better appreciate how Galen's decision to use these symbols allowed him to enter into further dialogue with his imagined interlocutors.

---

[42] Zanker (1995, 190–206), e.g., has argued that a beard marked its wearer's affinity for Greek philosophy, in particular. Smith and Borg, however, have objected that visual iconography is too polysemous to support such a specific correlation and instead see the beard, *himation,* etc. as attributes of the *pepaideumenos* more generally (R. R. R. Smith 1999; Borg 2004a).

[43] Borg (2004a, 167) seems to interpret the sundial as an indication that the subject practiced astronomy and/or astrology. Meanwhile, Marrou (1938, 30), Cumont (1942, 336), and Bonnin (2013, 482–3) have suggested that the sundial be considered the attribute of a philosopher, because philosophers often wrestled with the concept of time and the associated issues of birth, death, and change. Furthermore, Marrou (1938, 288) and Traversari (1991, 70–1) have suggested that, in these contexts, the sundial may have served as a symbol of the kind of immortality available to those who live in accordance with philosophical teachings.

[44] Recall, for instance, how Galen states that only a man who "is prudent by nature and who received the education (παιδείαν) that has been esteemed by the Greeks from the beginning" (*Aff. pecc. dig.* 51.24–52.2 DB = 5.75.13–15 K) can hope to master Galen's methodology.

In the *Epistles to Lucilius*, Seneca often employs metaphors to help his readers grasp the fundamentals of Stoic philosophy.[45] Such metaphors are drawn from diverse social spheres—law, finance, medicine, daily life—and serve to render familiar and concrete concepts that might otherwise appear foreign and abstract. One of these challenging concepts is the notion of "time." As M. Armisen-Marchetti has persuasively argued, a theme throughout the *Epistles* is the presentation of time as a tangible good with inherent value which one should not squander.[46] This serves one of Seneca's central paraenetic aims: to encourage Lucilius, and each future reader, to take ownership of the time allotted him upon this earth, for it is a finite and irreplaceable resource.[47]

Seneca often employs water metaphors to make the flow of time vivid to his reader's eyes. Most commonly, he presents time as a river that is turbulent and panic-inducing for those who have not mastered the Stoic precepts, but placid and pleasant for the lucky few who have.[48] In *Epistle* 24, however, Seneca includes a telling variation on this theme; he compares a human lifespan to the water draining from a water clock.[49]

> Just as it is not the final drop that empties the water clock, but whatever has flowed out before, so the final hour in which we cease to exist is not the only one to cause death; it is simply the only one to complete it. At that time, we arrive at death, but we have long been coming to it.

Seneca's particular concern in this letter is to assuage Lucilius's fear of death.[50] He does so by emphasizing the fact that we have all been dying steadily since the moment of our birth. Therefore, we should not consider our inevitable demise to be an aberration, an unnatural snatching-away of existence, but simply the final step in a continual, lifelong process of passing away. The water clock is an apt metaphor for a human lifetime viewed in this manner, for the dropwise

---

[45] On Senecan metaphors, see esp. Wilson 1987; Armisen-Marchetti 1989, 1995; Lotito 2001, 144–65; Bartsch and Wray 2009.

[46] Armisen-Marchetti 1995, 549–53. See, e.g., Sen. *Brev.* 8.1–4.

[47] On achieving mastery over time, see, e.g., Sen. *Brev.* 15.5: "Therefore life permits much more to the wise man. The boundary that shuts out others does not do the same to him. He alone is released from the laws of mankind. All the centuries serve him as if he were a god. What of the time that has passed? This he has grasped with his memory. What of the present? This he utilizes. What of the future? This he has anticipated." Cf. *Brev.* 7.5; *Otio* 4 and 5.

[48] E.g., *Clem.* 2.6.1; *Ira* 2.20.3; *Ep.* 68.13, 99.27; *Const.* 8.2; *Tranq.* 2.4. Armisen-Marchetti (1989) has collected Seneca's water metaphors at p. 108 (including the reference to the water clock), and his river metaphors, specifically, at pp. 121–2.

[49] *Ep.* 24.20: *Quemadmodum clepsydram non extremum stillicidium exhaurit, sed quicquid ante defluxit, sic ultima hora, qua esse desinimus, non sola mortem facit, sed sola consummat; tunc ad illam pervenimus, sed diu venimus.* Cf. *Ep.* 120.17–18, which makes the same argument without using the metaphor of a water clock. See also *Ep.* 1.2–3.

[50] On death in Seneca's writings and portrayals of Seneca's own death, see Ker 2012.

reduction in its water level neatly parallels the incremental shortening of a person's remaining lifetime as the moments pass by.

This metaphor is also consistent with Seneca's goal of reifying time. Seneca depicts each moment as a concrete entity (a water droplet) with a measurable value (a fraction of an hour). These fractions of hours are all the more valuable because, here, a single draining of a water clock (i.e., one day) represents the length of an entire human life. When one views the day as a microcosm of human existence, there really is no time to spare![51] Seneca explores this conception of time in even greater detail in his twelfth epistle:[52]

> A whole lifetime is made up of parts and has larger circles enclosing smaller ones.... For the longest span of time has that which you would find in a single day, namely light and darkness, and a day continually produces more of those changes; it is no different when the day is shorter or again when the day is longer. Therefore, every day ought to be regulated as if it contained the whole temporal series and both consumed and filled out a life.

Seneca goes on to tell the story of one Pacuvius, who held a burial sacrifice for himself every evening, complete with wine and feasting. Each night, when the party was over, he would have himself "carried out" (*ferebatur*) from the dining hall to his bed chamber, while his eunuchs sang, "He has lived! He has lived!"[53] While Seneca hastens to express his disapproval of Pacuvius's behavior, he encourages his readers to adopt a similar attitude mentally: "Let us go to sleep happy and cheerful, and let us say: I have lived and that which Fortune assigned me I have completed."[54] Petronius, too, exploits the funerary connotations of sundials when he introduces the character of Trimalchio as "a very well-to-do

---

[51] On the concept of the day as a metaphor for life in Seneca's writings, see Ker 2023, 235–8.

[52] *Ep.* 12.6–8: *Tota aetas partibus constat et orbes habet circumductos maiores minoribus... nihil enim habet longissimi temporis spatium, quod non et in uno die invenias, lucem et noctem, et in aeternum dies vices plures facit istas, non alias contractior, alias productior. Itaque sic ordinandus est dies omnis, tamquam cogat agmen et consummet atque expleat vitam.* Cf. *Brev.* 7.9; *Ep.* 61.1 and 101.10.

[53] *Ep.* 12.8: βεβίωται, βεβίωται.

[54] *Ep.* 2.9: *nos ex bona faciamus et in somnum ituri laeti hilaresque dicamus: vixi et quem dederat cursum fortuna, peregi.* A similar scene occurs in Petronius's *Satyricon*, wherein the extravagant host Trimalchio, after delivering what amounts to a eulogy for himself, demands the following of his slave Stichus: "Bring me the funeral garb in which I intend to be carried out. And some ointment and a mouthful from that jar which must be poured over my bones" (77). This pageantry is but the climax of a dinner party that is regularly punctuated with reminders of the diners' mortality. After pouring the wine, for example, Trimalchio calls for a little silver skeleton to be brought to him and, while playing with it, declares in verse: "Alas, we're sad creatures, poor man's all nothing. / Thus we'll all be, after Orcus takes us away. / Therefore let's enjoy life, while we can live well" (34, tr. Heseltine 1987 with modification; cf. 72). It is interesting to note that, whereas the narrator judges the mock funeral to be "utterly sickening" (34), he says that Trimalchio's performance with the skeleton was met with praise (35). That it was, in fact, appropriate to contemplate death at a dinner party in this way is suggested by the number of skeletons that appear in banqueting scenes on Roman-period sarcophagi and gemstones (e.g., the sarcophagus of Polybius from Heraclion [see Stylianos 1968] and the gem at Furtwängler

man, who has a clock in his *triclinium* along with a uniformed trumpeter, in order that he might know from them how much of his life he has lost."[55] This way of reading a clock—i.e., as something of a *memento mori*—is analogous to Seneca's metaphor of the water clock.

Although clocks do not reappear elsewhere in the *Epistles*, the unit which they alone can measure, the hour, pops up in a variety of contexts, usually to indicate the brevity of a timespan or the suddenness of an event. In *Epistle* 91, for example, during a discussion of Fate's fickleness, Seneca exclaims that the day is too large a unit for measuring the speed at which one's fortunes can go belly-up: "Whatever one has built over a long stretch of time, with many labors and through the indulgence of the gods, a single day scatters and dissipates. In fact, he who said 'a day' granted a long delay to evils that are, in fact, rapid. An hour, a moment of time, suffices for overturning empires."[56]

It is with this connotation that we see the hour appear in Epictetus's second *Discourse*, as transcribed by his pupil Arrian. In this text, Epictetus underlines the importance of distinguishing between the factors in life that are under the control of one's internal volition (*prohairesis*) and those things that fall outside of its purview (*ta exō*).[57] By concerning oneself solely with the former and cultivating indifference toward the latter, so Epictetus argues, one can lead a happy life. To model how one might go about identifying the elements that are and are not under one's control, Epictetus first offers the example of a sea voyage:[58]

> What can I do? Pick the captain, the ships, the day, the opportune moment. But then—a storm has come upon us! Still, what does it matter to me? My part has been fulfilled. The plan of action belongs to someone else, to the captain. But the ship is in fact sinking! What can I do? I do the only thing that I am able: I drown unafraid, having neither cried out nor called upon God, but knowing that that which has come into being must also perish.

1900, pls. 46, 26; see Bonnin 2013, 486–7, for discussion). On the theme of the skeleton in funerary art more generally, see Dunbabin 1986. It is especially interesting, for our purposes, that sundials also participate in this symbol system, appearing alongside skeletons in the examples listed above.

[55] Petr. *Sat.* 26: *Trimalchio, lautissimus homo, horologium in triclinio et bucinatorem habet subornatum, ut subinde sciat, quantum de vita perdiderit.*

[56] *Ep.* 91.6: *Quidquid longa series multis laboribus, multa deum indulgentia struxit, id unus dies spargit ac dissipat. Longam moram dedit malis properantibus, qui diem dixit; hora 1 momentumque temporis evertendis imperiis sufficit.* Cf. *Tranq.* 11.9: "But a fraction of an hour exists between being on the throne and at another's knees."

[57] Arr. *Epict.* 2.5.4.

[58] Arr. *Epict.* 2.5.10–12: τί μοι δύναται; τὸ ἐκλέξασθαι τὸν κυβερνήτην, τοὺς ναύτας, τὴν ἡμέραν, τὸν καιρόν. εἶτα χειμὼν ἐμπέπτωκεν. τί οὖν ἔτι μοι μέλει; τὰ γὰρ ἐμὰ ἐκπεπλήρωται. ἄλλου ἐστὶν ἡ ὑπόθεσις, τοῦ κυβερνήτου. ἀλλὰ καὶ ἡ ναῦς καταδύεται. τί οὖν ἔχω ποιῆσαι; ὃ δύναμαι, τοῦτο μόνον ποιῶ· μὴ φοβούμενος ἀποπνίγομαι οὐδὲ κεκραγὼς οὐδ' ἐγκαλῶν τῷ θεῷ, ἀλλ' εἰδώς, ὅτι τὸ γενόμενον καὶ φθαρῆναι δεῖ.

Epictetus anticipates the objection, however, that it is difficult to face the prospect of drowning with calm indifference. To mitigate the terror that such a violent death might instill, Epictetus, like Seneca, uses a horological metaphor to stress the fact that, one way or another, we all die in the end:[59]

> For I am not an ever-lasting creature, but a man, a part of all things as the hour is a part of the day. I must be present like the hour and pass like the hour. What difference does it make to me then how I pass, whether I do so by drowning or falling ill with fever? For I must pass in some such way.

Rather than establish the whole day as a microcosm of human life, as Seneca does, Epictetus increases the magnification by an additional step, seeing a metaphor for human life within a fraction of a day, a single hour.[60] He builds on this metaphor to establish an analogic relationship: just as an hour (*hōra*) is the smallest measurable fragment of eternity (*aiōn*),[61] so is a human being the smallest measurable fragment of the all-pervasive Stoic god.[62] Thus we see that, by the time Galen began to write *Affections and Errors*, a treatise that offers critiques of Stoicism and other sectarian philosophies, at least two eminent Stoics had already used horological metaphors to explain important aspects of their worldviews.

Affections and Errors is fundamentally a philosophical work, and Galen, like Seneca and Epictetus before him, is using horology to make a claim about how best to live one's life.[63] However, his concern is not with the span of that life, but with the *processes* that humans use to distinguish truth from falsehood and thereby make "correct" decisions about what to believe and how to act. Thus, *Affections and Errors* emphasizes the process by which a clock is made rather than the symbolism of the completed object. Galen does not expect the readers of *Affections and Errors* to go out and build a clock; he wants them to use the process of clock-making both as a blueprint for evaluating the truth of philosophical claims and, by extension, as a model for developing a habit of rational thinking within their own daily lives. "I have discovered," Galen tells us about

---

[59] Arr. *Epict.* 2.5.13–14: οὐ γάρ εἰμι αἰών, ἀλλ' ἄνθρωπος, μέρος τῶν πάντων ὡς ὥρα ἡμέρας. ἐνστῆναί με δεῖ ὡς τὴν ὥραν καὶ παρελθεῖν <ὡς> ὥραν. τί οὖν μοι διαφέρει πῶς παρέλθω, πότερον πνιγεὶς ἢ πυρέξας; διὰ γὰρ τοιούτου τινὸς δεῖ παρελθεῖν με.

[60] Lucretius, too, contrasts the "single hour" that a person spends on this earth with the "eternal time" that awaits him or her after death (3.1071–4).

[61] The concept of eternity was often personified by the figure Aiōn, a popular theme in Roman art. On the development of *aiōn* as a temporal concept, see Lackeit 1916; Zuntz 1989, 1991, 1992; Parrish 1993; Zaccaria Ruggiu 2006.

[62] For Epictetus's characterization of humans as part of the divine whole, see Arr. *Epict.* 1.12.26; 2.8.10–11; 2.10.3; 4.7.6–7. For discussion, see Long 2002, 233–4. Cf. M. Aur. *Med.* 2.3–4.

[63] Though Galen is surprisingly cagey about what the ultimate *goal* of life might be. See Donini 1988; Singer 2013, 230.

his scientific method, "not only the goal but also the *way of life* that conforms to the truth."[64]

Clocks in Architectural Writing: Symbols of the Human Body

"Clock as body" and "clock as lifespan" analogies were also well-established in the various genres of architectural writing that became popular during the Imperial period. As the built environments of many urban centers became ever more elaborate and impressive, architectural features came increasingly to be the subject of literary treatment. One device commonly used for this purpose was *ekphrasis*, in which an author vividly describes a technical or artistic object in order to amplify its meaning and meditate upon larger themes.[65] Other strategies included composing lists of monuments, such as the "Seven Wonders of the World;"[66] travel accounts, like Pausanias' *Description of Greece*; and epideictic oratory,[67] wherein rhetors delivered speeches designed to complement or compete with the artistry of well-known civic structures (including, often, the very buildings in which the speeches were being performed). Most relevant to our purposes, however, was the habit of comparing architectural features to parts of the human body. Allegories of this kind depended on multiple levels of resemblance, one of which was purely physical.

---

[64] *Aff. pecc. dig.* 52.19–53.1 DB = 5.77.3–4 K, emphasis added: τῆς ἀληθοῦς εὑρηκυίας οὐ μόνον τὸ τέλος, ἀλλὰ καὶ τὸν ἀκόλουθον αὐτῷ βίον. Galen expresses this sentiment elsewhere in his corpus, e.g.: "And, as all matters of enquiry require two tools, reasoning and experience, for any discovery in all the arts, and not less in our whole life, I consider it necessary to seek for discovery of the facts before us by using reasoning alone or experience alone, or both together" (*Cur. rat. ven. sect.* 11.255 K; cf. *MM.* 10.272 K). On the interrelation of βίος and μέθοδος in the Galenic corpus, see Boudon-Millot 2009.

[65] Examples of famous *ekphraseis* include: Verg. *Aen.* 1.418–93, 6.20–30; Catull. 64; Philostr. *Im.* For ancient descriptions of the genre of *ekphrasis*, see the various *Progymnasmata*. For modern discussions, see esp. Elsner 1996; 2007; Webb 2016. Mesomedes Fr. 8 includes an excellent *ekphrasis* of an anaphoric water clock, a kind of water clock that was often outfitted with rotating models of the celestial bodies and a bell or whistle to announce the hour. This second-century CE poet presents the clock as a technical marvel and uses it as a jumping-off point for contemplating the cosmic order: "*Another to the clock*: / Who produced, by the art of bronze-forging, / the course of the blessed ones in order to measure the day? / Who arranged the course of the stars in a circle, / an all-bronze likeness of the cosmos, / having divided the matrix of the well-running lines, / having designated the pure road among the paths, / the number of the [zodiac] animals, three times four? /.../ After the boundless battle of heaven, / a bronze delight rang out, / making clear to mortals the measure of the day (Fr. 8 Heitsch). Cf. Fr. 7, titled "To the clock."

[66] E.g., the papyrus from Ptolemaic Egypt discussed at Diels 1904, 1–16.

[67] E.g., Luc. *Hipp.*, in which the artistry of Lucian's rhetoric vies with the imagined artistry of the baths commissioned by Hippias. Lucian also incorporates clocks into his description, presenting them as useful embellishments: "Besides [the baths'] adornment with other subtleties, furnished with two bathrooms and many means of egress, you will see two clocks, one that works by means of water and a bell, and another which operates by tracking the path of the sun" (8). For a discussion of this text as a commentary on the character of Hippias, see Thomas 2007, 228.

Building façades, for example, were often called "faces" (Latin *facies* or Greek *prosōpon*) and their walls "eyebrows" (Greek *ophrus*).[68] Greek authors sometimes exploited the homophony between a *metōpē*, a rectangular element in Doric friezes, and a *metōpon*, the human "brow."[69] Sundials were also incorporated into this kind of rhetoric, as analogs for the human face or head.[70] An epigram attributed to Trajan, for instance, exploits this analogy to humorous effect: "If you raise your nose to the sun and open up wide / you will mark off the hours for all who pass by."[71]

As an extension of these analogies, buildings and individual architectural elements were also frequently said to reflect the moral dispositions of their owners. Plato may have been influential in this trope's development since, in the *Timaeus*, he uses architectural terms to situate the human soul within the body and to describe the functions of each ensouled organ.[72] Plato explains, for example, that "the mortal part of the soul" resides in the cavity of the thorax, which is divided into two sections "in the way that women's and men's apartments are divided within houses."[73] A few lines later, Plato likens the human head to an acropolis and the heart to a guardroom.[74] This attitude leads to a broader

---

[68] E.g., *SEG* 2: 545 (Mylasa).

[69] On the use of facial terminology to describe building façades in general, see Orlandos and Tavlos 1986, 224. For the application of the terms "head" and "arms" to an aqueduct, see, e.g., *CIL* 9.3018 = *ILS* 5761. Greeks also compared elements of their natural environment to human body parts, such as breasts, arms, and heads. See Finzenhagen 1939, 74. Similar practices occur in Latin literature. In Vitruvius's description of the columnar orders, e.g., he asserts that the Doric and Ionic styles acquired their "proportion, strength, and beauty from the human figure...the former being of a masculine character, without ornament, and the latter being of a character which resembles the delicacy, ornament, and proportion of a female" (4.1.6-7). Vitruvius goes on to describe how the Corinthian order is meant to evoke the grace of a young virgin (4.1.8). Cf. Cicero's comment that the city of Himera had the "form of a woman" (*Verr.* 2.2.87).

[70] This practice likely derived from the fact that many sundials were mounted on pillars and had a spherical or conical shape, causing them to look like heads perched atop torsos. It may also have resulted from the fact that, prior to the advent of sundials and water clocks, ancient Greeks and Romans often used their own bodies to measure time. This practice could take many forms and persisted alongside the use of sundials and water clocks. In Aristophanic comedy, e.g., characters often specify the time of day in terms of "feet," i.e., the length of one's shadow at that point in the day (*Ekk.* 651-2. Cf. Ath. *Deipn.* 1.8b-c, 6.243a). On this practice, see Hannah 2009, 75; 2020, 323–6. According to another trope—attested in both Plautus (preserved at Aul. Gell. *NA* 3.3.5) and Alciphron (*Ep.* 3.1)—a social parasite complains that life was much better when mealtimes were determined by one's belly instead of by a clock. For an analysis of this trope, see Gratwick 1979; Wolkenhauer 2011, 126–37.

[71] *Anth. Graec.* 11.418 Beckby (1965): Ἀντίον ἠελίου στήσας ῥίνα καὶ στόμα χάσκων/δείξεις τὰς ὥρας πᾶσι παρερχομένοις. Cf. Cratinus's joke comparing the bulging shape of the Odeion to the shape of Pericles's head (Storey [2011] Fr. 73 = Plut. *Per.* 13.9). In contrast to sundials, I have found no evidence of water clocks being used as analogs for the human form, even by comedians.

[72] Pl. *Ti.* 69c–72b; cf. Sen. *Ep.* 82.5.

[73] Pl. *Ti.* 69e6–70a2: ἐν δὴ τοῖς στήθεσιν καὶ τῷ καλουμένῳ θώρακι τὸ τῆς ψυχῆς θνητὸν γένος ἐνέδουν. καὶ ἐπειδὴ τὸ μὲν ἄμεινον αὐτῆς, τὸ δὲ χεῖρον ἐπεφύκει, διοικοδομοῦσι τοῦ θώρακος αὖ τὸ κύτος, διορίζοντες οἷον γυναικῶν, τὴν δὲ ἀνδρῶν χωρὶς οἴκησιν, τὰς φρένας διάφραγμα εἰς τὸ μέσον αὐτῶν τιθέντες.

[74] Pl. *Ti.* 70a2–b3. One of the more extended explorations of the moral similarities between men and buildings appears in Plautus's fragmentary comedy *Mostellaria* or "The Haunted House." The

trend of seeing certain private buildings and architectural features as extensions of their owners.[75] By Galen's time, sundials had become part of this conversation. Sundials symbolize their owners most frequently within the context of omens and portents. Valerius Maximus, writing in the mid-first century CE, offers an example. In his *Memorable Deeds and Sayings*, Valerius describes how Cicero's death was presaged when a raven collided with Cicero's sundial and knocked the gnomon out of true:[76]

> The imminent death of Cicero was predicted by an *auspicium*: for when he was in the villa at Caletana, a raven in his sight knocked out of place the iron gnomon by which the hours were distinguished and headed straight for him. It held the end of Cicero's toga continuously in its beak until the assassins came to kill him.

In order for this portent to make sense, we must infer the existence of a symbolic link between this particular sundial and its owner, Cicero.[77] A similar principle underlies Artemidorus's explanation of sundial symbolism in dreams. "The clock (*hōrologion*)," he says, "signifies deeds, undertakings, movements, and enterprises. For all actions men take they perform with an eye toward the hours. Hence, a clock that has collapsed or ceased to function would be bad, even fatal, especially to those who are ill."[78] The structural integrity of a given dream-clock, Artemidorus implies, is indicative of the dreamer's own structural integrity. Should a clock cease to function, as Cicero's does when the raven knocks its gnomon askew, then the owner (or dreamer) of the

---

character of Philolaches—a youth who, during the frequent absences of his father Theuropides, has become "corrupted" by the pleasures of city life—delivers a lengthy monologue in support of the claim that "a man is similar to a new house when he is born." Philolaches develops this theme over 74 lines, comparing a child's parents to those who construct a house, the child's education to a building's foundations, and the misfortunes that can lead a person into vice and destitution to the storms and floods which cause buildings to decay.

[75] See Bodel 1997, 5; Thomas 2007, 81. Not unlike today, Romans under the Empire were taken with the idea of visiting the houses of historical celebrities, such as Pindar and Simonides, thinking in this way to become more intimately acquainted with them. See Paus. 9.25.3 (Pindar) and Strabo 17.1.29 (Plato and Eudoxus). For discussion of the houses of Pindar and Simonides, see Slater 1971, 1972.

[76] V. Max. 1.4.6 Briscoe: *Ciceroni mors imminens auspicio praedicta est: cum enim in villa Caletana esset, corvus in conspectus eius horologii ferrum loco motum excussit, et protinus ad ipsum tetendit, ac laciniam togae eo usque morsu tenuit donec servus milites ad eum occidendum venisse nuntiaret.* This chapter, which belongs within book 1's extensive lacuna (1.1.ext.5–1.4.ext.1), has reached us only through the epitomes of Julius Paris and Nepotianus. It is discussed at Wolkenhauer 2011, 144–5; Murray 2022, 7–9.

[77] Davis also reads into this scene a pun on the word ζῆθι, which is both the Greek imperative "Live!" and the sequence of letters qua numbers that marks the seventh through tenth hours (1956, 70). However, as very few extant sundials from this period have labeled hour-lines, Davis's interpretation is speculative at best.

[78] *Oneir.* 3.66: Ὡρολόγιον πράξεις καὶ ὁρμὰς καὶ κινήσεις καὶ ἐπιβολὰς <τῶν> χρειῶν σημαίνει· πάντα γὰρ πρὸς τὰς ὥρας ἀποβλέποντες οἱ ἄνθρωποι πράσσουσιν. ὅθεν συμπίπτον ἢ κατεασσόμενον πονηρὸν ἂν εἴη καὶ ὀλέθριον, μάλιστα δὲ τοῖς νοσοῦσιν. Artemidorus does not make a clear distinction here between sundials and water clocks but uses the general term ὡρολόγιον which can refer to either.

clock will be unable to go about his or her business[79]—even, in some cases, the business of staying alive.[80]

As we will explore further below, Galen capitalizes on this close moral link between the clock and the individual. In *Affections and Errors*, however, he adjusts the metaphor: rather than employing sundials (and water clocks) as symbols of their owners, he makes the *process* of constructing a clock into a model for how an individual should conduct his or her life. To better understand this shift, and the cultural frameworks that enabled it, it will be helpful to consider the semiotic domain of clocks within horoscopic astrology.

Clocks and Horoscopic Astrology: Symbols of Death and the Afterlife

The development of a close conceptual link between sundials and the physical persons of their owners is itself related to the connection between hourly timekeeping and the increasingly popular practice of horoscopic astrology, which predicts certain aspects of an individual's life experience based on the alignment of significant asterisms at the time of his or her birth.[81] Despite periods of virulent imperial opposition,[82] astrology flourished under the Roman Empire.[83] By Galen's time it was not unusual to see funerary *stēlai* that recorded the deceased's lifespans, often down to the hour, for the probable purpose of casting horoscopes

---

[79] It is possible that etymological puns also facilitated the development of such clock symbolism. A common Latin term for a sundial's pointer, as we saw in the passage from Valerius Maximus, was *motum*. This term derives from the verb *moveo*, meaning "to move, to undertake." Although the *Oneirocritica* was written in Greek, Artemidorus was a citizen of the Roman Empire and surely familiar with the Latin term; perhaps the etymological root of *motum* influenced his assessment of the clock as something that "signifies deeds, undertakings, movements, and enterprises." The Greek term for a sundial's pointer also lends itself to punning, since γνώμων derives from the verb γιγνώσκω and translates literally to "one who knows." As the seat of knowing and rationality was understood by many (including Galen) to be the human brain, this etymology may have served to reinforce the relationship between the functioning of a sundial's pointer and a person's capacity for rational thought while alive. It is impossible to say which came first—the terminology or the symbolic connotations of sundials—but it is likely that the availability of these puns helped to propagate such symbolic readings of clocks.

[80] It is possible that the depictions of sundials in funerary art served a prescriptive purpose in this regard. In addition to potentially signifying the lifespan of the deceased and evoking the idea of eternity as the sum of all hours, the inclusion of erect, unbroken sundials on sarcophagi and *stēlai* might have been intended to persuade the gods to "resurrect" the deceased and grant him or her a "life" in the beyond.

[81] As A. Jones has noted (2003, 337-8), the horoscopic astrology of the Hellenistic and Roman periods should be distinguished from other types of astrology known in the ancient Mediterranean, such as the astrological weather predictions included within earlier Greek *parapēgmata* and the Near Eastern practice of observing the heavens for ominous signs.

[82] Astrologers were expelled from Rome on at least eight occasions between 139 BCE and 139 CE. See, e.g., V. Max. 1.3.2; Dio Cass. 49.43.5, 56.25.5, 57.15.8, 64.1.4, 65.9.2; Tac. *Ann.* 2.32, *Hist.* 2.62; Suet. *Tib.* 36; *Vit.* 14.4; Ulp. *Mos. et Rom. legum coll.* 15.2. For discussion, see Cramer 1954, 232-48; Ripat 2011.

[83] We recall that Augustus himself was said to have published his horoscope (Suet. *Aug.* 94.12). On the status and roles of astrology under the Empire, see esp. Barton 1994.

for the afterlife.[84] Knowledge of the exact hours at which a person was born and/or died was considered important for horoscopic astrology. It is no coincidence that the very word *horoscope*, in fact, means "hour-watcher" and that a common word for sundial or water clock in Greek was *hōroskopeion*.

Presumably because sundials and water clocks were the only instruments that could provide temporal measurements down to the hour, iconographic representations of clocks—specifically, sundials—began to crop up in scenes with astrological significance.[85] A common motif on sarcophagi of the second and third centuries CE involves the three Fates determining a person's horoscope, with Atropos pointing out the hour on a sundial, Lachesis spinning out the appropriate length of thread, and Klotho making the cut.[86] In a handful of cases, the Fates are part of a larger mythological scene in which Prometheus molds the human bodies to which the horoscopes will be assigned (Figure I).[87] The three Fates together, or simply Atropos alone with her sundial, also appear frequently on the sarcophagi of young children (perhaps emphasizing the brevity of the span allotted them) and on sarcophagi that depict scenes from daily life. The lid of the "Portonaccio Sarcophagus," for example, dated to c.190–200 CE and currently in the Palazzo Massimo of Rome, captures a series of moments from a deceased general's life, proceeding from his birth on the far left, supervised by Atropos with her sundial, to the erection of a military trophy on the far right.[88]

Iconography is, by nature, polyvalent, so it would be difficult, if not impossible, to attempt to pin down which precise interpretations were being evoked on the sarcophagi and funerary *stēlai* mentioned above.[89] However, it is clear that, by Galen's time, clocks had developed a series of associations with the human body.

---

[84] Ehrlich 2012. Some examples include: *IEph* 2268, 1636.6–17; *SEG* 9: 877, 18: 402; *IG* XIV 2308; *IGUR* II 389, 436, 703, 727, 799, 890, 903, 993, 1004, 1023, 1028, 1084, 1090, 1355.

[85] We have yet to identify a single iconographic representation of a water clock, despite the fact that this is the only tool capable of marking the hour of a night-time birth or death. This may have something to do with the fact that, unlike sundials, water clocks did not bear much resemblance to the human form.

[86] Persius describes a similar scene at 5.45–51. For a survey of the role of sundials in funerary art, see Bonnin 2013. On the iconography of Roman sarcophagi, see Turcan 1999.

[87] E.g., the Prometheus sarcophagus at the Museo Pio Clementino at the Vatican and one at the Capitoline Museum. For images of the former, see Robert 1897, 440–1, no. 354, pl. 116, and for discussion, see Turcan 1968. For images of the latter, see G. Koch and Sichtermann 1982, 183, no.41, fig. 215 and Robert 1897, 441–4, no. 355, pl. 117.

[88] We also have evidence to suggest that sundials were sometimes mounted atop funerary *stēlai* to encourage passersby to pause at the grave monument in order to check the time and, while doing so, invoke the memory of the deceased. See, e.g., Petron. *Sat.* 71; a lamp in the Kunsthistorische Museum of Vienna's Ephesus Museum, discussed by Veyne (1985); and two funerary *stēlai*, those of Izmit and Odessos. For the former *stēlē*, see Fıratlı et al. 1971, pls. 17 and 22. For the latter, see Tonchéva 1969, 17, no. 16, figs. 15 and 39.

[89] It is similarly difficult to clearly distinguish the semiotic fields of sundials and water clocks from one another. To my knowledge, water clocks are not represented iconographically on any extant artifact, and their use as literary symbols is rare. Yet, within these literary sources (such as Mesomedes's poem, the ambiguous reference in Artemidorus's dream book, and the *Epistles* of Seneca), water clocks take on a range of meanings, representing the cosmos, a human being, and a human's lifespan, respectively.

Figure I Detail of Prometheus sarcophagus. Capitoline Museum, Inv. 638. Image © Shutterstock.

A clock could represent a single moment—specifically, the hour of that body's birth or death—or it could stand for the length of time between the two: that body's entire lifespan. Analogously, clocks could signify the brevity of a human life, which lasts but a metaphoric hour, or conversely, the sum of all hours in existence, i.e., eternity.

One could argue that *Affections and Errors* participates in the broader trends of using clocks to represent either the brevity of life on earth or the eternity of life hereafter. In a passage examined in the previous chapter, Galen situates developments in clock technology within the context of continuous scientific progress.[90] Though each individual clock is the product of a particular human *arkhitekton* or team, Galen points out that the theoretical framework on which the act of clock-making relies has been built up over many generations and continues to mature. In this passage, Galen mocks the hubris of sectarian philosophers who think they can squeeze many lifetimes' worth of study, experimentation, and reasoning into their single, lonely lifespans.[91] Rather than engage in such navel-gazing, Galen advocates recognizing and embracing one's small part in the "eternal" progress of scientific thought. In short, for Galen in this text, it is the process of scientific

---

[90] *Aff. pecc. dig.* 59.3–8 DB = 5.86.15–87.5 K.
[91] On the degree to which experimentation can be said to have taken place in Greco-Roman antiquity, see von Staden 1975; Lloyd 1979.

reasoning that achieves a kind of immortality, not necessarily the individual human. However, by utilizing logic and empirical demonstration, the individual can participate in that kind of immortality, and be assured that he is leading a life of truth and virtue. By situating himself as the true proponent of this method, Galen also grants himself an association with the unfolding process of scientific reason and presents himself as the culmination of this tradition.

Conclusion: The "Clock-Construction" Lifestyle

Our investigation has revealed that writers and craftsmen of Galen's time used the symbol of the clock to represent a whole constellation of ideas, including Roman imperialism, Greek *paideia*, mathematical ingenuity, and the rhythms of the heavens and the human body.[92] Therefore when Galen, in *Affections and Errors*, uses the processes of sundial and water clock construction as his central paradigm of apodeictically informed scientific method, he can tap into this rich semiotic network, relying on the sociocultural knowledge of his educated, elite readers. In some instances, Galen seems to export these symbolic valences directly. If, for example, a reader happened to connect Galen's *paradeigmata* to themes of imperial authority, cosmic and political harmony, mathematical ingenuity, and Greek intellectual accomplishment, so much the better; these resonances directly supported Galen's own claims that his method was itself a medium for cultivating harmony, consensus, and impressive intellectual attainment, particularly in mathematics and the exact sciences.

In other ways, however, Galen seems to reconfigure contemporary tropes to suit his own purposes. For instance, I have suggested that Galen engaged with motifs that were employed by two eminent Stoic philosophers (Seneca and Epictetus) and are reflected more broadly in certain literary and iconographical contexts of the Imperial period: namely, the practice of using a clock (or the hour) as a symbol of the human body or its time on earth. Although Galen, in *Affections and Errors*, briefly addresses himself to the "body" (*sōma*) of the clock as a whole,[93] he is clearly less concerned with how these bodies are sculpted or used than with the ways in which the hour- and date-lines are marked upon their surfaces (*katagraphē*). As we saw in the previous chapter, Galen spends pages walking his reader through the basic processes of inscribing lines on sundials and water clocks and then checking them for accuracy. For Galen, the precise inscription of hour- and date-lines, achieved through a dialectic between logical reasoning and empirical observation, is what makes a clock a clock. This inscription

---

[92] Talbert (2017, 159–63) also observes that portable sundials, in particular, could act as symbols of their owners' cosmopolitanism and personal freedom to travel widely throughout the Empire.

[93] See *Aff. pecc. dig.* 55.5–17 DB = 5.81.3–15 K.

process allows one to transform simple shadows and water tanks into tools that transmit useful information. Thus, in *Affections and Errors*, Galen is more concerned with clock-construction as a *process* than with the clock as a finished *product*. Galen uses clocks not as metaphors for the human life*span*, but as metaphors for his ideal life*style*: one in which decisions are made through that consummate blend of reasoning and empiricism that characterizes his scientific method.[94]

In Chapters 3 and 4, we focused on how Galen employed sundials and water clocks as symbols to represent certain aspects of his signature scientific method. In the following chapters, we will shift our attention to the ways in which Galen used actual clocks and numbered hours as heuristics, to put his method into action in the sickroom. By examining case studies drawn from a variety of Galen's other works, we will develop a richer sense of the roles that hourly timekeeping played for Galen in diagnosis, prognosis, and therapeutic treatment, as well as in his larger program of self-presentation. We will also consider how his heuristic use of hours compares to that of certain of his medical rivals.

---

[94] Feke (2018) has explored how one of Galen's rough contemporaries, the astronomer Ptolemy, also viewed his scientific methodology and mathematics in general as guides to good living.

PART II
HOURS IN ACTION

5
From "Season" to "Hour"
Galenic Refinements of the Hippocratic *Hōra*

In the preceding discussion of *Affections and Errors*, we saw how, in Galen's view, the process of clock-construction exemplifies his ideal scientific methodology and serves as a model for unerring truth-assessment and decision-making.[1] At the same time, however, we also saw that Galen considered clocks to be tools that were not only good to think with representationally but were also "useful" (*khrēsimos*), on a practical level, for obtaining temporal measurements. This leads one to wonder just how an Imperial-period physician like Galen might have incorporated hourly timekeeping into his day-to-day medical practice. Or, to put it another way, in what contexts did ancient Greek physicians, like Galen, feel that numerical exactitude, down to the level of the hour, was important for successful diagnosis, prognosis, and treatment?

Chapters 5–7 explore Galen's engagement with this question across a variety of his medical works. Chapter 5 begins by investigating how Galen, in *On Critical Days*, uses hourly timekeeping to chart the course of febrile illnesses with greater precision and improve his ability to predict their outcomes. From this analysis, the picture that emerges of Galen himself is that of a doctor who advocates for numerically exact timekeeping. Chapter 6, however, offers an important corrective to the impression that Galen promoted such exactitude across the board. In that chapter, we will see that Galen actually sought to position himself as a moderate within an active, contemporary debate over the degree to which physicians should rely on numbers in their daily medical timekeeping. By examining passages from Galen's works on irregular intermittent fevers, medical definitions, and venesection, it will become clear that Galen held strong views about which diagnostic, prognostic, and therapeutic contexts were benefitted by close attention to hourly timekeeping and which could, instead, be harmed by it. Finally, Chapter 7, through close readings of passages from Galen's *On Hygiene*, presents and dissects some of the rationales that underlie Galen's decisions about when hourly timekeeping could be medically beneficial. Chapter 7 reveals that, for Galen, the utility of this kind of timekeeping is contingent not only upon the nature of the disease confronting the physician and the kind of procedure he is trying to perform, but

[1] Material from this chapter was previously published in K. J. Miller 2018.

also upon the very nature of the patient in question: specifically, his or her age, health, habits, preferences, and physical capacities. Taken together, Chapters 5–7 demonstrate that Galen's decisions about when and why to employ hourly timekeeping in the sickroom are intimately related (a) to his scientific methodology and admiration of the mathematical sciences; (b) to his theories about how to define health, disease, and a medical *kairos*; and (c) to his deep-seated desire to portray himself as the successor (and refiner) of that paragon of Classical Greek medicine, Hippocrates.

In *On Critical Days*, the focus of the present chapter, Galen seeks to assess and justify Hippocratic "critical-day" schemes—that is, schemes for anticipating important moments of change in the course of febrile diseases. This text, I argue, highlights how Galen uses the concept of hourly timekeeping to defend Hippocratic doctrine, while simultaneously bringing it into greater harmony with Imperial-period ideas and technologies. I suggest that this strategy permits Galen to present himself as a *pepaideumenos* whose medical theories and practices are backed by authoritative, Classical precedent and as the true intellectual successor of Hippocrates. It also enables him to showcase his appreciation for and familiarity with aspects of astronomy and, in so doing, to affiliate the *tekhnē* of medicine more closely with the mathematical sciences.

Galen's Hippocratism

Nowadays, one often hears Hippocrates lauded as the very "Father of Medicine."[2] During Galen's time, however, there were multiple contenders for this title, and to whom it was awarded depended largely on the medical sect of the physician doing the talking. The Methodists,[3] for example, who were so loathed by Galen, often called upon the authority of Thessalus of Tralles, while many Pneumatists followed in the footsteps of Agathinus and Athenaeus.[4] W. D. Smith, in his account of the development of "Hippocratism," has cogently argued that the campaign for Hippocrates as the Father of Medicine was first initiated by the Empiricists of the Hellenistic period.[5] Empiricist doctors were at pains to differentiate their medical ideology from that of the early Rationalists. Hippocrates was an excellent candidate for their sect's figurehead, because so many Hippocratic texts emphasize observation and engage in detailed descriptions of the symptoms and

---

[2] On the legendary status of Hippocrates in the present day, see King 2019.
[3] The extant writings of Methodist physicians have been collected by Tecusan (2007).
[4] On the distinctions between ancient medical sects and on their most prominent adherents, see, e.g., W. D. Smith 1979, 198–215 (specifically on the rivalry between Empiricist and Rationalist or Dogmatic sects); Gourevitch 1999; Nutton 1992, 2013.
[5] W. D. Smith 1979, 177–210. Smith cites the first-century CE author Celsus as our earliest coherent source for the conflict between the Empiricists and the Rationalists, but notes that Celsus's account is influenced by the earlier writings of Asclepiades (first century BCE).

environmental conditions associated with particular illnesses. By Galen's day, Hippocratic exegesis had also become a favored pastime among physicians like Marinus, Quintus, and Numesianus, whose affiliations are uncertain,[6] but whose views on physiology, pharmacology, and clinical medicine likely influenced Galen.[7]

Determining what the great Hippocrates actually said or intended, however, was—and still proves to be—something of a challenge. From the perspective of modern scholarship, "Hippocrates" is more of a legendary persona than a historical personage. We know very little about "Hippocrates" the man;[8] we do know, however, that the "Hippocratic Corpus" as we have it today was not composed by a single author but is rather a collection of texts by a variety of contributors.[9] These men wrote from different geographical locations and at different times within the late Classical and early Hellenistic periods. Even the theories of medicine that are articulated or implied within the Hippocratic texts differ from one another to varying degrees.[10] These inconsistencies were not lost upon readers from the late Hellenistic and Imperial periods and, from then on, there has been a culture of debate surrounding the proper attribution of individual Hippocratic treatises. In antiquity, however, it was assumed that there was a core of "genuine" texts composed by Hippocrates himself. More problematic texts were often attributed to one of Hippocrates' sons or grandsons (presumed to be writing from the great man's notes) or were rejected as spurious. Which works constituted the Hippocratic "canon" depended on the individual commentator; there was no strict consensus.

As scholars of recent decades have pointed out, the inconsistencies, ambiguities, and generalities within the Hippocratic writings left later commentators plenty of room to exercise interpretative license.[11] Galen himself regularly seized the opportunity to reinterpret Hippocratic doctrines in light of his own ideologies. One strategy that Galen frequently employed was to explain what Hippocrates had "really meant" by a particular word or phrase, using thick descriptions to expand upon pithy Hippocratic statements and thereby to refashion the great man's message.[12] At other times, Galen would read back into Hippocratic works

---

[6] Gourevitch calls theirs an "unnamed sect" (Gourevitch 1999, 118). On Marinus and Quintus, see Keyser and Irby-Massie 2008, 532 and 717, respectively.
[7] While in Pergamon, Galen studied under Satyrus, a pupil of Quintus, and attended lectures by Pelops, a pupil of Numesianus. On Pelops and Satyrus, see Keyser and Irby-Massie 2008, 634 and 728.
[8] Plato, for instance, mentions that Hippocrates charged money for his medical services (*Prot.* 311b-c).
[9] For an overview of the Hippocratic corpus and how Hippocratic medicine related to contemporary intellectual trends, see, e.g., Jouanna 1999.
[10] For a detailed discussion of the medical theories present within the Hippocratic Corpus, see Langholf 1990.
[11] See W. D. Smith 1979, 175–6; Lloyd 1988; von Staden 2002, 114–16; Yeo 2005; Flemming 2007a, 343–6; van der Eijk 2012, 33–4.
[12] For discussion of this phenomenon, see Manuli 1983; Asper 2013, 423–4. Manuli calls this practice "l'integrazione del significato" (472).

concepts and terms that were not there to begin with, either because they were anachronistic or simply because they did not pique the Hippocratic authors' interest.[13] This revisionist technique allowed Galen to claim for his own theories the full weight of Hippocratic tradition and authority.[14] In what follows, we will consider how Galen employs such tactics in *On Critical Days* to introduce the concept of hourly timekeeping into frameworks that are essentially Hippocratic.

Galen's *On Critical Days* and the Hippocratic *Epidemics*

Despite the recent upsurge in Galenic studies, *On Critical Days* has not received much scholarly attention, due perhaps in part to the dearth of critical editions and translations.[15] Nevertheless, it is an important work, in which Galen seeks to test a set of theories advanced by Hippocratic writers for anticipating when the course of a febrile illness might change. Central to these theories is the concept of a "crisis" (*krisis*), a decisive moment at which a fever either dissipates or begins a transition into a new phase (including, but not limited to, death).[16] Some Hippocratic writers had come to recognize through their own bedside experiences that, in illnesses lasting longer than a single day, such turning points tended to recur after certain numbers of days. The Hippocratic doctors sought to map these intervals and come up with schemes that would allow them to anticipate future crises. It was important that ancient physicians predict the outcome of an illness as accurately as

---

[13] Holmes (2012), e.g., has argued that Galen adapted the Stoic concept of "sympathy" (Greek *sympatheia*) for incorporation into his own theory about the interconnectivity of anatomical systems. In his liberal commentary on *Epidemics* 2, Galen manages to read the concept back into the Hippocratic text, despite the fact that the term *sympatheia* itself does not appear there. Manetti (2003) has observed that Galen also reads certain stylistic qualities back into the Hippocratic Corpus. By being selective about which texts and passages he cited, Galen could claim that Hippocrates' style was clear, concise, and properly Greek, and thus should act as a model to the loquacious, long-winded physicians of his own day. For further discussion of Galen's views on language usage, see Manetti and Roselli 1994; von Staden 2002; Manetti 2009.

[14] Galen makes no secret of his admiration for and reliance upon Hippocratic teachings, and many modern scholars have investigated the techniques, biases, and agendas with which Galen approached Hippocratic exegesis. A partial list includes W. D. Smith 1979; Manuli 1983; Nutton 1992; Manetti and Roselli 1994; Manetti 2003; Holmes 2012.

[15] The most recent edition of the original Greek is that published by C. G. Kühn in *Claudii Galeni opera omnia*, vol. 9. Leipzig: Knobloch, 1825 (repr. Hildesheim: Olms, 1965), 769–941. In 2011, G. Cooper published a much-needed critical edition, English translation, and commentary of Hunayn Ibn Ishaq's Arabic translation, penned in the ninth century CE (Cooper 2011b). For a critique of this edition, see Langermann 2012. On the later Alexandrian summaries of Galen's *On Critical Days*, see Bos and Langermann 2015.

[16] Galen often indicates a febrile crisis by using the term *kairos*, which includes within its range of meanings the notions of "right time" and "critical or decisive time." On the semantic field of the term *kairos*, see Sipiora and Baumlin 2002. On the notion of *krisis* within Greek epic, medicine, and Platonic philosophy, see Longhi 2020. On *kairos* and *krisis* in Imperial-period medicine, see Singer 2022, 102-122.

possible, so that they might recommend optimal therapies and avoid taking on terminal patients, whose deaths could damage their professional reputations.[17]

The Hippocratic authors of the *Epidemics* display a particular interest in constructing critical-day theories. The *Epidemics* is a collection of useful aphorisms, individual case histories, and "constitutions" (*katastaseis*), i.e., detailed accounts of the seasonal disease patterns produced in different geographic locations. The dating and authorship of each of the *Epidemics*' seven books has been controversial since antiquity.[18] There have been tendencies, however, in ancient as well as modern scholarship, to group certain of these books together chronologically according to the following scheme: (a) books 1 and 3, which are dated to c.410 BCE and have a more polished, coherent form; (b) books 2, 4, and 6, which have been dated to between 427/6 and 373/2 BCE, and are less polished and consistent in their style; and finally, (c) books 5 and 7 which, already in antiquity, were considered by many to be post-Hippocratic additions to the Corpus.[19] In short, the *Epidemics* was probably pieced together out of a variety of sources, whose authors were located primarily in Northern Greece and wrote, by-and-large, during the late fifth and early fourth centuries BCE.[20]

Different configurations of critical days appear throughout the Hippocratic Corpus. The author of *Epidemics* 1, for instance, proposes two different schemes, depending on whether a fever's paroxysms arrive on odd or even days.[21] The author of *Aphorisms* 4.36, however, asserts that crises can occur on the third, fifth, seventh, ninth, eleventh, fourteenth, seventeenth, twenty-first, twenty-seventh, and thirty-fourth days.[22] By Galen's time, it seems that the correct sequence and even the utility of critical days were open to debate. In *On Critical Days*, Galen informs us that many doctors of his acquaintance asked "what is meant by the

---

[17] E.g., *Adv. typ. scrip.* 7.479.16–7.480.2 K. Cf. Hippocrates at *Prog.* 1.2.110–12 L = 2.24 Jouanna 2013.
[18] Among the earliest attested Hippocratic commentators are Bacchius, Heraclides, and Zeuxis, each of whom wrote in the third century BCE. Our earliest extant Hippocratic commentary was penned by Apollonius of Citium. On these developments, see W. D. Smith 1979, 235–9.
[19] The dating of books 1 and 3 is based on corroborating inscriptions discovered at Thasos (Langholf 1990, 77). In fact, some scholars now consider them to be two halves of a single work produced by a single author. See, e.g., Álvarez Millàn 1999, 22. Wee, however, cautions against conflating the two, noting in particular that "variables of geography, time, and patient were defined and prioritised differently in Books 1 and 3" (2015, 145). Books 5 and 7 constitute two separate collections of case histories, although parallel versions of certain case histories appear in both. The extreme heterogeneity of *Epidemics* 7, along with the dramatic style of its storytelling, suggests that it was constructed from a wide variety of sources and edited to enhance its narrative effect. For more on the generic style of these two books, see W. D. Smith 1981.
[20] On the history, dating, and style of the *Epidemics*, see Deichgräber 1971; Langholf 1990, 77–242; W. D. Smith 1979, 237; 1981; Álvarez Millàn 1999, 21–7.
[21] Hippoc. *Epid.* 1.12, 2.678–82 L = 37.6–39.2 Jouanna.
[22] Hippoc. *Aph.* 4.514–16 L. For other critical day lists within the Hippocratic Corpus and discussion, see Langholf 1990, 78–127. For overviews of critical-day theories, see also Lloyd 1979, 154–68; Cooper 2011b, 127–8.

term *crisis*" and "whether such a thing is even possible."[23] Galen aimed to resolve this controversy by means of his own scientific method. In the first two books, he proceeds to test the Hippocratic systems against the rubric of his own experiences and the observational data aggregated in the *Epidemics* itself.[24] Then, in the third book, Galen reasons through the theoretical underpinnings of his preferred predictive scheme. Galen explains his method thus: "Since everything to do with the medical art is discovered and tested sometimes through experience, sometimes through reasoning, and sometimes through the two together, one must therefore attempt, by means of both tools, to refute what is erroneous and to commend and accept what is correct."[25] This chapter focuses on two aspects of Galen's approach: first, the temporal structure of the fever case histories that Galen offers as empirical evidence in books 1 and 2; second, the links that he forges between medical and astronomical periodicities during book 3's theoretical discussion. In both instances, I examine how Galen uses hourly timekeeping to build upon Hippocratic precedents.

Temporality in Galen's Fever Case Histories

Interestingly, recent scholarship on Galen's patient case histories has emphasized the many ways in which they tend to depart from, rather than adhere to, Hippocratic models. Álvarez Millán, for example, has asserted that, when it came to clinical narratives, "Galen did not follow the Hippocratic pattern."[26] In the same vein, Lloyd's contribution to the volume *Galen and the World of Knowledge* is titled "Galen's Un-Hippocratic Case Histories."[27] These scholars' views are based primarily on Galen's work *On Prognosis*, which compiles most of his case histories. Yet medical historians have often overlooked the various case histories embedded within Galen's fever treatises, such as *On Crises, On Critical Days,* and *On the Distinct Types of Fever.*[28] While references to hours do not appear in the case histories of *On Prognosis*, hourly timekeeping is an important structuring device in Galen's fever narratives. As I demonstrate, the latter bear a

---

[23] *Di dec.* 9.772.11–13 K: τὸ σημαινόμενον ὑπὸ τοῦ τῆς κρίσεως ὀνόματος and εἰ δυνατὸν ὑπάρχει τὸ πρᾶγμα.

[24] Cooper has written extensively on Galen's methodology in *On Critical Days*. See esp. Cooper 2004, 2011a, 2011b.

[25] *Di dec.* 9.841.9–9.842.4 K: ἐπειδὴ πάντα τὰ κατὰ τὴν ἰατρικὴν τέχνην εὑρίσκεταί τε καὶ δοκιμάζεται, τὰ μὲν ἐμπειρίᾳ, τὰ δὲ λόγῳ, τὰ δὲ συναμφοτέρῳ, πειρατέον κἀνταῦθα δι' ἀμφοτέρων ὀργάνων ἐξελέγξαι μὲν τὸ ἡμαρτημένον, ἐπαινέσαι δὲ καὶ προσίεσθαι τὸ κατωρθωμένον.

[26] Álvarez Millàn 1999, 32.

[27] Lloyd 2009. For an overview of how this genre of medical writing developed over time and across cultures, see Pomata 2014.

[28] E.g., *Cris.* 9.680.11–9.683.2 K; *Di. dec.* 9.800.1–9.802.6 K; *MM.* 10.608.5–10.615.15 K; *Diff. feb.* 7.351.15–7.354.13, 355.12–357.17, and 359.9–363.3 K. Hours play an important role in structuring each of these narratives.

close resemblance to the case histories included within the *Epidemics* (especially books 1 and 3).²⁹ Furthermore, Galen's narratives introduce even greater temporal precision.

As Langholf has pointed out, critical-day schemes serve as the dominant organizing principle in the case histories of *Epidemics* 1 and 3. Some of these case histories are laconic and offer no additional temporal markers beyond the potentially critical days. Case 10 of *Epidemics* 3.2, for example, proceeds as follows (with day-references emphasized):³⁰

> A woman from among those around Pantimides, just out of childbirth, was seized *on the first* [*day*] by fever. Tongue very dry. Thirsty. Nauseous. Sleepless. Disturbed bowels, with slender, frequent, undigested [stools]. *On the second* [*day*], much shivering. Acute fever. Much discharge from the bowels. Didn't sleep. *On the third* [*day*], sufferings were greater. *On the fourth* [*day*], she was delirious. *On the seventh* [*day*], she died. Bowels moist throughout, with frequent, slender, undigested stools. Urine small, thin. Causic fever.

We can see here both how the patient's symptoms are grouped according to the day on which they manifested and how the author has been selective about which days to include in his account.³¹ Yet, while critical days create the main framework within *Epidemics* 1 and 3, many case histories also take an interest in the timing of symptoms *within* the day—what I will refer to as "intra-day" timing. Case 3 from *Epidemics* 3.17 offers a representative example. In addition to putting the references to day-units in italic, I have underlined the intra-day time markers and references to temporal durations.³²

---

[29] These are the only books of the *Epidemics* that Galen consistently attributes to Hippocrates himself. Elsewhere, Galen argues that books 2, 4, and 6 were composed by Hippocrates' son Thessalus and that books 5 and 7 are spurious. This appears in Hunayn Ibn Ishaq's Arabic translation of Galen's commentary on the *Epidemics*: Escorial, MS. 805, fol. 1v (according to Western pagination). For discussion, see Lloyd 2009, 116.

[30] *Epid.* 3.2, 3.60 L = 75.7–76.2 Jouanna: Γυναῖκα ἐξ ἀποφθορῆς νηπίου, τῶν περὶ Παντιμίδην, τῇ πρώτῃ πῦρ ἔλαβεν· γλῶσσα ἐπίξηρος· διψώδης· ἀσώδης· ἄγρυπνος· κοιλίη ταραχώδης λεπτοῖσι, πολλοῖσιν, ὠμοῖσιν. Δευτέρῃ, ἐπερρίγωσεν· πυρετὸς ὀξύς· ἀπὸ κοιλίης πουλλά· οὐχ ὕπνωσεν. Τρίτῃ, μείζους οἱ πόνοι. Τετάρτῃ, παρέκρουσεν. Ἑβδόμῃ, ἀπέθανεν. Κοιλίη διὰ παντὸς ὑγρὴ διαχωρήμασι πολλοῖσι, λεπτοῖσιν, ὠμοῖσιν· οὖρα ὀλίγα, λεπτά. Καῦσος.

[31] It is unclear whether practice preceded theory in the development of critical-day systems or vice versa. Many scholars follow Langholf (1990, 115) in supposing that, while critical-day theories may have found their inspiration in actual experience, it is likely that many physicians simply placed their trust in a specific critical-day system and only visited patients at times that the system anticipated would be decisive. For physicians, such a practice could help with self-advertisement, as patients were bound to be impressed by doctors who appeared ready to perform at precisely the right moment.

[32] *Epid.* 3.17, 3.112–16 L = 96.1–97.20 Jouanna: Ἐν Θάσῳ Πυθίωνα, ὃς κατέκειτο ὑπεράνω τοῦ Ἡρακλείου,...πυρετὸς ὀξὺς ἔλαβε·...Δευτέρῃ, περὶ μέσον ἡμέρης ψῦξις ἀκρέων, τὰ περὶ χεῖρας καὶ κεφαλὴν μᾶλλον· ἄναυδος, ἄφωνος, βραχύπνοος ἐπὶ χρόνον πολύν· ἀνεθερμάνθη· δίψα· νύκτα <δι'> ἡσυχίης· ἵδρωσε περὶ κεφαλὴν σμικρά. Τρίτῃ, ἡμέρην δι' ἡσυχίης· ὀψὲ δὲ περὶ ἡλίου δυσμὰς ὑπεψύχθη σμικρά·...Ἐν ἀρτίῃσιν οἱ πόνοι τούτῳ.

> In Thasos, Pythion, who was living above the Temple of Heracles, was seized... by a powerful shiver and acute fever.... *On the second [day]*, around the middle of the day, a sense of cold in the extremities around the hands and head. Speechless, voiceless, short of breath for a long time. He heated up again. Thirst. Had a quiet night. Sweated a little around the head. *On the third [day]*, had a quiet day. Late, around sunset, he became a little chilled.... The sufferings of this man were *on the even days*.

Although this doctor concludes by identifying a critical pattern in the patient's illness (i.e., suffering increased on the even days), he is interested in charting the *whole* temporal progression of the disease. He records his patient's condition not only on the critical even days, but on *all* of the days between the onset of the sickness and the patient's death. Furthermore, this doctor wants to explore the variations in the patient's condition *within* each given day. Thus, his time-keeping is more meticulous than the previous author's. Here, the physician's intra-day time descriptors are approximate and cued primarily by the position of the sun (e.g., *peri hēliou dusmas*). He occasionally alludes to symptom durations but does not record them with any specificity (e.g., *epi khronon poulun*). Hours make no appearance in this account, nor do we see any intra-day time markers that are derived from social, rather than celestial, cycles.

By the time we get to the "later" books of the *Epidemics*, however, the situation has changed. On the one hand, we begin to see some socially based time markers, derived primarily from the cycles of activity in the agora. Case 92 in *Epidemics* 7, for instance, describes sweats that come "on the third [day], when the agora fills,"[33] and Case 62 in *Epidemics* 5 informs us that the patient "died before the opening of the agora, coincident with daybreak."[34] As discussed earlier, in Chapter 2, *Epidemics* 4 may even contain an isolated reference to a numbered hour: "He felt the same amount of pain later in the third hour."[35] Yet, as Álvarez Millán has pointed out, the authors of these later case histories seem less interested in temporal patterns, and increasingly prioritize self-advertisement and dramatic narrative.[36]

---

[33] *Epid.* 7.92, 5.448 L = 104.9–10 Jouanna: τρίτῃ δὲ, ἀγορῆς πληθούσης. Cf. *Epid.* 5.88.3, 5.252 L = 40.8 Jouanna and 7.25.5.394 L = 66.23 Jouanna. The LSJ lexicon lists comparanda from other well-known literary sources, including Herodotus (2.173, 4.181, 7.223), Xenophon (*Mem.* 1.1.10; *An.* 1.8.1 and 2.1.7), and Plato (*Gorg.* 469d).

[34] *Epid.* 5.62, 5.242 L = 28.8–9 Jouanna: ἔθανε πρὶν ἀγορὴν λυθῆναι, ἅμ᾽ ἡμέρῃ πληγείς.

[35] *Epid.* 4.12, 5.150 L: τρίτην ὥρην ἴσως ὠδυνήθη ὕστερον. τρίτην mss. and edd.; αὐτὴν Smith. As discussed in Chapter 2, one other reference to numbered hours appears in the Hippocratic Corpus, at *Int.* 27.7.238 L. For discussion of the fever theory implied in *Epidemics* 5 and 7, see W. D. Smith 1981.

[36] Álvarez Millàn 1999, 24–7. Some later physicians abandoned critical-day theories all together. One such was Asclepiades, whom later Methodists claimed to be the founder of their school (W. D. Smith 1979, 228). Pearcy makes the interesting observation that when Aelius Aristides contrasts divine and human medicine (*Or.* 61–8), he presents the god Asclepius's medicine as "perplexing, fey, and ambiguous," while the medicine of the human doctors is "chronological, rational, [and] particularized" (1992, 606–9). However, Aristides does not use hours in this passage.

There are many ways in which Galen's fever case histories can be said to imitate the temporal patterns and principles found in *Epidemics* 1 and 3. A representative case history occurs in the first book of *On Critical Days*. In this instance, Galen presents a clinical narrative that he marks as hypothetical, but which he recounts with such specificity that a reader might be inclined to assume that Galen had experienced one or more similar cases in real life. He begins in the following manner:[37]

> Let some such patient be set before us, as an example for the clarity of our teaching, a patient who began to experience fever acutely in the tenth hour of the day…. We ourselves, during the second day, will closely observe whether, on that day, another paroxysm makes another beginning, which is perceptible and clear. Then we will do the same on the third day, in order that we might know whether the paroxysms occur every third day or every day and, furthermore, whether the paroxysms are more robust on the odd or even days.

Galen's narrative is clearly structured according to a critical-day format. But in specifying the precise hour at which this hypothetical fever comes on, Galen signals that his temporal framework may deviate somewhat from the Hippocratic. As the case history progresses, it maintains this extra level of temporal precision:[38]

> [Let] the paroxysms come every third day. And let there be a paroxysm in the eleventh hour of the day, and another on the fifth day in the first hour of the night, and another on the seventh day in the third hour of the night. For let us always assume that the paroxysm comes two hours later, with the result that there will be a paroxysm on the ninth day during the fifth hour of the night, as well as on the eleventh day during the seventh hour.

Galen's temporal map of this imagined illness is plainly at a higher resolution than those of the Hippocratic writers. Galen organizes his account according to a scheme that indicates not only the critical *days* upon which a paroxysm occurred, but also what one might call the "critical *hour*" within each day. He asserts that, ideally, such precision will help the physician "to be able to say at times, with

---

[37] *Di. dec.* 9.800.1–9 K: ἔστω δή τις τοιοῦτος ἡμῖν ἄρρωστος ἐξ ὑποθέσεως εἰς σαφήνειαν τῆς διδασκαλίας προκείμενος, τῇ δεκάτῃ τῆς ἡμέρας ὥρᾳ πυρέττειν ὀξέως ἀρξάμενος…. αὐτοὶ τὴν δευτέραν ἡμέραν, ἢ εἴ τιν' ἀρχὴν ἑτέραν ἐν αὐτῇ ποιεῖται παροξυσμὸς ἕτερος, αἰσθητὴν καὶ σαφῆ παραφυλάξομεν· εἶθ' οὕτω καὶ τὴν τρίτην, ἵν' εἴτε διὰ τρίτης, εἴτε καὶ καθ' ἑκάστην ἡμέραν οἱ παροξυσμοὶ γίνονται, γινώσκωμεν· ἔτι δὲ τούτου μᾶλλον, εἴτ' ἐν ταῖς ἀρτίαις, εἴτ' ἐν ταῖς περιτταῖς ἡμέραις παροξύνεται σφοδρότερον.

[38] *Di. dec.* 9.800.11–16 K: διὰ τρίτης δ' οἱ παροξυσμοί· καὶ τῇ τρίτῃ μὲν τῶν ἡμερῶν ἑνδεκάτης ὥρας, τῇ πέμπτῃ δὲ νυκτὸς ὥρας πρώτης παροξυνέσθω, τῇ δὲ ἑβδόμῃ νυκτὸς ὥρας τρίτης· ἀεὶ γὰρ ὑποκείσθω δυοῖν ὥραιν ὑστερίζειν τὸν παροξυσμόν, ὥστε καὶ τῆς ἐννάτης νυκτὸς ὥρᾳ πέμπτῃ παροξυνθήσεται καὶ τῆς ἑνδεκάτης ἑβδόμῃ.

certainty, not only the day but also *the very hour* in which one of his patients will experience a crisis—or must die."[39]

Unfortunately, due to the paucity of extant medical writings by Galen's contemporaries, it is difficult to say how unique Galen was in privileging hourly schemes within his clinical narratives. The best available source for comparison is a collection of case histories, preserved only in Arabic, that claims to reproduce faithfully texts written by "Rufus of Ephesus and other ancient and recent doctors."[40] M. Ullmann, who edited the compilation and provided a German translation, has argued on the basis of syntax, diction, and the case histories' internal references to one another that the Arabic texts not only mask authentic Greek originals, but also can be attributed, as a collection, to Rufus or to members of his school.[41] Ullmann's second claim has since been contested, but the former is still accepted widely.[42] Thus, with due caution, let us examine a few case histories from this collection to see how their temporal structures compare with those we have already seen in Galen's *On Critical Days* and in the Hippocratic *Epidemics*.

Of the twenty-one case histories included in this compilation, only one has fever as its focus.[43] Ullmann's Case 5, addressing a "Quartan Fever with Melancholic Symptoms," reads:[44]

> Another man lay for a long time with quartan fever. He was thereby an ascetic; he suppressed his appetites and fasted for a long time. Thence impairment overtook his thoughts, and he developed bad ideas about himself. When I saw the sign of coction in his urine, and when, at the evacuation, black bile came out of him, I could hope that he would get better, since the humor had come out cooked. However, that was not at the beginning, but he lacked coction, until he was <...> Then I made his body moist and reestablished his vigor. Then he recovered without having had a noteworthy evacuation. I have cured many patients of this disease by bringing their temperament into balance, without evacuation.

---

[39] Di. dec. 9.831.7–9 K, emphasis added: μὴ μόνον ἡμέραν ἔχειν εἰπεῖν βεβαίως ἐνίοτε, ἀλλὰ καὶ τὴν ὥραν αὐτὴν ἐν ᾗ κριθῆναί τινα τῶν νοσούντων, ἢ ἀποθανεῖν ἀναγκαῖον. Cf. Cris. 9.674.16–9.675.4 K: "At this point, one must pay attention to the hours of the paroxysms, in order that one might predict something about the third day. For if, by chance, the paroxysm of the tertian is going to occur around the first hour, and the paroxysm of the daily fever around the eleventh, the beginnings of each will be clear." See also *Cris.* 9.749.8–10 K.

[40] The manuscript, at the Bodleian Library, Oxford, is MS. Hunt. 461, fols. 38b–50a.

[41] Ullmann 1978, 16.

[42] One contestation appears at Mattern 2008, 33. Álvarez Millàn (1999), however, accepts both claims.

[43] Ullmann suggests that the extant group is only the beginning of a more extensive compilation (1978, 16).

[44] This represents the author's English translation of Ullmann's German translation of the Arabic translation of the (lost) Greek. Proceed with caution.

There are no time markers in this passage whatsoever, not even a consideration of critical days. This physician is concerned less with the temporal cycles of the quartan fever than with the patient's dietary regimen and humoral balance. The majority of the other case histories in the collection share this focus, though some reveal slightly more interest in the timing of a patient's symptoms.[45]

As in the *Epidemics*, the most common temporal indicators in the case histories attributed to Rufus are approximate and pegged to solar rhythms. Ullmann's Case 1, for instance, describes a patient who "experienced a fever in the evening and melancholia the next morning, in which, however, he did not persist for long."[46] References to numbered hours do occur, but they are rare and diffuse. I have counted only seven, extracted from four different case histories.[47] Only in Case 21, an account of angina, are hours used as an organizing principle, and this only for the brief section of the case history in which the physician records the frequency with which he lets the patient's blood.[48] In this instance, hours are used to describe the behavior patterns of the physician, not of the disease.[49]

If these Arabic case histories are indeed translations of Greek originals composed under the Empire (whether or not they should be attributed to Rufus himself), they illustrate that, around Galen's time, case histories could take many forms. Although critical-day doctrines seem to have circulated widely, these were not always used to structure case histories, even those concerned with fevers.[50] In fact, the Arabic examples suggest that case histories need not have discussed the temporal patterns of fevers at all, much less at Galen's level of horological precision. Thus, in imitating the temporal structures of early Hippocratic case histories but adapting them to emphasize the role of what I am calling critical hours, it becomes clear that Galen made a deliberate choice, one available to his contemporaries but not always adopted. I suggest that Galen's decision was based, at least partially, on his desire to present himself as Hippocrates' intellectual descendent. By recalling the systems of critical days

---

[45] For further discussion of Rufus's case histories, see Swain 2008. On Rufus's approach to medicine generally, see Thomssen 1994.

[46] Line 12.   [47] Ullmann 1978, Cases 14.9, 15.9, 18.6, and 21.7 and 10–12.

[48] Lines 10–12. We will see in Chapter 6 that hourly timekeeping often crops up in discussions of bloodletting.

[49] This relative disinterest in the temporal cycles of disease seems to persist into the Islamic period. Álvarez Millán, e.g., has shown that the case histories of the tenth-century physician Abū Bakr Muhammad ibn Zakariyā al-Rāzī (which were collected posthumously by his students in 2 vols: the *Kitāb al-Tajārib* and the *Kitāb al-Hāwī*) are not structured according to any critical temporal system (1999, 33–42). Álvarez Millán has also demonstrated that the case histories of Avicenna (Ibn Sīnā), whose *Canon of Medicine* (*al-Qānūn fi l-ṭibb*) was considered an essential medical textbook through the eighteenth century, do not seem to have been based on personal experience and hardly employ temporal markers at all (2010, 209–13). It seems that the case histories from Rufus's collection, the Hippocratic *Epidemics*, and Galen's *On Prognosis* (as opposed to his fever case histories) were the models most commonly used by Islamic writers.

[50] This is suggested, e.g., by *Di. dec.* 9.934–6 K, where Galen addresses alternative explanations for critical-day periods.

that pervade *Epidemics* 1 and 3, Galen establishes continuity between those systems and his own temporal framework for analyzing fevers, complete with its emphasis on critical hours.

We recall that, while Galen's contemporaries acknowledged Hippocrates as a founding father of the art of medicine, many contested the claim that he should still be considered the ultimate medical authority. Thus, physicians like Galen, who advocated for Hippocrates's medical supremacy, were under constant pressure to demonstrate the coherence, sagacity, and continued relevance of the Hippocratic texts. By adjusting the critical-days concept to allow for developments in timekeeping that had occurred between the Classical and Imperial periods, Galen defends Hippocrates against the charge of being outdated and, in fact, asserts that the seeds of these developments were present in the Hippocratic writings all along.[51] In what follows, we will see more explicit examples of how Galen uses revisionist techniques to read hourly timekeeping and Imperial-period astronomical knowledge back into Hippocratic critical-day theories.

Hours and Astronomy I: The Period of the Moon

Let us turn now to the third book of *On Critical Days*, where Galen transitions from his more "empirical" discussion of how and when febrile crises manifest in the sickroom to a more "rational" meditation on how the cycles of human biology relate to the cycles of the heavenly spheres. I propose that Galen's interest in such cycles was motivated, at least in part, by a desire to adhere to the Hippocratic dictum that good doctors incorporate astronomical principles into their theories of medicine.

Galen often asserts that astronomical knowledge is a critical component of medicine, and he traces this view directly back to Hippocrates.[52] In *That the Best Doctor is also a Philosopher*, for example, Galen criticizes other physicians who praise and seek to assimilate themselves to Hippocrates while, at the same time, refusing to take the renowned physician's advice about integrating geometry and astronomy into their medical studies: "For [Hippocrates] says that astronomy contributes no small part to medicine, and it is clear that geometry is antecedent to this. Yet they themselves [i.e., other doctors] not only take no part in these things,

---

[51] Tieleman observes that Galen "tends to stress his independence from his contemporaries, while representing himself as conversing directly with the classical authors.... And of course he understands the great past thinkers much better than their self-styled followers do" (1996b, p. xxii). Tieleman goes on to discuss how Galen updates Hippocratic anatomy and physiology to account for Hellenistic developments and his own discoveries (p. xxix). See also Dillon 1977, 289; Mansfeld 1991, 137 n. 78.

[52] On Galen's relationships with astronomy and its practitioners, see Toomer 1985; Strohmaier 1997.

but even censure those who do."⁵³ Galen, on the other hand, is eager to avoid this mistake. He has chosen to heed Hippocrates's advice and to follow him in his interdisciplinary scientific pursuits. For if, as Galen cautions in *On the Method of Healing*, physicians ignore astronomy (and related disciplines, like geometry), "it will soon be permitted to everyone to become a doctor easily."⁵⁴ Galen claims to have learned from Hippocrates that what distinguishes a true doctor from a lay healer is not simply the doctor's cache of medical experiences, but also his grounding in mathematical arts like astronomy.

In book 3 of *On Critical Days*, Galen draws upon his mathematical and astronomical knowledge in order to explain and defend the theoretical bases of Hippocratic critical-day models. He concludes, on the basis of his examinations in books 1 and 2, that there are two kinds of critical-day periods: a primary cycle of one week, and a secondary cycle of four days.⁵⁵ Galen's data, however, present him with some problems. First, he would like the primary cycle to be divisible into secondary-cycle units, but the number of days in a week, seven, is not neatly divisible by four. In book 2, Galen uses the common Greek practice of inclusive counting to sidestep this problem.⁵⁶ If one divides the seven-day week in half, he points out, the midpoint of the week will fall on the fourth day. If you then count this day twice—both as the last day of the first four-day period and as the first day of the second four-day period—you will wind up with a seven-day primary period that is indeed divisible into two four-day periods.⁵⁷ Galen adopts a similar strategy to resolve a second conundrum; his empirical testing has indicated that the clearest and most decisive crises occur on the seventh, fourteenth, and *twentieth* day—not on the seventh, fourteenth, and twenty-first.⁵⁸ This is all right, Galen explains, because the second and third weeks can be said to "overlap," so that one day is shared between them.⁵⁹

In book 3, however, Galen offers a different explanation for critical-day patterns, one that is more mathematically complex and appeals to astronomical principles. He may have been motivated to alter his approach in response to a change in intended audience. Galen tells us that books 1 and 2 were written for medical students, whose astronomical knowledge could perhaps not be assumed.⁶⁰ He composed book 3, on the other hand, for a small group of professional colleagues demanding a technical account of why critical days exist.⁶¹ To this

---

⁵³ *Opt. med.* 1.53.5–1.54.2 K = 284.9–13 Boudon-Millot (2000): ὁ μὲν γὰρ οὐ σμικρὰν μοῖραν εἰς ἰατρικήν φησι συμβάλλεσθαι τὴν ἀστρονομίαν καὶ δηλονότι τὴν ταύτης ἡγουμένην ἐξ ἀνάγκης γεωμετρίαν· οἱ δ' οὐ μόνον αὐτοὶ μετέρχονται τούτων οὐδέτερον ἀλλὰ καὶ τοῖς μετιοῦσι μέμφονται.

⁵⁴ *MM.* 10.5.8–9 K: ἕτοιμον ἤδη προσιέναι παντὶ γενησομένῳ ῥᾳδίως ἰατρῷ.

⁵⁵ *Di. dec.* 9.900.5–9.901.8 K. He also acknowledges the possibility of less-decisive crises occurring on days that do not correspond to either cycle.

⁵⁶ For discussion of this strategy, see Cooper 2004, 48–9; 2011b, 448.

⁵⁷ *Di. dec.* 9.845.16–9.846.6 K.

⁵⁸ On the significance of the twentieth day, see, e.g., *Di. dec.* 9.851.11–9.853.6 K.

⁵⁹ *Di. dec.* 9.850.16–9.851.11 K.   ⁶⁰ *Di. dec.* 9.789.17–9.790.10 K.

⁶¹ *Di. dec.* 9.934.1–9 K. See also Cooper 2011b, 61.

group, Galen explains that the critical days result from the changing position of the moon and the waxing and waning of its influence over earthly matters during the course of a month. G. Cooper and others have analyzed this argument in detail.[62] I will offer only a brief outline here, paying particular attention to Galen's use of hours.

Galen's argument in book 3 stems from the observation (already articulated in the Hippocratic writings) that changes in environmental conditions and human health correlate with changes in solar and lunar "seasons."[63] Just as the sun passes through four seasons in the course of a year (each approximately three months long), the moon passes through four "seasons," or phases, in the course of a month (each approximately seven days long). Galen sees that each step in the primary sequence of critical days corresponds to one lunar phase of approximately seven days, and that each step in the secondary sequence corresponds to one half-phase (approximately 3.5 days, which Galen elsewhere rounds up to four). Thus, he concludes that critical-day patterns must be tied to the motion and influence of the moon.

But how does this help to explain the primary-sequence progression from seven to fourteen to *twenty* days, rather than twenty-one? The key, Galen asserts, is that, while in common parlance people often say that a year is three hundred sixty-five days or a month thirty, these are only rough approximations; neither solar nor lunar periods can be expressed accurately in whole days. In support of this fact, he cites not only Hippocrates,[64] but also the Hellenistic astronomer Hipparchus.[65] Therefore, Galen continues, in order to anticipate when a febrile paroxysm will actually occur, one must use a more precise value for the length of an average "medical week." To calculate this, one must first discover the more precise length of an average month and then divide that value by four.[66]

Galen proceeds to calculate as follows. First, he makes a distinction between *synodic* and *sidereal* months; the former represent the time the moon takes to return to its starting phase, the latter the time it takes to return to a certain position in the zodiac.[67] Because one year is approximately 365¼ days long, Galen calculates that each synodic month is, on average, approximately 29½ days long, while the average sidereal month lasts only a little over 27 days.[68] Galen subtracts three days from the average synodic month (yielding 26½) to reflect the period

---

[62] See esp. Cooper 2011b, 61–76, and Langermann's (2012) critique.   [63] *Di. dec.* 9.908.4–12 K.
[64] *Di. dec.* 9.928.16–9.929.3 K. Galen is here quoting Hippoc. *Prog.* 20.2.170 L = 58.9–59.2 Jouanna. On fever theories in the Hippocratic Corpus, see Sticker 1928, 1929, 1930.
[65] *Di. dec.* 9.907.14–16 K. Hipparchus's name occurs six other times in the Galenic corpus: *Di. dec.* 9.907.16 K; *Hipp. Epid.* 1.17a.23.13 K; *Hipp. Prog.* 18b.240.14 K; *MM* 10.12.10 K; *Comp. med. loc.* 13.353.14 K; *Sept. part.* line 19 Schöne (1933); *UP* 4.359.11 K. In this last instance, Galen locates Hipparchus in the same intellectual stratum as Plato, Aristotle, and Archimedes.
[66] On Galen's process of defining "medical" weeks and months, see Garofalo 2003, 52–3; Cooper 2004, 53–5; 2011a, 129; 2011b, 75; Heilen 2018; Singer 2022, 117–22.
[67] *Di. dec.* 9.907.8–9.908.1 K.
[68] The discussion that follows can be found at *Di. dec.* 9.930.15–9.933.4 K.

when the moon is not visible in the sky, and its influences are therefore negligible. Then, because he has reasoned that both synodic and sidereal months must be influential for determining clinical crises, Galen averages their two lengths to arrive at 26¹¹⁄₁₂ days.[69] Finally, he divides that number into fourths to produce an average "medical week" of 6³⁵⁄₄₈ days. Three such weeks comes to 20⁹⁄₄₈ days, which is closer to twenty days than twenty-one. In this way the dilemma of the seven-fourteen-twenty progression is resolved.

Here, Galen expresses his calculations as a combination of integers and unit-fractions.[70] Elsewhere in his corpus, however, Galen chooses to translate these day-fractions into hours and fractions of hours. The relevant passage occurs in *On Seven-Month Children*, which Galen composed in response to the Hippocratic texts *On the Seven-Month Child* and *On the Eight-Month Child*.[71] *On Seven-Month Children* has come down to us both in a fragmentary Greek version and in a complete Arabic translation by Hunayn Ibn Ishaq.[72] In this text, Galen participates in a lively medical debate, ongoing since the time of Hippocrates, over the appropriate length of human fetal gestation.[73] Galen observes that at the heart of this debate lies a dispute over the definition of a month, so he explains to his readers how trained astronomers measure its length. Preserved among the Greek fragments of *On Seven-Month Children* is the following passage:[74]

> Hipparchus demonstrated that, of a whole day, one thirtieth and one twentieth and one twenty-seven-thousandth are added, and another small fraction again in

---

[69] He asserts that the synodic month influences the general atmosphere, while the sidereal month influences the particular changes that affect an individual.

[70] *Di. dec.* 9.932.13–9.933.1 K. The ancient Greeks and Romans, like the Egyptians, expressed the remainder of a quotient as the sum of a series of "unit fractions," which had the number one as numerator. On fractions in ancient Greece and Egypt, see Knorr 1982.

[71] The relationship between these two texts is controversial. *On the Eight-Month Child* appears in a different sequence in manuscripts M and V. As Potter, editor and translator of the Loeb Classical Library edition, explains, "V presents the whole of chs. 10-13 and 1-9 in succession under the title *Eight Months' Child*, and then another short spurious text under the title *Seven Months' Child*.... To avoid the unnecessary confusion a departure from Littré's chapter numbering would entail, I have kept his and M's order of the text, but adopted Joly's and V's title *Eight Months' Child* for the whole work" (2010, 73–4).

[72] The incomplete Greek text is based on the Greek codex Laurentianus LXXIV 3 (L), fol. 104r–105v and Laurentianus Gr. LXXIV 2 (l). The complete Arabic translation appears in Codex 3725 of the Library of the Aya Sofia in Istanbul (fol. 127b–134b). See Walzer 1935.

[73] Mention of this debate also occurs at, e.g., Plin. *HN* 7.4 and Aul. Gell. *NA* 3.10 and 16.

[74] *Sept. part.* 19–38 Schöne: ἄλλο τι μόριον Ἵππαρχος ἀπέδειξε τῆς ὅλης ἡμέρας εἰς τριακοστὸν τὸν καὶ εἰκοστὸν καὶ δισμυριεπτακισχιλιοστὸν ἄλλο τέ τι πρὸς τούτῳ πάλιν σμικρόν, οὗ περιττὸν εἰς τὰ παρόντα μεμνῆσθαι. τὸ γάρ τοι προκείμενον ἤδη πέρας ἔχει, κατὰ μὲν τὸ δεύτερον τῶν Ἐπιδημιῶν.... "Οἱ δ' ἑπτάμηνοι, φησίν, γίνονται ἐκ τῶν ἑκατὸν ἡμερῶν καὶ ὀγδοήκοντα καὶ δύο καὶ προσεόντος μορίου." ἑπταμήνους ἤτοι παιδία ἢ τόκους, τὸ δ' ἐπὶ ταῖς ἑκατὸν ὀγδοήκοντα δύο ἡμέραις μόριον τὰς πεντεκαίδεκα ὥρας λέγει μετά τινος, ὡς ἔμπροσθεν ἔφην, μορίου σμικροῦ, ὃ καὶ αὐτὸ μιᾶς ὥρας ἐστὶν εἰκοστὸν τέταρτον ἔγγιστα. δῆλον γὰρ ὅτι τὰς ἰσημερινὰς ὥρας λέγομεν ἐν πᾶσι τοῖς τοιούτοις, ὧν ἐστι καὶ τὸ νυχθήμερον ὀνομαζόμενον ὑπὸ τῶν ἀστρονομικῶν εἴκοσι καὶ τεττάρων ὡς ἁπάντων αὐτῶν ἴσων ὄντων, ἐπειδὴ τὸ παραλλάττον ἐλάχιστόν ἐστι, ὡς τινας νομίζειν ὅλως αὐτὸ μηδ' εἶναι.

addition to this, which it is superfluous to mention at present.[75] For the point is that [the month] has the aforementioned boundary, according to the second book of the *Epidemics*.... [*Galen quotes the Hippocratic* On the Eight-Month Child]: "The seven-monthers," [Hippocrates] says, "are born after one hundred eighty-two days and an additional fraction." With regard to seven-month infants or children, he means that the portion added to the one hundred eighty-two days is fifteen hours, with, as I said before, some small fraction close to one twenty-fourth of an hour. It is clear that we are talking about equinoctial hours in all these cases, a twenty-four-hour span of which is called the "night-and-day" by astronomers as if all of the hours were equal, since their deviation is very slight, with the result that some think it should be entirely ignored.[76]

Galen's central aim in *On Seven-Month Children* is to vindicate Hippocrates from accusations of self-contradiction. In two texts, *On the Seven-Month Child* and *On the Eight-Month Child,* Hippocrates correctly states that the length of an average synodic month is 29½ days. In two other texts, however, *On Nourishment* and *Epidemics* 2, Hippocrates gives a less precise month-length of thirty days. To resolve this apparent contradiction and to demonstrate that Hippocrates was not, in fact, in error, Galen argues that *On Nourishment* and *Epidemics* 2 were written early in Hippocrates' career, whereas *On the Seven-Month Child* and *On the Eight-Month Child* were composed later, after Hippocrates had learned the more precise value. To support this interpretation, Galen even cites the same passage from *Prognostic* that he cites in *On Critical Days*, about the lengths of solar and lunar periods being immeasurable in whole days.[77]

It is notable, however, that no Hippocratic writer expresses the length of an average month with precision down to the hour. While the author of *Epidemics* 2 states that "seven-monthers... are born after one hundred eighty-two days and an additional fraction," it is Galen himself who redefines that fraction as fifteen hours and one twenty-fourth. Galen explicitly reads this figure into the subtext of *On the Eight-Month Child* when he asserts, "[Hippocrates] *means that* the portion added to the one hundred eighty-two days is fifteen hours" (emphasis added). Galen thereby attributes to the Hippocratic author the same concern with temporal precision that he himself shares, despite the fact that the word *hōra*, in the sense of "hour," only appears in the Hippocratic corpus on a handful of occasions, as

---

[75] For an explanation of this calculation and its relationship to Hipparchus's data, see Neugebauer 1983.

[76] For the corresponding passage in Hunayn's Arabic, see 126–37, pp. 339–40 Walzer 1935 (German tr. at pp. 349–50).

[77] 114–124, p. 347 Walzer. The text in the Greek differs somewhat from the Arabic and can be found at 21–7, p. 354 Walzer.

discussed in Chapter 2.[78] It seems that, while at least some of the Hippocratic writers had exposure to hourly timekeeping, they elected not to use hours in their calculations (or, as we have seen, in structuring their case histories). This contribution is wholly Galen's, and it serves him in a variety of ways: it helps to exonerate Hippocrates from charges of error and contradiction, to strengthen the link between Hippocrates and Galen, and to highlight Galen's own astronomical proficiency.

Hours and Astronomy II: The Periods of Planets

Returning to book 3 of *On Critical Days*, it is worthwhile to consider one other passage where hourly timekeeping, though not mentioned explicitly, is fundamental to Galen's argument. Here, Galen seeks to explain, by recourse to contemporary astrological theory, how it is that the moon's quarterly position can affect human health.[79] He refers to a form of astrology, handed down "by the Egyptian astronomers," dedicated to determining whether an event or undertaking will be auspicious.[80] According to these unnamed astronomers, a day is auspicious if the moon, on that day, is in an astrologically significant aspect (e.g., quartile, trine, etc.) with "the well-disposed planets, which they call benefic," based on its position at the time of the person's birth.[81] If, on the other hand, the moon is in astrological aspect with "ill-disposed" or "malefic planets," the day is inauspicious.[82] By extension, if, on a critical day, the moon is in quartile with a malefic planet, it is likely that the patient will experience a crisis that is severe and even fatal. If, however, it is in quartile with a benefic planet, the patient's condition will likely improve.

Since the moon can move from one zodiacal sign to another at any time of day or night, the astrological character of a day can change from one hour to the next.[83] Thus, it was important for ancient astrologers to be able to track the movements of celestial bodies over time with as much accuracy and precision as possible—ideally, down to the hour or fraction of an hour.[84] As we have seen, the technology

---

[78] *Epid.* 4.12, 5.150 L and *Int.* 27.7.238 L. The earliest example of their popular (as opposed to technical) use is in Callim. Fr. 550 Pfeiffer (1924). See Langholf 1973; Hannah 2009, 73–5. Sattler (2020) argues that their first appearance in a philosophical work is in Plato's *Laws*.

[79] The full passage can be found at *Di. dec.* 9.911.14–9.913.10 K. I am indebted to Alexander Jones (pers. comm.) for his assistance with the following translation and discussion.

[80] *Di. dec.* 9.911.15–16 K: πρὸς τῶν Αἰγυπτίων ἀστρονόμων.

[81] *Di. dec.* 9.911.18–9.912.1 K: τοὺς εὐκράτους ... τῶν πλανητῶν, οὓς δὴ καὶ ἀγαθοποιοὺς ὀνομάζουσιν.

[82] *Di. dec.* 9.912.2 K: τοὺς δυσκράτους.

[83] This is evident, e.g., in *P.Oxy.* 65.4483 (194 CE), where the author recommends that his addressee meet a friend while the moon is in Sagittarius, which he claims will be from the fourth hour on Thoth 12 until the seventh hour on Thoth 14.

[84] On the extent to which ancient Greek scientists concerned themselves with precise measurement, see Lloyd 1987, 215–84.

for tracking and modeling celestial cycles improved dramatically between Hippocrates's day and Galen's. The late Classical and early Hellenistic periods, when the Hippocratic writers were active, constituted an important transitional time for Greek astronomy. It was already recognized in the Archaic period that close observation of celestial movements could enable one to make predictions about related, personally relevant matters. Farmers and merchants were better off if they could anticipate the likelihood of rain or high winds, and if they knew when the period of daylight would be longer or shorter. Since both the weather and the length of daylight correlate with the time of year, farmers and merchants discovered that they could get a handle on these factors by pegging their own enterprises to recurring celestial events. The poet Hesiod offers our earliest literary testimony of what seems like a Farmer's Almanac in *Works and Days*.[85]

By the end of the Classical period, however, new technologies and new data enabled Greeks to develop predictive models of greater complexity. One such tool was the *parapēgma* (from *parapēgnumi*, "to fix beside"), which coordinated the risings and settings of fixed stars with weather predictions for given dates.[86] Another was the practice of astronomical table-making.[87] Babylonian priests, eager to predict the occurrences of celestial omens, such as lunar and solar eclipses, had maintained meticulous records of astral, planetary, and lunar positions with which they could calculate impending syzygies.[88] At some point during the Hellenistic period, these tables made it into the hands of Greek astronomers, providing them with raw data and models[89] for organizing their own records and observations in the future.[90]

---

[85] Hesiod recommends, for instance, that his layabout brother Perses gather the grapes from his vines "whenever Orion and Sirius reach mid-heaven / and rosy-fingered Dawn sees Arcturus" (*WD* 609–10).

[86] *Parapēgmata* came in a variety of forms, both literary and inscriptional. Inscriptional varieties use a hole-and-peg system to keep track of the date, and it is from this practice of "fixing" the peg in each successive hole that the term *parapēgma* derives. For an in-depth history of Greek and Roman *parapēgmata*, see Lehoux 2007. He notes (2007, 14, 24) that the Greek *parapēgmata* of the Hellenistic period show a greater interest in astrometeorology than do the Latin *parapēgmata* of the Roman period. See also Lehoux 2005; Hannah 2009, 53–4.

[87] On the history of mathematical table-making, see Campbell-Kelly 2003.

[88] I use the term "syzygy" here in the astrological sense of "alignment," not in the modern astronomical sense of three celestial bodies that appear in a line.

[89] Lehoux has pointed out that the producers of *parapēgmata* often preferred to collate data from preexisting models rather than make new observations of their own (2007, 69). The Babylonian system was also based primarily on mathematics, not observation (see Neugebauer 1975, 363–8).

[90] These developments in astronomical data organization also contributed toward the invention of more complex predictive devices, like the so-called Antikythera Mechanism, which was salvaged from an ancient shipwreck off the coast of Antikythera and has been dated to c.100 BCE. The Antikythera Mechanism (National Archaeological Museum of Athens, Inv. 15087) is a geared, computational device that allows the user to input a date and determine where that date falls within a number of calendrical cycles (both astronomical and political) and the positions of the celestial bodies at that time. On the mechanism and its functions, see, e.g., Freeth et al. 2008; Carman et al. 2012; Lin and Yan 2016; A. Jones 2012, 2017.

As sundials and water clocks became more widely available over the Hellenistic and Imperial periods, a variety of tables began to appear in the papyrological record that record the hour of a celestial event's occurrence alongside other factors.[91] Ephemerides, for example, list the daily positions of celestial bodies and often include the hour of zodiac sign-entry as well as the longitude and time of particular syzygies.[92] Sign-almanacs from the third century CE onward also tend to include a column for the seasonal hour of sign-entry.[93] Thus, by Galen's day, astronomers were in the habit of recording the exact hours at which periodic events occurred, and then using that observational data (whether their own or someone else's) to construct mathematical models for anticipating future events. In turn, catarchic and horoscopic astrologers, whose trades became wildly popular under the Roman Empire, often attempted to impress their clients by drawing upon the latest developments in astronomical table-making, timekeeping, and celestial trigonometry.[94]

Galen appeals to such developments in order to provide a more nuanced explanation of why Hippocratic critical-day theories work. Since these astronomical and astrological trends, in full force during Galen's time, were only just beginning to take off during the period of the Hippocratic writers, it seems likely that these writers did not share the same degree of exposure to or interest in this mode of interpretation. Many of the Hippocratic works certainly display an appreciation for astronomy.[95] The *Epidemics*, for example, contains many references to seasonal astronomical events, such as equinoxes, solstices, and the rising and setting of stars and constellations.[96] The author(s) of *Epidemics* 1 and 3, whose case histories were examined above, could even be said to map the progression of diseases over time for predictive purposes in a manner analogous to the astronomer tracking the movements of celestial bodies. Galen, however, seems to read back into these texts a capacity for and familiarity with hourly timekeeping (and its astronomical and astrological applications) which was not appropriate for the late Classical and early Hellenistic eras. In so doing, Galen appears once again to present himself as a follower and refiner of Hippocratic doctrine.

---

[91] See esp. A. Jones 1999, 2009.

[92] Some ephemerides were used for catarchic astrology (e.g., A. Jones 1999 no. 4180), as evidenced by the fact that they evaluate each day as "good" or "bad" for certain kinds of endeavors. Our extant ephemerides date from the first century BCE to the late fifth century CE. The most extensive collection and commentary to date is A. Jones 1999.

[93] E.g., A. Jones 1999 nos. 4192 and 4194–4196a.

[94] For Greek and Roman horoscope collections, see, e.g., Vettius Valens, Ptolemy's *Tetrabiblos*, and the astronomical papyri compiled by Neugebauer and Van Hoesen (1957) and Jones (1999, 371–450). On the data-processing and marketing strategies of ancient astrologers, see Barton 1994, 27–94. On their tools, see Evans 2004.

[95] See, e.g., Hippoc. *Aer.* 2.2.14 L = 189.4–6 Jouanna; *Vict.* 2.25, 6.470 L. On Hippocratic engagement with meteorological medicine, see Liewert 2015.

[96] For discussion, see Phillips 1983.

Conclusion

This chapter investigated the roles of hourly timekeeping in Galen's *On Critical Days*. Galen introduced greater temporal precision into clinical narratives that are otherwise modeled on Hippocratic critical-day structures. Galen's account of the theory behind critical days and the astronomical and astrological arguments he musters imply the use of hourly timekeeping technology. Engaging with hourly timekeeping in these ways enabled Galen not only to weigh in on the debate surrounding the utility of critical-day systems, but also to support a network of interrelated claims. One of these claims is that the genuine Hippocratic writings are internally consistent and contain within them the seeds of important medical advancements, including the use of hourly timekeeping. Another is that Galen's own medical theories represent truthful and logical extensions of Hippocratic doctrine. The refinements that Galen introduces (such as precision down to the hour) simply bring Hippocratic teachings into closer alignment with contemporary developments and technologies. Ultimately, Galen claims that he, like Hippocrates before him, practices a form of medicine which is grounded in demonstrative method and incorporates data and strategies from mathematical arts like astronomy. Thus, Galen seems to have considered hourly timekeeping to be both a practical and a rhetorical tool for promoting himself, his scientific method, and his professional hero, Hippocrates.

6
When Is Temporal Exactitude Desirable?

In the previous chapter, we saw how Galen used hourly timekeeping to intervene in medical debates, ongoing since the Hellenistic period, over which times would prove critical during a patient's febrile illness.[1] The present chapter will investigate the roles of hourly timekeeping within other active debates of the Imperial period—debates over how to diagnose and classify intermittent fevers with irregular periodicities, how to distinguish between the essential features of a disease and its epiphenomena, and the appropriate circumstances under which to let a patient's blood. Underlying all of these discussions are larger, more general disputes over the relative values of personalization and standardization in patient care and over the extent to which medicine should be considered an "exact science," i.e., one whose fundamental rules and principles can be expressed quantitatively. In the face of these controversies, which still fuel medical debates in the present day, how did ancient Greek physicians decide when numerical exactitude was desirable, particularly in their daily timekeeping?

In what follows, we will examine, as case studies, passages drawn from a series of Galenic texts: first, a suite of treatises on intermittent fevers, including *On Crises, On Types, On the Distinct Types of Fevers,* and *On Periods*; second, *On the Method of Healing*, Galen's most popular and exhaustive medical handbook, in which we will be particularly concerned with his critique of the Methodist distinction between a "symptom" and an "affection"; finally, *On Treatment by Venesection*, one of three texts attributed to Galen that are devoted entirely to the art of bloodletting or phlebotomy. I propose that these texts can help us to sketch the outlines of a larger debate among Imperial-period physicians over the importance of numerical timekeeping within the day. Taken together, these passages offer insights into how this debate played out both in the diagnosis and classification of diseases and in the administration of particular treatments. Ultimately, I argue that Galen sought to portray his own attitude toward quantitative timekeeping as the harmonious mean between two extreme positions: one that favored excessive and programmatic numerical precision, and another that overlooked quantitative timekeeping all together. In Galen's view, members of this latter group fail to recognize the important temporal patterns that could help them diagnose conditions more accurately and treat them more effectively. On the other

---

[1] Some material from this chapter was previously published in K. J. Miller 2020.

hand, members of the former group fail to acknowledge the diversity inherent among patients, diseases, and their environmental conditions, and as a result, their diagnoses and treatment plans lose touch with reality. Only Galen himself, he would have us believe, knows when numerically exact timekeeping is both useful and appropriate.

Insufficient Exactitude: Fevers and Symptoms

Hourly timekeeping played an important role in Imperial-period debates over how to describe and classify types of intermittent fevers—that is, fevers whose intensities ebb and flow according to recognizable temporal schemes. Such periodicity is now recognized as the signature feature of malaria, three strains of which are endemic to the Mediterranean: *Plasmodium* (= *P.*) *falciparum*, *P. vivax*, and *P. malariae*.[2] From the late Classical period onward, physicians competed to identify the temporal patterns of these diseases in order to better predict the outcomes of individual cases. It was important that a doctor learn to prognosticate as accurately as possible, for he could endanger his medical reputation by taking on incurable patients.[3]

In addition to tracking and theorizing about the days on which periodic fevers would reach their crisis points, the Hippocratic writers also classified such fevers according to the number of whole days that passed between each paroxysm of fever and chills.[4] A tertian (*tritaios*) fever, for example, could be recognized by the fact that it recurred every two days (or every third day, counting inclusively). Troublingly, however, some intermittent fevers recurred at irregular intervals,

---

[2] All malarial species belong to the genus *Plasmodium* and are transmitted by mosquito vectors. *P. falciparum* produces a malignant tertian fever (whose intensities peak every forty-eight hours, or third day inclusive), *P. vivax* a benign tertian, and *P. malariae* a quartan (whose intensities peak every fourth day inclusive). These periodicities result from the fact that, at the end of synchronized cycles of spore-generation, malarial protozoa rupture host blood cells, inducing a paroxysm in the patient. See Scheidel 2001, 75; Sallares 2002, 9–11; Harper 2017, 84–8.

[3] The Hippocratic author of *Prognostic* summarizes this state of affairs thus: "It seems to me to be an excellent thing that the physician practice prognosis. For if he should prognosticate and foretell, in the presence of patients, things present, things past, and what will be in the future, and if he explains however many things the patients have neglected, he would be more trusted to know the circumstances of those who are sick, with the result that men will dare to entrust themselves to this physician.... [A]nd by knowing and declaring in advance who will die and who will be saved, he can be blameless" (2.110–112 L = 1.1–3.11 Jouanna). Galen, likewise, declares in *On Periods*: "Nothing is so useful—and especially in the case of fevers—as prognosticating the beginnings of the paroxysms yet to come. If someone is able straightaway, from the beginning, to accomplish this by means of a technical estimate, he will benefit the patient very greatly throughout his illness" (7.479–80 K).

[4] See, e.g., *Nat. hom.* 15.1–29 L. The Hippocratic authors of the *Epidemics*, e.g., organized their patient case histories according to the ordinal day (counting from the onset of the illness) on which various symptoms occurred. These physicians were particularly interested in developing systems to predict the day on which a disease might undergo a crisis and whether that crisis would be good or bad. A crisis, we recall, was a decisive moment in the course of a disease at which the illness would either be cured or begin a transition into a new phase.

which could not be neatly measured in whole numbers of days.[5] These fevers defied reliable prognosis and could make fools of physicians at the bedside.

This unpredictability was a problem, and one that stimulated debate into the Roman Imperial period. By that time, however, as we have seen, more sophisticated tools had become available to physicians interested in tracking febrile paroxysms. Sundials and water clocks had become increasingly widespread, as had the process of scientific table-making, which made its way to the Greco-Roman world from Babylonia in the Hellenistic period. While, at first, these technologies belonged almost exclusively to the purview of astronomers, they were soon adopted by authors writing on a variety of subjects, including, of course, Galen himself. This section will explore how Galen used these tools to propose a system for diagnosing and classifying irregular fevers that, in his opinion, walked the delicate line between being overly and insufficiently exact. In so doing, Galen also presents himself, in contrast to rival doctors, as an astute observer and a logical, systematic thinker: in other words, someone who puts into practice his preferred scientific method.

In order to appreciate the nuances of Galen's strategies, it is important first to become familiar with the principles of Galen's fever classification system, which he develops most extensively in the treatises *On Crises, On Types,* and *On the Distinct Types of Fever*. For Galen, fevers ultimately arise because one's internal, life-giving fire has been excessively stoked.[6] This stoking could be caused by a variety of factors, one of which was that certain bodily humors (*khymoi*: specifically, yellow bile, black bile, and phlegm) could become clogged in the arteries and start to ferment.[7] Fevers brought about in this way Galen calls septic (*sēpomenoi*).[8]

Different putrid humors yielded septic fevers with different characteristics. Galen famously associated each humor with two of four qualities (*poiotētes*): hot, cold, moist, and dry. Yellow bile was hot and dry, black bile cold and dry, phlegm cold and moist.[9] Thus, in Galen's view, a fever caused by an excess

---

[5] The Hippocratic author of *Prognostic*, e.g., describes a number of acute fevers, none of whose periods can be calculated exactly using whole days (2.169 L = 58.8–9 Jouanna).

[6] As Wittern notes (1989, 4), Galen is inconsistent in his details. In *On the Distinct Types of Fever*, he claims that fevers should be categorized among "external" or "unnatural" fires (e.g., at 7.374 K), but in his commentary on the Hippocratic *Aphorisms*, he defines fevers as transformations of the inherent fire within living creatures (e.g., at 17B 414 and 426 K).

[7] Galen also recognized a fourth humor, blood, but did not believe it could ferment to produce an intermittent fever. Instead, he considered fevers that originated in the blood to be "ephemeral" (*Diff. feb.* 7.374–7 K).

[8] As Lonie (1981, 29) observes, "The basic feature of [Galen's] explanation was that the temporal discontinuity of the fever is correlated with a spatial discontinuity of the humors in the body. Both the fits of shivering and the period of the onsets were accounted for by *quanta* of putrid humors moving from one part of the body to another. These *quanta* accumulated, putrefied, and were expelled in periods of time which differed according to the quantity and quality of the particular humor." Lonie goes on to note that sixteenth-century writers found this component of Galen's fever theory to be unpersuasive.

[9] *Nat. fac.* 2.129–31 K.

of yellow bile would overheat and desiccate a patient. To alleviate this condition, Galen recommends forbidding the patient from having contact with other "hot" or "dry" substances and, instead, treating the patient allopathically (in this case, with substances considered "cold" and "moist").[10] For Galen, it was important that a doctor know which humor was to blame for a given fever, so that he could not only predict the course of the illness, but also design a treatment plan for the patient tailored to that humor's qualities.

Another important aspect of septic fevers was their temporality. Galen divided such fevers into two broad categories: the continuous (*synekheis*) and the intermittent (*dialeipontes*). Intermittent fevers went through cycles of paroxysm and remission, and the lengths of those periods depended upon the type of humor that had been corrupted. Phlegm, for instance, if fermented, produced a fever that underwent a full cycle of paroxysm and remission every twenty-four hours, earning it the name quotidian (*aphēmerinos*). Fermented yellow bile caused a fever with a forty-eight-hour cycle, which was called tertian (*tritaios*) because it arrived every third day, inclusive of the day from which the counting began. Septic black bile produced a fever with a seventy-two-hour cycle, called quartan (*tetartaios*).[11]

"Quotidian," "tertian," and "quartan" were, for Galen, the three simple fever types (*typoi haploi*). To account for so-called wandering or irregular (*planētai*) fevers, which deviated from these types, Galen also incorporated into his system mechanisms for combining and adjusting the three primary types to produce a wider range of periodicities.[12] He made a distinction between stationary fever types (*typoi hestōtes*), whose paroxysms arrived at precisely the same hour every day, and moving fever types (*typoi kinoumenoi*), whose paroxysms tended to arrive two to three hours early or late.[13] This allowed Galen to include under the heading of a "moving quotidian," for instance, fevers with a wide range of paroxysmal cycles, extending from roughly twenty-one to twenty-seven hours. Galen also dealt with irregular fevers by suggesting that the three simple fever types could synthesize to produce compound fever types (*typoi synthetoi*). Such compounds could either be homogeneous (*homogeneis*), i.e., composed of multiple fevers of the same simple type (such as three tertians or two quartans), or heterogeneous (*heterogeneis*), i.e., formed by a mixture of different simple types (such as one tertian and one quartan).[14]

---

[10] *MM.* 10.590 K.

[11] These terms—and the recognition of one-day, two-day, and three-day fever periods—did not originate with Galen. We see them already in the Hippocratic works of the late Classical period.

[12] *Cris.* 9.680 K.   [13] *Typ.* 7.464 K.

[14] *Typ.* 7.464.18–7.465.8 K. Galen also speaks often of the "semitertian" fever, though he does not classify it among the simple types. For Galen, a semitertian was the deadly product of a continuous quotidian fever and a tertian, and he distinguishes between multiple kinds (*Diff. feb.* 7.358–9 K). Another Imperial-period physician, Celsus, describes the semitertian as a tertian that never undergoes complete remission and whose paroxysms last roughly thirty-six hours (3.3.2).

Galen's classification system proved so attractive to later physicians that it became the foundation of fever theory first in the medieval Arabic world and then in the European West, up to the end of the fifteenth century.[15] Galen's reputation began to crumble during the Enlightenment, but vestiges of his fever classification system remain embedded in modern medical terminology.[16] To this day, fevers are still categorized as "continuous" or "intermittent," and while the terms "quotidian," "tertian," and "quartan" have recently gone out of fashion in the medical world, they are still used on occasion to describe strains of malaria.[17]

An extended passage from *On Crises* offers a glimpse of Galen's classification system in action. Galen's aim in this passage is to demystify an irregular fever by interpreting it as a homogeneous compound of three tertians. Here, Galen's references to hourly timekeeping help him to illustrate the temporal patterns that underlie the illness in question, and to promote the image of himself as an exemplary scientist. The passage is quoted in full to better demonstrate the shape of Galen's argument and the temporal landscape he depicts:[18]

> And yet frequently [fevers] have a clear mathematical proportion, just like in the case of the young man who began to have a fever in autumn, on the first day and around the fifth hour together with short shivering spells.... To me, who

---

[15] During the medieval period, most of Galen's writings were only indirectly available in the Latin West, whether through compilations that included works by other medical authorities (e.g., the Articella) or because Galenic material was incorporated into the writings of other major figures like the Arabic philosopher Ibn-Sīnā or Avicenna (980–1037 CE).

[16] Galen's reputation was shaken in the sixteenth and seventeenth centuries, initially under the pressure of attacks launched by the mystic Paracelsus and his followers, but ultimately also due to scientific discoveries, such as those made by Vesalius and Harvey, which disproved central elements of Galen's medical theories. Vesalius's *Fabrica* (1543), e.g., demonstrated that Galen's was not the last word on human anatomy, while Harvey's (1628) work on the circulatory system revealed that Galen had been wrong in his interpretations of veins and arteries. On the reception of Galenic medicine from the Renaissance onward, see Temkin 1973; W. D. Smith 1979, 1–44; Bates 1981, 45–70; Lonie 1981; Yeo 2005, 435; Nutton 2008, 355–90; 2020, 132–56.

[17] On the modern medical definitions and treatments of malaria, see, e.g., Pampana 1963 and the Center for Disease Control's entry for "malaria." On the roles that malaria may have played in the ancient world, see W. H.S. Jones 1907, 1909; de Zulueta 1973, 1–15; Grmek 1989, 245–83; Burke 1996, 2252–81; Scheidel 2001, 76–89; Sallares 2002; van der Eijk 2014, 112–17.

[18] *Cris.* 9.680.13–9.683 K: καίτοι πολλάκις γε σαφεστάτην ἔχουσι τὴν ἀναλογίαν, ὥσπερ ὁ νεανίσκος ὁ περὶ τὰ τέλη τοῦ φθινοπώρου τῇ πρώτῃ τῶν ἡμερῶν ὥρας που πέμπτης ἀρξάμενος πυρέττειν ἅμα ῥίγει βραχεῖ.... ἡμῖν μὲν οὖν ἅπασαν τοῦ πυρετοῦ τὴν ἰδέαν ἐπισκεψαμένοις ἐπιπλοκὴ τριταίων ἐφαίνετο τριῶν.... τριῶν μὲν δὴ τριταίων ἐπιπλοκὴ κατὰ τὴν δευτέραν εὐθὺς ἡμῖν ἡμέραν ἐφαίνετο σαφῶς, τὸ δ᾿ ὅτι καὶ προληπτικῶς παροξυνόντων ἐπὶ τῆς τρίτης πρῶτον ἐγνώσθη. δευτέρας μὲν γὰρ ὥρας ὁ παροξυσμὸς εἰσέβαλλεν, οὐ πέμπτης ὥσπερ ἐν τῇ πρώτῃ, τὰ δὲ τῆς τριταίας περιόδου γνωρίσματα πάντ᾿ ἐναργέστερα τῶν ἐπὶ τῆς πρώτης ἡμέρας ἐκόμισεν. ἐντεῦθεν δὲ παραφυλάττοντες ἅπαντας εὕρομεν τοὺς παροξυσμοὺς ἢ δυοῖν ἢ τρισὶν ὥραις προλαμβάνοντας ὥστε καὶ τὸν ἀνάλογον τῷ πρώτῳ κατὰ τὴν περίοδον ἀπαντῶντα τῇ μὲν πέμπτῃ τῶν ἡμερῶν ἀνίσχοντος ἡλίου γενέσθαι, τῇ δ᾿ ἑβδόμῃ πολὺ θᾶττον ὡς περὶ ὥραν που τῆς νυκτὸς ἐνάτην, εἶθ᾿ ὁ μετὰ τοῦτον ἀνάλογον ὥρας νυκτὸς ἑβδόμης, ὁ δ᾿ ἐφεξῆς τετάρτης ἐγένετο, τοὐντεῦθεν δὲ πάλιν ὁ μὲν ἐφεξῆς τετάρτης, ὁ δ᾿ ἐπὶ τούτῳ πέμπτης, εἶθ᾿ ἕκτης αὖθις, εἶτ᾿ ὀγδόης ὁ δέκατος ἀπὸ τῆς ἀρχῆς ἐγένετο παροξυσμός. ἀνάλογον δὲ τούτῳ καὶ οἱ ἄλλοι δύο μετὰ τὴν ἑβδόμην περίοδον οὐ μόνον οὐ προελάμβανον ἀλλὰ καὶ ὑστέριζον, ὅτε δὴ καὶ σαφέστατ πᾶσιν ἐφάνη καὶ τοῖς ἡμιτριταῖον εἶναι νομίζουσιν αὐτόν, ὡς ἐσφάλλοντο μέγιστα.

examined the whole form of the fever, it seemed to be a combination of three tertians.... [This fact] appeared clear to me immediately upon the second day. The first [tertian] was recognized to have also had an early paroxysm on the third [day], for the paroxysm came on in the second hour—not the fifth as on the first [day]—and conveyed all the signs of the tertian period more clearly than on the first day. From there, keeping an eye on everything, we found that the paroxysms were coming on two or three hours early, with the result that the mathematical proportion corresponding with the first with regard to its period came into being on the fifth day during sunrise, then much more rapidly on the seventh day, at around the ninth hour of the night. Then the next one according to this mathematical proportion [arose] in the seventh hour of the night. The next one in succession arose in the fourth hour, and thence again the next one at the fourth hour, and the one after this during the fifth hour, and then again during the sixth. Then the tenth paroxysm from the beginning came about during the eighth hour. Analogously to this one, the other two after the seventh period not only did not come early but they also arrived late, which thing is very clear to all, even to those who think that [this fever] is a semitertian, although they are very much mistaken.

In the opening of this passage, Galen makes it clear that he is participating in an ongoing terminological debate over how to define atypical fevers. Other doctors, he claims, find themselves entirely nonplussed by fever periods that do not fit their standard simple types. Galen, on the other hand, asserts that he knows better and can even support his diagnosis by offering physical and temporal evidence for his interpretation. By indicating, in the subsequent case history, the exact days *and hours* at which the patient experienced paroxysms, Galen attempts to demonstrate that, in spite of the apparent irregularity of these paroxysms, an underlying pattern is still discernible. He claims that this pattern can actually be explained as the result of three tertian fevers with different start-dates, whose paroxysmal periods overlap (see Table 1). Galen explicates the unusual periods of these tertians by calling upon his distinction between fixed and moving fever-types (see Table 2). These tertians, he argues, are moving, and thus one should expect each paroxysm to arrive two to three hours later or earlier than the one that preceded it.[19] Characteristically for Galen, the resulting argument seems designed to strike the reader as both straightforward and mathematically complex at the same time.[20]

---

[19] Galen seems untroubled by the nine-hour jump in the timing of the paroxysms on Day 9 and Day 11.
[20] Barton has observed that, "despite [Galen's] idealization of the autodidactic doctor and his implication that anyone who follows his example can enjoy similar fame and success, he appears to relish the mastery of such knowledge more than he does attempting any real communication of it" (1994, 137).

Table 1. The pattern of paroxysms experienced by the patient within the first five days of his illness, and Galen's attribution of them to particular tertians. Shaded areas indicate seasonal hours of the night, unshaded areas seasonal hours of the day

Hour	Day 1	Day 2	Day 3	Day 4	Day 5
4				(Tertian 2)	
5					
6					
7		Tertian 2			
8					
9					(Tertian 3)
10			Tertian 3		
11					
12					Tertian 1
1					
2			Tertian 1		
3					
4					
5	Tertian 1				

Table 2. The paroxysms of what Galen identifies as Tertian 1. In this table, the days advance in units of two because the paroxysmal period of a tertian is approximately two days

Hour	Day 1	Day 3	Day 5	Day 7	Day 9	Day 11	Day 13	Day 15	Day 17	Day 19
6										
7					x					
8										
9				x						
10										
11										
12			x							
1										
2		x								
3										
4						x	x			
5	x							x		
6									x	
7										
8										x

Galen's decision to use hours as a structuring tool in this clinical narrative permits him to accomplish a number of things. First, Galen can produce a temporal map of these fevers that is so detailed as to reinforce the impression that Galen acquired his temporal data points through genuine autopsy. The reader pictures Galen by the sickbed, with one eye on his patient and the other on a clock.

Galen enhances this image later in the passage with additional references to his own personal experience: e.g., "to me, who examined the whole form of the fever"[21] or "from there, keeping an eye on everything."[22] Thus, Galen presents himself as a physician who grounds his theories in empirical research.[23] Meanwhile, by mapping out the progress of the triple tertian in a schematic way, Galen suggests to his readers that he has subjected his empirical data to sophisticated and systematic analysis. This exercise helps Galen to support the claim that a coherent, logical framework underlies even the most erratic fever periods. Such a framework promises to enable doctors to identify and make predictions about any kind of febrile behavior. The balance between logic and empiricism that Galen displays here is, of course, the hallmark of his own scientific method. Thus, by using hours to structure his clinical narrative in this context, Galen can bolster his persistent claim to be a meticulous practitioner of an apodeictically informed scientific method.[24]

Secondly, this passage makes it clear that, for Galen, hourly timekeeping offers a way to describe the behavior of an intermittent fever that is more precise and information-rich than the terminology commonly employed during his day. Galen makes this claim explicitly in another of his fever treatises, *On the Distinct Types of Fever*, during a discussion about how best to refer to tertians of varying paroxysmal lengths. Galen describes how other physicians often resort to vague adjectival modifiers to differentiate between tertians whose paroxysmal periods are exactly twenty-four hours and those that are longer by various degrees. These physicians call the former sort of tertian "strict" or "simply lengthened" and divide the latter category into the "sufficiently lengthened," "very lengthened," and "the most lengthened of all."[25] Galen observes that such labels are relative terms, at best, and do not necessarily help a doctor to predict a fever's specific behavior. Instead, he recommends a different method: "But if one should want to explain to someone else what sort of fever a patient has, it will be more precisely clear if one mentions the length of the paroxysm and of the interval, or if one seeks a term that can indicate the same thing clearly and definitively."[26] In the example that Galen offers, he states the durations of both the paroxysm and remission period specifically in hours.[27]

---

[21] *Cris.* 9.681 K: ἡμῖν μὲν οὖν ἅπασαν τοῦ πυρετοῦ τὴν ἰδέαν ἐπισκεψαμένοις.

[22] *Cris.* 9.682 K: ἐντεῦθεν δὲ παραφυλάττοντες ἅπαντας.

[23] Cf. (as one *comparandum* among many) *Di. dec.* 9.772 K, where Galen accuses "some doctors" of weighing in on medical discussions "despite never actually having attended" those who suffered from the illness at issue.

[24] Cf., e.g., *Diff. feb.* 7.370 K: "I have tried, as best I can, to express every fever type accurately, finding out the useful differences among them through both experience and logic over a long period of time."

[25] *Diff. feb.* 7.372 K. The Greek phrases used here are ἁπλῶς ... μεμηκυσμένον, μεμηκυσμένον ἱκανῶς, ἐπιπλέον μεμηκυσμένον, and ἐπιπλεῖστον [μεμηκυσμένον].

[26] *Diff. feb.* 7.373 K: ἀλλὰ καὶ δηλῶσαι βουλόμενος ἑτέρῳ τὸν τοῦ κάμνοντος πυρετὸν ὁποῖός τίς ἐστιν, ἀκριβέστερον δηλώσει τό τε τοῦ παροξυσμοῦ καὶ τὸ τοῦ διαλείμματος εἰπὼν μῆκος, ἢ ζητῶν ὄνομα σαφῶς καὶ ἀφωρισμένως ἐνδείξασθαι, ταὐτὸν δυνάμενον.

[27] *Diff. feb.* 7.372 K: "Let us assume that this particular man experiences fever for fifteen hours and becomes fever-free for thirty-three, and that this state of affairs persists for him in the successive days according to this progression."

Galen is concerned that disputes over fever terminology can lead to the propagation of harmful errors, both among students building up the foundations of their medical knowledge and among actual physicians attempting to treat patients. At this point in *On the Distinct Types of Fever*, Galen essentially asks: if one physician cannot understand what another physician means by "tertian fever," then what good is that term? And how is medical progress ever to be achieved, if physicians are unable even to agree on the technical language common to their art? As we saw in Chapter 3, Galen avidly promoted the development of a universal "scientific language" which would allow practitioners to communicate with one another without any loss or garbling of information. In the realm of fever typology, he seems to see the hour-unit as offering physicians a way to describe and identify fevers precisely, without resorting to more approximate or controversial terms.[28]

Thus Galen seems to have viewed the numbered hour as a tool that could help physicians to define certain medical concepts more clearly and to create more precise divisions within their classification systems. A passage from *On the Method of Healing* which is part of an extended diatribe against physicians of the Methodist sect reveals that Galen extended this idea beyond the realm of fever theory. Galen clearly felt threatened by this popular sect, and in his writings he frequently accuses Methodists of creating a slapdash and intellectually bankrupt system of medicine that could be mastered in just a few months. In the lead-up to the passage in question, Galen dissects and then lambasts the ways in which many Methodist physicians distinguished between two types of physical conditions: a "symptom" (*symptōma*) and an "affection" (*pathos*). Timing appears to have been an important factor in the Methodists' definitions of these terms: an "affection," for them, was an affliction that was "persisting" (*epimonos*) throughout the disease, while a "symptom" was an affliction that arrived sometime after the disease had begun.[29] In Galen's view, the temporal imprecision in these Methodist definitions made the very categories of "affection" and "symptom" totally unhelpful to practicing physicians. For where exactly, Galen demands, should one draw the temporal boundary between the two?[30]

---

[28] Galen does warn his readers that fever periodicities are not always sufficient, on their own, for producing accurate diagnoses and prognoses. A good doctor must consider the whole picture: the quality of the patient's internal heat; the presence or absence of other symptoms, such as shivering or inflammation; the characteristics of the pulse, urine, and stool. We see this in our *On Crises* passage, e.g., when Galen observes that the patient's pulse, temperature, shivering, and sweating all corroborate the conclusion that Galen had tentatively drawn from the fever's paroxysmal periods—namely, that he is faced with a triple tertian. Yet, among these diagnostic features, Galen seems to give pride of place to fever periodicity, as measured specifically in hours.

[29] *MM*. 10.67–8 K.

[30] *MM*. 10.72.16–10.74.7 K: ὁ κρίνων τὸ ἐπίμονον καὶ τὸ μὴ τοιοῦτον οὐδεὶς αὐτῶν ὥρισεν, ἆρά γε ἡμερῶν τις ἀριθμός, ἢ μηνῶν, ἢ ὡρῶν... πάλιν οὖν αὐτὸν ἐχρῆν εἰρηκέναι σαφῶς ὁπόσαις ὥραις ὁρίζεται καὶ διακρίνει τοῦ συνεισβάλλοντος τὸ ἐπιγιγνόμενον. Tr. Johnston and Horsley (2011) with modifications.

[N]ot one of them [i.e., these Methodist physicians] has defined what this time is that determines whether something is persisting or not—whether it is a number of days, months, or hours.... And furthermore, it behooves him[31] to say clearly by how many hours he defines [a symptom], and how he distinguishes the epiphenomenon from what appears together with it [i.e., with the disease].

A thought experiment can help us to understand Galen's criticisms here. Let us imagine, for example, that a patient presents with a fever and then, one hour later, begins to complain of a constant headache. Because the headache arrived a *whole* hour later than the initial fever, should it be considered a "symptom" of the patient's disease, and therefore merely an epiphenomenon? Or because the headache arrived *only* one hour after the fever began, should it actually count as a "persistent affection," and thus as an essential feature of the disease? This distinction is important, because a physician's diagnosis, and therefore his proposal for treatment, will differ depending on whether he sees the headache as being essential or auxiliary to the underlying disease. Therefore, Galen insists that these Methodist physicians should either abandon the categories of "symptom" and "affection" entirely or else redefine them with greater temporal precision.

While this portrayal of the Methodists, as physicians who utterly disregard temporal exactitude, serves Galen's argument nicely, it is actually quite misleading. It is true that, as C. Webster has demonstrated, the Methodists' attempts to create a non-causal, heuristic-based form of medicine often led to confusion and "ontological slippage" between categories like "symptom" and "affection."[32] Nevertheless, precise hourly timekeeping was a vital component of one of the Methodists' central therapeutic concepts: the *diatritos*. D. Leith—who has aggregated and analyzed testimonies of the *diatritos* not only from non-Methodist observers like Galen, but also from actual Methodist writers, including Soranus (late first/early second century CE) and Caelius Aurelianus (fifth century CE)—defines the concept thus (emphasis added):[33]

[The therapeutic system of the *diatritos*] describes a chronology which highlights the importance of the recurring third day in illness, i.e., counting inclusively, the third, fifth, seventh, ninth days, etc., limited by the duration of the disease. This is furthermore linked with a specific period of time on the recurring third day *which directly corresponds to the hour* on the first [day] at which the disease originally presented itself. It is not always clear to what degree there is an overlap between these two senses, or precisely how it was understood by various authors, and it seems best to judge from the context of each passage.

---

[31] Presumably Olympicus, although the referent is ambiguous.   [32] Webster 2015.
[33] Leith 2008, 594.

Methodist physicians used this framework to determine exactly when (i.e., after which *diatritos* interval) to give a patient a meal, bath, or another therapy. According to Galen, for example, Methodists often made their patients fast until the conclusion of the first *diatritos*, i.e., forty-eight hours after the onset of the illness.[34] Such a system required the physician to pay careful attention not only to the day but also to the *time of day* at which a patient fell ill, so that the physician could determine not only the day but also the *time of day* at which to transition from one phase of therapy to the next. Caelius Aurelianus describes this process: "The onset of illness should be judged to occur at the time when the patients fall over during it, from which we can discover from what point the time of the *diatritos* is to be calculated, so that the remedies can be correctly assigned to their own times."[35] This practice closely resembles a type of astrology attested in the Imperial and late antique periods (though it probably developed earlier) that is called "decumbiture" in English (*kataklisis* in Greek, *decubitus* in Latin). Decumbiture uses a patient's natal chart and the day and hour at which he or she took to their bed in order to predict the course and outcome of their disease.[36] The main differences between the concepts of *diatritos* and decumbiture seem to be that the former did not posit any astrological resonances and was not, to our knowledge, used for prognostic purposes. The Methodist *diatritos* was first and foremost a therapeutic tool, one that allowed physicians to create modular treatment plans with each module lasting (and perhaps being reevaluated after) one forty-eight-hour period.

It does not seem to have been the case, therefore, that Methodist physicians were uninterested in using clocks and precise hourly timekeeping for medical purposes. In fact, they paid very close attention to the time of day when it was a matter of determining the timing of the *diatritos*. Elsewhere in *On the Method of Healing*, Galen himself makes fun of the Methodists by describing a patient, forced to fast for the first forty-eight hours of his illness, as lying down "for two days without food, dry and full of distress, *staring at the clock* according to the orders of the diatritarians."[37] Where the Methodists parted ways with Galen and with other neo-Hippocratic physicians was instead over the level of personal tailoring that therapeutic timelines required. As we have seen, physicians in the Hippocratic tradition believed that treatment plans had to account for a wide variety of factors in each patient's life (e.g., their age, humoral balance, personal habits, place of origin, occupation, and of course, the *hōra*—whether season or time of day—at

---

[34] See, e.g., *MM.* 10.673 K and *Praen.* 14.664 K.

[35] *Tard. pass.* 1.105: *sed tunc erit accessio iudicanda, quoties ceciderint in ea aegrotantes, quo discere poterimus, quo tempore diatriti tempus sit numerandum, ut recte possint curationes suis reddi temporibus.* Tr. Leith (2008) with modification.

[36] For an overview of decumbiture, see Greenbaum 2020a; 2020b, 374–6.

[37] *MM.* 10.582.10–12 K: τὸν ἐμημεκότα τρίτης ἑσπέρας ὡς ἐχρῆν αὐτὸν ἄρτι δυοῖν ἡμερῶν ἄσιτόν τε καὶ ξηρὸν καὶ ἄσης μεστὸν κατακεῖσθαι, εἰς τὰς ὥρας ἀποβλέποντα κατὰ τὴν τῶν διατριταρίων πρόσταξιν. Tr. Leith (2008), emphasis added.

which their symptoms appeared). The Methodists, in contrast, considered a very limited number of factors (such as whether a condition was "constricted," "lax," or "mixed") and applied the same therapeutic framework, including its temporality, to all. The popularity of the Methodist sect, which attempted to dramatically simplify and standardize the medical art, indicates that there was some pushback within the medical community of the Imperial period (including both physicians and their patients) against the trends toward complex *diairesis* and hyper-mathematization that we are about to explore.

Excessive Exactitude: Fevers and Phlebotomy

The texts we have considered so far might leave us with the impression that Galen was a dyed-in-the-wool promoter of temporal exactitude. However, other works reveal that Galen did not, in fact, see quantitative timekeeping as a tool that should be used across the board. To him, it was but one of the many temporal tools available to the physician, and the best doctors knew not only when to use it, but also when to refrain. This becomes most apparent in *Against Those Who Have Written on Types* or *On Periods*, a highly technical and rarely discussed treatise of which no modern translation has yet been published. This text is a polemic directed against physicians whose fever-classification systems Galen considered to be excessively quantitative and programmatic. Galen's main targets throughout this work are physicians who are unwaveringly committed to two axioms: first, that all fever types must be "defined in complete days and nights"—i.e., in exact multiples of twenty-four hours.[38] There is no margin of error here, whereby a fever with a period of forty-six hours, for example, could still count as a tertian (whose typical period is forty-eight hours). Their second axiom is that there is no limit to the number of fever types. You can have not only quotidians, tertians, and quartans, but also quintans, sextans, and all the way up to quinquagintans, fevers whose periods, counting inclusively, are fifty days long. Fever classification systems of this ilk explain "irregular" fevers as combinations of various fever types and rely on mathematical tables and formulae in order to determine the exact kinds and quantities of fevers in given cases.

Galen finds this sort of system to be utterly impractical. Imagine, he exclaims, that physicians who adhere to these axioms are faced with a fever whose paroxysms recur every twenty-two hours:[39]

---

[38] *Adv. typ. scr.* 7.476.12 K: ὁλοκλήροις ἡμέραις τε καὶ νυξὶ περιγράφηται.
[39] *Adv. typ. scr.* 7.477.4–12 K: οἱ δὲ ταῖς εἰρημέναις ὑποθέσεσι δουλεύοντες ... εἴκοσι μὲν εἶναι καὶ δυοῖν ὥραιν φασὶ τὴν περίοδον, τὸν δ' ὅλον τύπον οὐχ ἁπλοῦν, ἀλλά τινα σύνθετον ὑπάρχειν νομίζουσιν· εἶτα ζητήσαντες ἱκανῶς καὶ λογισάμενοι κατὰ σφᾶς αὐτούς, οἱ μὲν ἐπὶ δακτύλων, οἱ δὲ καὶ ἐπὶ διαγράμματος ἔν τινι βιβλίῳ γεγραμμένου, δώδεκά φασιν εἶναι δωδεκαταίους τὸν τοιοῦτον τύπον. γελώντων δὲ ἐπὶ τούτοις τῶν ἀκουσάντων, ἀγανακτοῦσιν.

Those who are slaves to the aforementioned hypotheses... would say that [the fever's] period is twenty-two hours, and they would think that the whole type is not simple [i.e., not just a straightforward quotidian variant], but some sort of compound. Then, having investigated sufficiently and calculated on their own—some on their fingers, some with a diagram drawn in a book—they would say that such a fever type is twelve "dodecan" fevers [i.e., twelve fevers whose paroxysms arrive on the twelfth day, counting inclusively]. Then, when those listening laugh at them, they are irritated.

In Galen's view, these doctors misidentify a simple quotidian fever with a period of approximately twenty-four hours as an entirely different illness with a strict twenty-two-hour period. To explain this strange pattern, these doctors must then resort to an absurd combination of twelve so-called "dodecan" fevers, a diagnosis that fails to correspond to a recognizable clinical condition or suggest a course of treatment for the patient.

The failure of these physicians to acknowledge sickroom realities and practicalities is a theme throughout *On Periods*. At one point, for instance, Galen recounts a debate between himself and a rival physician over how to characterize a fever whose paroxysms last five hours, and whose periods of abatement last two hours (yielding a total cycle of seven hours). Galen says, "I replied straight away that this is not even possible. For paroxysms that are of such short duration produce abatements that are long and, on the whole, they result in a fever-free state."[40] He then presents his interlocutor as someone who "does not... even consider whether a hypothesis is impossible or possible but concedes that such things are sought simply for the sake of exercise."[41]

What can we learn about the physicians who promoted these convoluted, mathematical theories? Although Galen never names them or associates them with a particular medical sect, he does refer to these doctors as *hoi neōteroi* (literally, "the youngsters").[42] The phrase *hoi neōteroi* (often used in contrast to *hoi palaioi* or *hoi arkhaioi*, "the ancient ones") is quite slippery in its usage; Imperial-period authors seem to have applied it flexibly, its definition varying according to the context at hand.[43] The phrase could, for instance, refer simply to the author's younger contemporaries within a given field (such as poetry, philosophy, or medicine), regardless of the particular schools of thought or practice to

---

[40] *Adv. typ. scr.* 7.479.2–5 K: ἀπεκρινάμην οὖν εὐθέως αὐτῷ, μηδ' οἷόν τε γενέσθαι τοῦτο. τοὺς γὰρ ὀλιγοχρονίους οὕτω παροξυσμοὺς μακρὰς ποιεῖσθαι τὰς παρακμὰς, καὶ τοὐπίπαν γε καὶ εἰς ἀπυρεξίαν ἀφικνεῖσθαι.

[41] *Adv. typ. scr.* 7.479.7–9 K: μηδέν, ἔφη, σκόπει τό γε νῦν, πότερον ἀδύνατός ἐστιν ἡ ὑπόθεσις, ἢ δυνατή, γυμνασίας δ' ἕνεκα τὰ τοιαῦτα συγχωρεῖ ζητεῖσθαι.

[42] *Adv. typ. scr.* 7.476, 488 K.

[43] On the semantic range of this term in Galen's writings, see Singer 2022, 74–82. Singer also illustrates how Galen uses the terms *palaioi* and *neōteroi* to position his own theories as, in some ways, outside of or transcending time.

which they adhered.[44] On the other hand, *hoi neōteroi* could also refer to adherents of a particular school of thought that was considered "more recent" than others under discussion. Some scholars, for instance, have argued that Alexander, in his commentaries on Aristotle, uses the phrase in this way to contrast the beliefs of the Peripatetics (identified in this reading as *"hoi arkhaioi"*) with those of the Stoics (identified as *"hoi neōteroi"*).[45] Finally, *hoi neōteroi* could be used to distinguish more recent from more ancient adherents of the same school of thought, which Sextus Empiricus seems to do when he contrasts the views of "the more ancient" Stoics with those of "the more recent" ones (using *hoi arkhaioteroi* and *hoi neōteroi*, respectively).[46]

The wide semantic range of *hoi neōteroi* makes it difficult to identify the group that Galen is targeting in *On Periods*. The text, however, offers us a clue. Galen tells us that his debate with the rival physician took place during the reign of Septimius Severus—i.e., toward the end of Galen's own life and around the time at which Galen was composing *On Periods* itself.[47] This suggests that Galen was pointing a finger at young up-and-comers among his own contemporaries toward the very end of the second century CE—and not targeting, as he so often does, the writings of his influential predecessors of the Hellenistic and early Roman periods. It is not clear whether Galen had a specific school of medical thought in mind here, but these passages seem to imply that, at the height of the Roman Empire, the hyper-mathematization of medical knowledge was something of a trend, at least in certain contexts and among certain kinds of practitioners.[48]

This trend also manifests in Galen's writings on venesection, where he takes special aim at Imperial-period followers of the Hellenistic-period physician Erasistratus. We see this, for example, in *On Venesection, Against Erasistratus* at a point in the text that deals with the right time of day at which a physician should let a patient's blood. Galen says:[49]

> I think it is clear that on the particular day on which we are to phlebotomize we ought to watch out for the abatement of the fever; but this is not clear to some people, the sort who order phlebotomy only at daybreak, or up to the fifth and sixth hour at the latest. If, however, anyone calls to mind what has been said

---

[44] For examples and discussion, see Kieffer 1964, 130–1. [45] Kieffer 1964, 131–2.
[46] *Math.* 253.   [47] *Adv. typ. scr.* 7.478 K.
[48] Perhaps these physicians were responding to the same kinds of forces that led Ptolemy to structure his own philosophy around mathematics (see Feke 2018).
[49] *Ven. sect. Er.* 11.310.12–11.311.11 K: Ὅτι δ' ἐν αὐτῇ πάλιν ἐκείνῃ τῇ ἡμέρᾳ, καθ' ἥν φλεβοτομοῦμεν, ἐπιτηρῆσαι χρὴ τὴν παρακμὴν τοῦ πυρετοῦ, πρόδηλον εἶναι νομίζω. καίτοι γ' ἐνίοις ἐστὶν οὐδὲ τοῦτο πρόδηλον, ὅσοι κελεύουσιν ἕωθεν μόνον φλεβοτομεῖν, ἢ τὸ μακρότατον ἄχρι πέντε ἢ ἒξ ὡρῶν. ἀλλ' εἰ τῶν ἔμπροσθεν εἰρημένων ἐν ὅλῳ τῷ γράμματι μνημονεύει τις, οὐδὲν τοιούτων σφαλήσεται φλεβοτομῶν ἐν ἁπάσῃ μὲν ἡμέρας ὥρᾳ, πάσης δὲ νυκτός, σκοπὸν ἔχων ἐπὶ μὲν τῶν πυρεττόντων τὴν παρακμὴν τῶν κατὰ μέρος παροξυσμῶν. ... μηδενὸς δὲ τοιούτου κατεπείγοντος ἢ κωλύοντος ἄμεινόν ἐστιν ἕωθεν φλεβοτομεῖν, οὐκ εὐθέως ἅμα τῷ τῶν ὕπνων ἐξαναστῆναι, προγρηγορήσαντες δὲ χρόνον ὡς ὥρας μιᾶς... Tr. Brain 1986.

previously in the whole of this work, he will make no such mistake, but will phlebotomize at any hour of the day and throughout the night, taking as his indication the decline in the individual paroxysms in patients with fever, and the need for the remedy.... But where nothing of this kind either urges or forbids us, it is *best* to phlebotomize at daybreak, not immediately after patients have risen from sleep, but after they have been awake for about an hour....

Galen acknowledges that, all else being equal, there is a preferred window of time in which to let blood—namely, close to daybreak, about an hour after the patient has woken up. However, Galen is critical of physicians who delineate this window of time too sharply—who say, for example, that it can only extend from the first to the fifth hour of the day. For Galen, the right time to perform a venesection is really indicated by a specific medical event—namely, the abatement of the patient's fever—and not by a specific numbered hour. That it is generally preferable to phlebotomize in the morning is, for Galen, a good rule of thumb, but ultimately, doctors should be willing to abandon it, if a patient's fever should abate at another time of day or night.[50] In Galen's view, the doctor who is overly programmatic in timing his treatments by the clock will often find himself out of sync with the specific rhythms of the case at hand, and his caregiving will therefore prove ineffective.

But who, we might ask, are *these* temporally precise physicians, and are they likely to belong to the same group that Galen chastised in *On Periods*? Here, too, it is difficult to say. Once again, Galen refrains from naming the specific individuals or medical sects he has in mind (though it may be relevant that, as mentioned earlier, his works on venesection generally target followers of Erasistratus active in Galen's own day). However, a later passage in *On Treatment by Venesection* may help us to sketch a slightly clearer picture. Just prior to this passage, Galen stresses the importance of immediate venesection in cases where the patient seems to have "an overabundance of seething blood."[51] In the following passage, he criticizes physicians who refuse to phlebotomize in such cases if they occur outside of a predetermined temporal window:[52]

---

[50] One of the case histories attributed to Rufus of Ephesus suggests that this physician might have agreed with Galen. In Case 21 (lines 10–12), he describes phlebotomizing his patient in the ninth hour of the day, the first hour of the night, and then again at dawn and the third hour of the day (Ullmann 1978).

[51] *Cur. rat. ven. sect.* 11.287 K: ζέοντος αἵματος πλῆθος.

[52] *Cur. rat. ven. sect.* 11.287.18–11.288.9 K: γελοῖον γὰρ ὅπερ οἱ πολλοὶ πράττουσιν ἀπὸ δευτέρας ὥρας ἡμέρας ἄχρι ε΄ ἢ στ΄ ἀφαιροῦντες μόνον αἵματος, ἐν ἄλλῳ δ᾽ οὐδενὶ χρόνῳ οὓς εἰ μὴ καὶ κλυστῆρσι καὶ τροφῆς προσφορᾷ καὶ τοῖς ἄλλοις βοηθήμασιν ἑώρων χρωμένους ἐν ἅπαντι καιρῷ τῆς νυκτός, ἐνῆν ἄν μοι χαλεπὸς ὁ πρὸς αὐτοὺς λόγος. ἐπεὶ δὲ πάντα πράττοντες οὐ τὸν ὡρισμένον ἀριθμὸν ὡρῶν ἕνα καὶ κοινὸν ἐπὶ πάντων τῶν νοσούντων, ἀλλ᾽ ὡς ἂν ὑπαγορεύῃ τὸ πάθος ἐπὶ μόνης τῆς φλεβοτομίας τὸν εἰρημένον ἀρτίως ἀεὶ περιμένουσι καιρόν, εὐφωρότατον ἴσχουσι τὸ σφάλμα.

What most people do, letting blood only between the second hour of the day and the fourth or fifth, is laughable; if it were not that I have seen them giving enemas, food, and other remedies at any time of the night, I should have some hard things to say to them. Since, however, they do not observe one and the same restricted period of hours for everything they do for all their patients, but act as the disease requires, awaiting the time just mentioned only in the case of phlebotomy, their error is easier to excuse.

Three elements of this passage are important for our purposes. First, by using the phrase "what most people do," Galen suggests that the practice of phlebotomizing only within specific windows was actually common to the *majority* of his contemporaries. Second, the precise temporal boundaries of those windows seem to vary from one physician to another. Here, Galen describes windows that extend from the second to the fourth or fifth hour of the day, but in the passage we examined earlier, he mentioned physicians who phlebotomized "only at daybreak, or up to the fifth and sixth hour." Thus, Galen does not seem to be critiquing members of a homogeneous group who all follow the same program. Instead, once again, he seems to be putting his finger on a general interest among his contemporaries in expressing certain types of knowledge schematically and quantitatively.

Finally, Galen notes that, while these physicians time their phlebotomies by the clock, they decide when to administer other treatments using more qualitative time-indicators. Hence, venesection emerges from this discussion as a special therapy that stood out to "most doctors" as requiring exact numerical timekeeping. But why? What was it about venesection that made it such a good candidate for temporal exactitude? Galen does not offer a direct explanation, but certain of his comments, as well as those of his near-contemporaries Athenaeus and Antyllus, can help us to appreciate the kind of logic that may have been at work here. In *On Treatment by Venesection*, Galen tells us that he often practices prophylactic phlebotomy specifically during the season of spring:[53]

> And as for those who go down every year in summer with plethoric diseases, they too should be evacuated at the onset of spring. Similarly, those who are seized in

---

[53] *Cur. rat. ven. sect.* 11.271.11–11.272.8 K: καὶ ὅσοι δὲ καθ' ἕκαστον ἔτος ἐν θέρει νοσοῦσιν νοσήματα πληθωρικά, καὶ τούτους χρὴ κενοῦν εἰσβάλλοντος ἦρος. ὡσαύτως καὶ ὅσοι κατ' αὐτὸ τὸ ἔαρ ἁλίσκονται τοῖς τοιούτοις, ἔνιοι ὀφθαλμοὺς ἔχοντες ἀσθενεῖς, ἢ τοῖς ὀνομαζομένοις σκοτωματικοῖς πάθεσιν εὐάλωτοι καὶ αὐτοὶ κατὰ τὴν ἀρχὴν τοῦ ἦρος δέονται κενοῦσθαι, προσδιασκεψαμένων ἡμῶν ὁποῖόν τι τὸ ἀθροιζόμενον αὐτοῖς εἴη. τινὲς μὲν γὰρ τὸν πικρόχολον ἀθροίζουσι χυμὸν πλέονα τῶν ἄλλων, τινὲς τὸν μελαγχολικὸν ἢ φλεγματικόν, ἔνιοι δὲ ὁμοτίμως ἅπαντας ἐφ' ὧν αἷμα πλεονάζειν λέγεται. τούτους οὖν ἅπαντας κενώσεις, ὥσπερ καὶ τοὺς ποδαγρικούς τε καὶ ἀρθριτικοὺς ἐν ἀρχῇ τοῦ ἦρος, ἀλλ' ἤτοι φαρμακεύων ἢ φλεβοτομῶν. ἐγὼ γοῦν πολλοὺς ἐτῶν ἤδη τριῶν καὶ τεττάρων ἐνοχλουμένους ἐκ διαλειμμάτων ἀλγήμασι ποδῶν ἰασάμην, ἤτοι καθαίρων τὸν πλεονάζοντα χυμὸν ἐν ἀρχῇ τοῦ ἦρος ἢ αἵματος ἀφαιρῶν. Tr. Brain 1986.

spring itself with such diseases, some having weak eyes, or being subject to the diseases called scotomatic—these also need to be evacuated at the beginning of spring, after we have first considered what sort of concourse of humors they have. This is because some accumulate the bilious humor more than the rest, others the melancholic or phlegmatic variety, while others again accumulate all of them equally; in these, blood is said to be in excess. You will evacuate all these, as you will also your gouty and arthritic patients, at the beginning of spring, either by purging or by phlebotomizing. I have cured many who had already been troubled on and off for two or three years with pains in the feet, either purging away the excessive humor at the beginning of spring or removing blood.

What makes spring so conducive to phlebotomy? While Galen does not fully address this question in On Treatment by Venesection, he does so in another treatise, On Temperaments.[54] There, Galen outlines a controversy over how to apply Hippocratic humoral theory to the season of spring. The few who have "got it right," according to Galen, see spring as conducive to phlebotomy because it is an inherently *eukratic* season, i.e., one in which all of the humors (and, by extension, all of their "qualities") are in equilibrium. By associative thinking, a *eukratic* season would seem to be the ideal time to restore equilibrium to individual patients' bodies by removing excess humors of any kind—whether phlegm, yellow bile, black bile, or blood. Galen complains, however, that other physicians adopt a different, wrong-headed view. According to this view, none of the four seasons are *eukratic*; instead, each is characterized by a single pair of qualities (hot or cold, moist or dry) and linked to a single humor that shares those qualities. Spring, to these physicians, is both hot and moist and associated with an excess of the single humor blood.[55] The logic underlying this view seems to follow that of the Hippocratic author of *Human Nature*:[56]

> During the spring, although the phlegm remains strong in the body, the quantity of blood increases. Then, as the cold becomes less intense and the rainy season comes on, the wet and warm days increase further the quantity of blood. This part of the year is most in keeping with blood because it is wet and hot. That this

---

[54] *Temp.* 1.4 = 1.433–4 K.

[55] Galen, as our earlier passage (11.271.11–11.272.8 K) suggests, also disagrees with the way that these physicians characterize blood. While these doctors understand blood as "hot and moist," Galen understands blood as itself a *eukratic* mixture of the other three humors.

[56] *Nat. hom.* 7.12–23: Τοῦ δὲ ἦρος τὸ φλέγμα ἔτι μένει ἰσχυρὸν ἐν τῷ σώματι, καὶ τὸ αἷμα αὔξεται· τά τε γὰρ ψύχεα ἐξανίει, καὶ τὰ ὕδατα ἐπιγίνεται, τὸ δὲ αἷμα κατὰ ταῦτα αὔξεται ὑπό τε τῶν ὄμβρων καὶ τῶν θερμημεριῶν· κατὰ φύσιν γὰρ αὐτέῳ ταῦτά ἐστι μάλιστα τοῦ ἐνιαυτοῦ· ὑγρόν τε γάρ ἐστι καὶ θερμόν. Γνοίης δ' ἂν τοῖσδε· οἱ ἄνθρωποι τοῦ ἦρος καὶ τοῦ θέρεος μάλιστα ὑπό τε τῶν δυσεντεριῶν ἁλίσκονται, καὶ ἐκ τῶν ῥινέων τὸ αἷμα ῥεῖ αὐτέοισι, καὶ θερμότατοί εἰσι καὶ ἐρυθροί· τοῦ δὲ θέρεος τό τε αἷμα ἰσχύει ἔτι, καὶ ἡ χολὴ ἀείρεται ἐν τῷ σώματι καὶ παρατείνει ἐς τὸ φθινόπωρον· ἐν δὲ τῷ φθινοπώρῳ τὸ μὲν αἷμα ὀλίγον γίνεται, ἐναντίον γὰρ αὐτέου τὸ φθινόπωρον τῇ φύσει ἐστίν· Tr. Chadwick and Mann (Lloyd 1978, 264–5).

is so, you can judge by these signs: it is in spring and summer that people are particularly liable to dysentery and to epistaxis, and these are the seasons too at which people are warmest and their complexions are ruddiest. During the summer, the blood is still strong, but the bile gradually increases, and this change continues into the autumn, when the blood decreases since autumn is contrary to it.

According to the Hippocratic author's framework, both the season of spring and the humor blood share the qualities "hot and moist." Thus, the theory would suggest that blood is particularly abundant in the body during spring and that, in order to treat this excess allopathically, one should also let patients' blood during this season.

But how do we get from the idea that venesection is best performed during the spring to the idea, expressed by Galen's interlocutors, that it ought to be performed in the morning? A fragment of the physician Athenaeus of Attalia, who was active in the first century CE and considered by Galen to be a pillar of the Pneumatist school of medical thought,[57] can help us to bridge this gap:[58]

> During each day, there are differences of the air. For dawn is wet and hot like the spring. For this reason, the bodies of both healthy and sick persons loosen, so that this period of time is most manageable even for those who are feverish. The middle times of the day are likened to summer, those during the afternoon autumn, and around evening winter. And of night, the first resembles evening, the middle seems like winter, and correspondingly the others.

Athenaeus, in a move reminiscent of our Stoic authors Seneca and Epictetus from Chapter 4, proposes that each period of daylight or night-time should be seen as a synecdoche of the entire solar year. Hence, the "qualities" that correspond to each subsection of the day or night are analogous to those that correspond to each of the four seasons. In the model that Athenaeus outlines here, the morning is the time of day that is most like spring in that it is both hot and moist. Though Athenaeus does not come right out and say so, we might assume that, for him, the morning, like spring, was also a time in which the hot, moist humor of blood was most abundant and therefore most ripe for the letting.

This idea is reprised in the writings of another Imperial-period physician associated with the Pneumatist school of thought, Antyllus. The exact dates of Antyllus's life are uncertain; S. Coughlin offers a *terminus post quem* of c.100 CE,

---

[57] *Diff. feb.* 8.749.11 K.
[58] Aët. Amid. *Libr. med.* 3.162.29–35: καὶ καθ' ἑκάστην δὲ ἡμέραν διαφοραὶ τοῦ ἀέρος γίγνονται· ὁ μὲν γὰρ ὄρθρος ὑγρὸς καὶ θερμὸς ὡς τὸ ἔαρ· διὰ τοῦτο καὶ τὰ σώματα ἀνίεται καὶ τῶν ὑγιαινόντων καὶ τῶν νοσούντων, ὥστε καὶ τοῖς πυρέσσουσιν ὁ καιρὸς οὗτος εὐφορώτατος· τὰ δὲ μέσα τῆς ἡμέρας θέρει παρείκασται, τὰ δὲ κατὰ τὴν δείλην φθινοπώρῳ, τὰ δὲ περὶ ἑσπέραν χειμῶνι καὶ τῆς νυκτὸς δὲ τὰ πρῶτα τῇ ἑσπέρᾳ παρείκασται· τὰ δὲ μέσα χειμῶνι παρέοικε καὶ τὰ ἄλλα ἀκολούθως. Tr. Coughlin n.d.

since Archigenes's name is mentioned in our extant fragments, and a *terminus ante quem* of the 350s CE, since Oribasius quotes him by name.[59] Thus, Antyllus seems to have been either a close contemporary of Galen's or a later author, writing in the hundred years or so after Galen's death. In either case, the following fragment from Antyllus's writings attests to the longevity of Athenaeus's theory of daytime and night-time as synecdoches of the year:[60]

> With "day" we include the night. We claim that it has the kind of relation to the year that Hippocrates indicated in the second *Epidemics*. The dawn is wet and warm, like spring. For this reason, sleep promotes crises and bodies relax, those of both the healthy and the sick. Hence, this time is also easiest to bear for those in fever. At dawn, exhalations rise, moist breezes of rivers blow, dew falls, flowers blossom, and everything that grows from the earth sprouts forth, the condition of the morning resembling spring. Middays are like summer, late afternoons like autumn, which is why they are productive of heaviness and malaise, the afternoon being unhealthy proportionally to autumn. Of night, the first part, around evening time, is like the later afternoon. Hence, we do not order patients to sleep at this time, nor do we offer them drink, nor do we administer other remedies, unless something else engages us, looking at this time with suspicion, just like the evening. The middle of the night resembles winter. For then the sun is furthest away from us, just as it is in the winter. Therefore, it is reasonable that the pains of the sick are stronger during the night, the [pores] on the surface of the body contracting because of the chill and the imperceptible transpiration shut up. And discharges from those with ophthalmia, dysentery, and coeliac, as well as bloody ones, supervene for the same reason, since everything that is usually discharged through imperceptible pores is concentrated into one place, that of discharge. The end of the night, because of its proximity to the dawn, shares in the same mixture as it.

---

[59] Coughlin n.d.
[60] Antyl. ap. Stob. *Ecl*, 4.37.15: Ἡμέραν λαμβάνομεν σὺν τῇ νυκτί· ἀναλογίαν δ' αὐτὴν ἔφαμεν ἔχειν τινὰ πρὸς τὸν ἐνιαυτόν, καθάπερ Ἱπποκράτης ἐν τῇ δευτέρᾳ τῶν Ἐπιδημιῶν ἐπεσημήνατο. ἔστι δὲ ὁ μὲν ὄρθρος ὑγρὸς καὶ θερμός, ἔαρι ἐοικώς. διὰ τοῦτο οἵ τε ὕπνοι εὐκρινεῖς καὶ τὰ σώματα ἀνεῖται, καὶ τὰ τῶν ὑγιαινόντων καὶ τὰ τῶν νοσούντων· ὥστε καὶ τοῖς πυρέσσουσιν εὐφορώτατον εἶναι τόνδε τὸν καιρόν. ἀναθυμιάσεις δὲ ἀνίασι περὶ τὸν ὄρθρον καὶ αὖραι ποταμῶν ἀποπνέουσιν ὑγραί, καὶ δρόσος ἐπιπίπτει, ἄνθη ἀναβαίνει, καὶ πάντα τὰ ἐκ τῆς γῆς φυόμενα ἄνεισιν, ἔαρι τῆς καταστάσεως τῆς ὀρθρινῆς ἐοικυίας. τὰ δὲ μέσα τῆς ἡμέρας θέρει παρείκασται, τὰ δὲ κατὰ τὴν δείλην φθινοπώρῳ, διὰ τοῦτο βαρῶν καὶ δυσαρεστημάτων οἰστικά, νοσερᾶς οὔσης τῆς δείλης ἀνάλογον φθινοπώρῳ. τῆς δὲ νυκτὸς τὰ μὲν πρῶτα καὶ τὰ περὶ τὴν ἑσπέραν ὅμοια τῇ δείλῃ· ἐντεῦθεν οὔτε ὑπνοῦν ἐπιτρέπομεν τοῖς κάμνουσι περὶ τήνδε τὴν ὥραν οὔτε ποτὸν προσφέρομεν οὔτε ἄλλο προσάγομεν βοήθημα, εἰ μή τι ἕτερον προκαλοῖτο, ὑφορώμενοι τὸν καιρὸν τοῦτον ὥσπερ καὶ τὸν δειλινόν. τὰ δὲ μέσα τῆς νυκτὸς χειμῶνι ἐξείκασται· πλεῖστον γὰρ τότε ὁ ἥλιος καθάπερ καὶ ἐν τῷ χειμῶνι ἀφέστηκεν ἡμῶν. αἵ τε οὖν ὀδύναι τοῖς νοσοῦσιν ἰσχυρότεραι τῆς νυκτὸς εἰκότως, διὰ τὴν ψῦξιν πυκνουμένης τῆς ἐπιφανείας καὶ ἀπολαμβανομένης τῆς κατὰ τὸ ἄδηλον διαπνοῆς. καὶ τὰ ῥεύματα δὲ καὶ τὰ τῶν ὀφθαλμιώντων καὶ τὰ τῶν δυσεντερικῶν καὶ κοιλιακῶν καὶ τὰ αἱματικὰ ἀπὸ τῆς αὐτῆς προφάσεως ἐπιγίνεται, παντὸς τοῦ εἰωθότος ἀπιέναι κατὰ τὸ ἀφανὲς εἰς ἕνα τόπον συνδιδομένου τὸν ῥευματιζόμενον. τὰ δὲ τελευταῖα τῆς νυκτὸς διὰ τὴν πρὸς τὸν ὄρθρον γειτνίασιν τῆς αὐτῆς ἐκείνῳ κράσεως μεταλαμβάνει. Tr. Coughlin n.d.

It seems likely, then, that this theory was circulating among Galen's contemporaries, and indeed, his interlocutors in *On Treatment by Venesection* appear to apply a similar logic. Broadly speaking, they prefer to phlebotomize patients between dawn and midday—i.e., during the morning. A belief that this time of day was warm and wet, like spring, and therefore tended to produce in patients an excess of blood, would help to explain the unwillingness of these doctors to consider venesecting at other times of day. If this was, in fact, the reasoning that motivated such doctors, it remains for us to ask whether they should be considered followers of Athenaeus (and hence, in Galen's terms, "Pneumatists") specifically, or whether the idea that the day was a humoral synecdoche of the year had become so widespread that it cannot help us to establish these doctors' affiliations.

Conclusion

This chapter has explored how Galen used hour-units to participate in ongoing debates about medical diagnosis, prognosis, and treatment and about the larger role of quantitative exactitude within Greco-Roman medicine. On the one hand, for Galen, numbered hours offered physicians a lexicon for describing periodic fevers that is more quantitative and broadly comprehensible than the language typically employed. Furthermore, Galen's willingness to chart the exact hours at which a fever's paroxysms arrived allowed him to distinguish between stationary and moving fevers, a distinction that he considered useful for explaining seemingly irregular patterns. By incorporating systematic hourly timekeeping into his fever theory, Galen could demonstrate to his readers that he, in contrast to many of his contemporaries, possessed the qualities of an exemplary scientist. He was meticulous about recording and organizing observational data in keeping with specific temporal schemes; he was adept at interpreting this data with the aid of logical reasoning and mathematical formulae; and, thanks to his scientific method, he could catch and correct potential errors.

Galen, however, did not see hourly timekeeping as a one-size-fits-all approach. There are some contexts in which he saw such temporal exactitude as overly programmatic, causing physicians to pay more attention to clocks or mathematical models than to the realities and idiosyncrasies of the cases in front of them. Through the tinctured lens of Galen's writing and perspective, we were able to sketch the broad outlines of a wider debate over quantitative timekeeping in the Roman Imperial period. This debate seems to have been catalyzed by certain questions—such as how to define the periods of irregular intermittent fevers, how to decide when to perform a venesection, or how to measure a *diatritos* interval—and connected to wider debates about the degree to which medical knowledge should be mathematized and about the relative value of personalized versus standardized medicine. Galen presents himself, of course, as occupying a sensible middle ground between physicians who are insufficiently exact, like the Methodists (at least, in certain contexts), and those who, like the elusive *neōteroi*, are overly committed to exactitude. In the following chapter, we will probe more deeply into the reasons behind Galen's own decisions about when physicians should and should not privilege quantitative over qualitative modes of daily time-reckoning.

7

"Right Timing" in Sickness and in Health

Hourly Timekeeping and *Kairos* in Galen's *On Hygiene*

When it comes to therapeutics, modern biomedicine focuses primarily on *what* we do to our bodies—which medicines we ingest at which doses, what surgeries we might require, etc. However, we have seen that ancient physicians, from the Egyptian authors of the New Kingdom medical papyri to Galen and his rivals, were deeply curious about *when* to perform medical interventions. What is the "right time" to feed, bathe, massage, or apply a dressing to a patient? As discussed briefly in Chapter 2, ancient Greek had a special term for the "right," "critical," or "opportune" time: *kairos*, a term with a broad semantic field that could also include "symmetry, propriety, occasion, due measure, fitness, tact, decorum, convenience, proportion, fruit, profit, and wise moderation."[1] The question at the heart of the present chapter is: how can a close reading of one of Galen's works help us understand how individual physicians thought about the relationship between the concept of the numbered hour and the concept of *kairos*? We will focus on how this relationship plays out in Galen's treatise *On Hygiene*, a text concerned less with the healing of sick bodies than with the maintenance of healthy ones.[2] As such, this treatise provides an instructive contrast to the texts examined in the previous two chapters, which showcased how Galen employed hourly timekeeping in the context of diagnosing and treating diseases. It will reveal that Galen conceptualized the relationship between *kairoi* and numbered hours very differently depending on the age and health of the patient.

We will begin by reviewing some of the ways in which Galen's intellectual role models understood the concept of *kairos* (particularly in relation to medicine) and proceed to identify ways in which Galen's understanding aligned with and differed from those of his predecessors.[3] We will then briefly revisit one of Galen's fever

---

[1] Sipiora 2002, 1. For a full bibliography on the concept (up to 2002), see Zhelezcheva and Baumlin 2002, 237–45. More recently, see Kim 2017; Vadan 2018, 124–70.
[2] All references to *On Hygiene* will refer to the editions of Kühn and Koch: 6.1–452 K and Koch et al. 1923, 3–198. An excellent translation and commentary has just been produced by Singer (2023).
[3] *Kairos* was clearly a very important concept for Galen. According to a lemma search of the *TLG* database, the closely related words *kairos* and *eukairos* ("particularly opportune moment") appear a total of 1,248 times within the Galenic corpus, while their antonym *akairos* appears sixty-five times.

Time and Ancient Medicine: How Sundials and Water Clocks Changed Medical Science. Kassandra J. Miller, Oxford University Press. © Kassandra J. Miller 2023. DOI: 10.1093/oso/9780198885177.003.0008

texts, *On Crises*, for the sake of comparison, in order to see how Galen defines a *kairos* in the context of illness. Finally, we will turn to *On Hygiene* to explore how Galen's understanding of *kairos*—and its relationship to hourly timekeeping— changes when he is discussing healthy individuals, and shifts again when he considers the elderly, a group that Galen understood to occupy a precarious position between sickness and health.

I will demonstrate not only that Galen consistently employs hourly timekeeping in order to answer the question, "What is the *right moment* for this specific patient, illness, or physician to engage in a particular behavior?" but I will further argue that Galen envisions the relationship between hourly timekeeping and *kairoi* to differ depending on the context. In *On Crises*, hours act as indices that are *predictive* of febrile *kairoi*. In *On Hygiene*, on the other hand, hours can also be variables that *influence kairoi* as antecedent causes. Throughout his corpus,[4] Galen consistently defends the notion that "antecedent, external factors operating upon a patient's body are causally relevant to that patient's subsequent condition."[5] These antecedent causes (*aitia prokatarktika*) are central to Galen's theory of medicine, which is founded on the premise that a patient's present condition results from the combination and interaction of a variety of causes, ones that are internal and external to the patient, occur prior to or coterminous with the patient's symptoms, and are within and outside of the patient's control.[6] In *On Hygiene*, Galen posits that the number of hours left before sunset, for example, or the precise time of day at which a patient performed a therapeutic action can be counted among the external, antecedent causes of a patient's present condition and, therefore, are relevant for determining the right times for future treatments.[7] Finally, we will also see how, for Galen, the kairotic window of opportunity differs in aperture depending on whether a patient is sick, healthy, or aged, and how the pursuit of this "right moment" requires Galen's scientific method. Ultimately, this investigation should help us to better appreciate the roles that Galen envisions for quantitative temporal exactitude within the often-messy, human science of medicine.

Kairos and Galen's Predecessors

By the Hellenistic period, writers across a variety of genres were praising *kairos* as an essential virtue. Among these authors were some of Galen's favorites, including

---

[4] Though most especially in the treatise *On Antecedent Causes*, preserved only in Latin. See Hankinson 1998.
[5] Hankinson 1998, 45. [6] On Galen's theories of medical causality, see Hankinson 1998.
[7] While Galen does not use the specific phrase *aitia prokatarktika* to describe hours or *kairoi*, I believe his treatment of the concept in *On Antecedent Causes* warrants applying the phrase here.

Plato, Aristotle, and of course, Hippocrates. Over the course of this chapter, I will argue that, while Galen's understanding of *kairos* closely resembles those of his intellectual heroes, it also differs in a few significant ways. In order to best appreciate both these similarities and these differences, we will first identify some of the essential features of the Classical and Hellenistic *kairos*, as presented by these authors. We will then discuss how Galen interprets this concept first in the context of febrile illness and then in the contexts of good health and old age.

Kairos appears in the famous opening line of *Aphorisms*, where the Hippocratic author opines, "Life is short. The art [i.e., of medicine] is long. Opportunity is sharp."[8] The idea that a kairotic moment is "sharper than any edge" led many ancient writers to contrast *kairos* with another temporal concept, *khronos*.[9] This term differs from *kairos* in two primary ways. First, whereas *kairos* indicates a moment with no temporal duration, *khronos* indicates an interval of time, whether definite (e.g., "for a time of six months") or indefinite (e.g., "for a time"). Second, whereas the term *kairos* is morally charged, in that a kairotic moment calls for decisive action, *khronos* carries no moral freight. The Hippocratic author of *Precepts* describes the relationship between these two terms in the following manner, "*Khronos* is that within which *kairos* exists; *kairos* is that in which there is not much *khronos*."[10] Fleeting kairotic moments punctuate the neutral timeline of *khronos*; they take place in time, but scarcely take up time.

How ought one to identify a *kairos* and determine how best to act at that decisive moment? Plato—inspired, no doubt, by figures such as Gorgias and perhaps even Pythagoras himself[11]—explains in his *Phaedrus* how a student might begin to answer these questions within the domain of rhetoric.[12] Plato points out that different people have different kinds of souls (*psykhai*), and that this variation will lead one man to be more easily persuaded by one style of oration, another man by another.[13] The would-be orator must first learn "how

---

[8] 1.1.1–2, 4.458 L: Ὁ βίος βραχὺς, ἡ δὲ τέχνη μακρὴ, ὁ δὲ καιρὸς ὀξύς.... Cf. *De morb.* 1.5, 6.146 L.

[9] The Hellenistic poet Posidippus composed an epigram in which a casual passerby has a dialogue with a statue of the divinely personified Kairos. In response to the passerby's query as to why Kairos carries a razor in his right hand, Kairos declares, "As an example to men that I am sharper than any edge!" (*Ep.* 16.275.5–6 Beckby).

[10] 1.1.1–2, 9.250 L: χρόνος ἐστὶν ἐν ᾧ καιρός, καὶ καιρὸς ἐν ᾧ χρόνος οὐ πολύς. On the role of *kairos* in the Hippocratic corpus, see Eskin 2002.

[11] The "Pythagorean" references to *kairos* come from the *Carmen aureum* (27–39), which some scholars (e.g., Thom 2001) argue should be dated to the Hellenistic rather than the Imperial period. On *kairos* in the writings of Gorgias and Pythagoras, see Kucharski 1963; Poulakos and Whitson 2002; Rostagni 2002.

[12] For further discussion of how *kairos* is used in this passage, see J. E. Smith 2002, 52–3. On *kairos* in Plato's writings, see Levi 1924. On the history of *kairos* in Classical and later rhetoric, see Baumlin 1984; Enos 1995; Sipiora and Baumlin 2002. On its usefulness in modern scientific rhetoric, see C. R. Miller 1992.

[13] *Phaedr.* 271d.

many forms the soul has," so that he may develop a theoretical basis for classifying the souls he will encounter. But theory alone will not suffice:[14]

> Once he [i.e., the student] has understood these things sufficiently, afterward he must see that they are so in actual affairs and events, and he must be able to verify them sharply by means of his senses. Otherwise, the majority of the relevant theory he heard back then will be no good to him. But whenever he can say sufficiently which sort of man is persuaded by which sorts of things, and is able to indicate to himself, perceiving fully and attended by others, that *this* is the man, and *this* is the nature... to which he must apply *these* sorts of arguments in *this* way in order to achieve persuasion for *these* purposes—for the man who grasps all of these things in addition to the right moments for speaking and for being silent, for the man who distinguishes the best occasion and worst occasion for using brachylogies, piteous appeals, and hyperbole... only for this man is the art [of speaking] well and completely perfected.

We observe that for Plato, as for Hippocrates, *kairos* is a moment that is "sharply" (*oxeōs*) perceived by the senses; it occupies little to no *khronos*. Furthermore, Plato's *kairos* is utterly contingent. Plato's repeated use of the deictic pronoun "this/those" (*houtos*, etc.) hammers home his point that the "right moment" to use a particular rhetorical strategy depends upon a variety of factors. To whom is the orator speaking? What kinds of souls or characters do his listeners have? What is the ultimate goal of his persuasion? Which rhetorical forms has he been taught, and which are likely to be of the greatest use in the present context? In order to answer this barrage of questions, the orator must synthesize his theoretical knowledge with both his prior personal experiences and his current, real-time observations. Thus, for Plato, these decisive, kairotic moments are dynamic and require a process not unlike Galen's scientific method.[15] We will revisit this observation below.

Aristotle, in his *Nicomachean Ethics*, echoes Plato's sentiments, but goes on to provide two examples of arts other than rhetoric that are especially dependent upon *kairoi*: navigation and medicine. Aristotle's treatment of medical *kairos* takes place in the midst of a larger discussion about how general ethical systems

---

[14] Phaedr. 271c10-272a8: Ἐπειδὴ λόγου δύναμις τυγχάνει ψυχαγωγία οὖσα, τὸν μέλλοντα ῥητορικὸν ἔσεσθαι ἀνάγκη εἰδέναι ψυχὴ ὅσα εἴδη ἔχει...δεῖ δὴ ταῦτα ἱκανῶς νοήσαντα, μετὰ ταῦτα θεώμενον αὐτὰ ἐν ταῖς πράξεσιν ὄντα τε καὶ πραττόμενα, ὀξέως τῇ αἰσθήσει δύνασθαι ἐπακολουθεῖν, ἢ μηδὲν εἶναί πω πλέον αὐτῷ ὧν τότε ἤκουεν λόγων συνών. ὅταν δὲ εἰπεῖν τε ἱκανῶς ἔχῃ οἷος ὑφ' οἵων πείθεται, παραγιγνόμενόν τε δυνατὸς ᾖ διαισθανόμενος ἑαυτῷ ἐνδείκνυσθαι ὅτι οὗτός ἐστι καὶ αὕτη ἡ φύσις... ᾗ προσοιστέον τούσδε ὧδε τοὺς λόγους ἐπὶ τὴν τῶνδε πειθώ, ταῦτα δ' ἤδη πάντα ἔχοντι, προσλαβόντι καιροὺς τοῦ πότε λεκτέον καὶ ἐπισχετέον, βραχυλογίας τε αὖ καὶ ἐλεινολογίας... τούτων τὴν εὐκαιρίαν τε καὶ ἀκαιρίαν διαγνόντι, καλῶς τε καὶ τελέως ἐστὶν ἡ τέχνη ἀπειργασμένη....

[15] On the differences between Plato's and Galen's approaches to scientific investigation, see Cooper 2011a, 161-3.

should not be overly rigid because so many ethical decisions are predicated on situational circumstances. "Matters pertaining to conduct and expediency," Aristotle asserts, "are in no way fixed, just like matters pertaining to health. If such is the case for the general theory [of ethics], still more is precision lacking in the theory of specific cases, for it falls under no established art or set of rules. Rather, the people who perform actions must always consider for themselves those things that are opportune, just as is the case in both the medical and the navigational arts."[16] Aristotle sees the sickroom as a paradigmatic example of a domain in which contingency reigns.[17] The physician must know the principles of his art, of course, but he must also recognize that, when it comes to human health, many variables are in play. He must therefore pay close attention to the circumstances of each individual case, if he is going to correctly identify and respond to kairotic moments and thereby excel at his craft.

Galen's beloved "Hippocrates" also emphasizes the contingency of medical *kairoi*. Directly after specifying the relationship between *kairos* and *khronos*, as we saw above, the Hippocratic author of *Precepts* goes on to add, "Healing happens in *khronos* when it is also in *kairos*."[18] The Hippocratic writings stress the importance of tailoring medical treatments to suit the specific characteristics and environmental conditions of particular patients.[19] They take a wide variety of factors into consideration: the patient's age, physical constitution, and daily habits; the climate to which he is accustomed and the current season of the year.[20] The Hippocratic authors recognized that all of these factors could affect the ways in which health or disease manifest in individuals, and could similarly influence the kairotic timing of that individual's treatments.

To review, we have seen how in the writings of Galen's favorite authors—Plato, Aristotle, Hippocrates—the concept of *kairos* often has the following features. It is both "sharp" (or "momentary") and morally charged, in contrast to *khronos* which is neither. To identify a *kairos*, one must assess the circumstances of the specific situation, a process which involves a combination of theory, experience, and

---

[16] *Nic. eth.* 1104a3–10: τὰ δ᾽ ἐν ταῖς πράξεσι καὶ τὰ συμφέροντα οὐδὲν ἑστηκὸς ἔχει, ὥσπερ οὐδὲ τὰ ὑγιεινά. τοιούτου δ᾽ ὄντος τοῦ καθόλου λόγου, ἔτι μᾶλλον ὁ περὶ τῶν καθ᾽ ἕκαστα λόγος οὐκ ἔχει τἀκριβές· οὔτε γὰρ ὑπὸ τέχνην οὔθ᾽ ὑπὸ παραγγελίαν οὐδεμίαν πίπτει, δεῖ δ᾽ αὐτοὺς ἀεὶ τοὺς πράττοντας τὰ πρὸς τὸν καιρὸν σκοπεῖν, ὥσπερ καὶ ἐπὶ τῆς ἰατρικῆς ἔχει καὶ τῆς κυβερνητικῆς.
[17] On *kairos* in Aristotelian rhetoric, see Kinneavy and Eskin 2000.
[18] 1.1.3, 9.250 L: ἄκεσις χρόνῳ, ἔστι δὲ ἡνίκα καὶ καιρῷ.
[19] Craik outlines some of the "profound perceptions" regarding health, disease, and diet that are put forward in the Hippocratic writings: namely, "that different individuals have different needs; that the same individual has different needs at different ages and in different seasons; that to understand the patient in illness it is useful to understand the patient in health; that what we eat affects how we function; that some illnesses affect all inhabitants of a region while others are peculiar to individuals in it; that environment cannot be disregarded as a factor in human well-being. These perceptions are in line with such modern catch-phrases as social geography, environmental health, and preventive medicine, and they are at the root of the homeopathic principles formulated under their influence in the nineteenth century" (1995, 398).
[20] See, e.g., the "constitutions" and case histories in the *Epidemics*.

improvisation. Furthermore, by the Hellenistic period, the concept of *kairos* was already seen as important particularly for arts like rhetoric, navigation, and medicine.[21] In what follows, we will see that Galen's understanding of *kairos* often cleaves closely to those of his predecessors. However, we will also discover that he parts ways with them in certain respects. First, he incorporates hourly timekeeping into the discussion, which his predecessors did not.[22] Second, he distinguishes between different kinds of *kairoi* for the sick, the healthy, and the aged.

Kairos and Febrile Disease

Let us now return to a text that we discussed in the previous chapter, the fever treatise *On Crises*, in order to understand how Galen defines a *kairos* within a febrile context. On the first page of this text, Galen tells us that every febrile illness can be divided into four "parts" (*moria*): the "beginning" (*arkhē*), the "increase" or "growth" period (*auxēsis*), the "prime" or "acme" (*akmē*), and the "past-prime" or "abatement" phase (*parakmē*).[23] These correspond to terms that Galen often uses to denote the stages of a human or animal life, which suggests that Galen viewed the disease process as having a microcosmic "life" of its own.[24] He goes on to say that "some men call these parts the '*kairoi* of the disease as a whole;' for this same reason, they also call them the 'general *kairoi*,' in order that they might be differentiated clearly from the 'specific *kairoi*,' where the paroxysms vary."[25] In other words, Galen is saying (a) that all fevers pass through these four general phases (though, as he later points out, the precise timing and character of each differs from one fever to another);[26] and (b) that each paroxysm goes through its own set of four phases, making it something of a microcosm of the disease as a whole. By referring to these phases specifically as *kairoi*, the physicians (and Galen, too, as he adopts their practice) direct the reader's attention to the critical decision-making peculiar to each phase; an action performed during the wrong phase could, at best, prove ineffective and, at worst, cause the patient serious harm.

---

[21] Cooper (2011a) highlights how the methodologies of such sciences, particularly medicine, differ from the ultra-rational approaches characteristic of ancient astronomy.

[22] The Hippocratic author of *On Diseases* 1 does recognize that some treatments are best administered during a particular part of the day (i.e., morning, evening, night), but he does not translate these times into hours (1 5, 6.148–50 L).

[23] *Cris.* 9.551.1–2 K.

[24] For *auxēsis*, see e.g., *San. tu.* 6.5.9, 386.10 K. For *akmē*, 6.346.8, 386.12, 387.2 K. For *parakmē*, 6.387.3 and 12, 397.12 K.

[25] *Cris.* 9.551.5–9 K: τὰ δ' οὖν τέτταρα μόρια ταῦτα καλοῦσι μὲν ἔνιοι καιροὺς ὅλου τοῦ νοσήματος (δι' αὐτὸ δὲ τοῦτο καὶ καθόλου καιροὺς ὀνομάζουσιν ὡς ἂν διαφέροντας δηλονότι τῶν κατὰ μέρος εἰς οὓς ἕκαστος τῶν παροξυσμῶν διαιρεῖται).

[26] *Cris.* 9.551.9–12 K.

Physicians of Galen's time seem to have quarreled over where to set the boundaries of these stages for any given illness. "Some," Galen says, "ask whether the paroxysms come early or late; others look at their length, others consider their magnitude (which they call the 'severity'), and some examine these things along with the abatements.... Scarcely anyone can thus make an accurate guess as to the designated *kairos*...."[27] The doctors to whom Galen refers here clearly detect a link between the temporal patterns of an illness and the timing of a kairotic phase-change. There was no consensus, however, as to which indicators were more reliable or how the data from multiple indicators should be synthesized.

Galen, as we saw in the preceding chapter, presents himself as the only physician capable of resolving this thorny problem. He advocates comparing a series of two or more paroxysms with regard not only to their length, duration, magnitude, and intervals, but also to their attendant symptoms and the characteristics of certain indicator substances, like urine, sweat, and stool.[28] He then offers these rules of thumb for determining whether a febrile illness is currently in the *kairos* of "growth" or "abatement," or whether it has actually reached its "acme:"[29]

> For let the paroxysm arrive earlier than the customary hour, and let it be longer, more violent, of worse character, and accompanied by many bad symptoms. And let the time of the abatement be short and not precisely easy to bear, nor free of all the symptoms of the paroxysm.... Then, in another case, let everything be opposite to this, with the onset later, the length and magnitude of the paroxysm shortened, the bad character mitigated, and the attendant symptoms fewer and simpler or not entirely visible. And let the interval or abatement that follows be easy to bear and long, and let it completely dispel the paroxysmal symptoms. These last, in particular, are the clearest indicators of abatement, just as the ones mentioned first are the clearest indicators of growth. Whenever the features of both paroxysms are particularly equal, it is clear that the disease is at its acme.[30]

---

[27] Cris. 9.552.1–7 K: ἀλλ' οἱ μὲν εἰ προλαμβάνουσιν ἢ ὑστερίζουσιν οἱ παροξυσμοὶ σκοποῦσιν, οἱ δ' εἰς τὸ μῆκος αὐτῶν ἀποβλέπουσιν... μόλις γὰρ ἂν οὕτως τις ἀκριβῆ στοχασμὸν ποιήσαιτο τοῦ καθεστῶτος καιροῦ....

[28] See, e.g., Cris. 9.562.6–9.563.3 K, where Galen attributes this insight to Hippocrates.

[29] Cris. 9.554.13–9.555.13 K: προλαμβανέτω γὰρ ὁ παροξυσμὸς τῆς συνήθους ὥρας καὶ μακρότερος γινέσθω καὶ σφοδρότερος καὶ κακοηθέστερος καὶ σὺν πολλοῖς καὶ μοχθηροῖς συμπτώμασιν, γινέσθω δὲ καὶ ὁ τῆς παρακμῆς χρόνος βραχύς θ' ἅμα καὶ οὐκ ἀκριβῶς εὔφορος οὐδ' ἐλεύθερος ἁπάντων τῶν ἐκ τοῦ παροξυσμοῦ συμπτωμάτων.... Αὖθις δὲ τἀναντία πάντα τούτοις γινέσθω, τῆς μὲν εἰσβολῆς ὑστεριζούσης, τοῦ μήκους δὲ καὶ τοῦ μεγέθους τοῦ κατὰ τὸν παροξυσμὸν ἐλαττουμένου καὶ τῆς κακοηθείας πραϋνομένης καὶ τῶν ἐπιγινομένων συμπτωμάτων ἐλαττόνων τε καὶ ἁπλουστέρων γινομένων ἢ οὐδ' ὅλως ἐπιφαινομένων καὶ τοῦ διαλείμματος ἢ τῆς παρακμῆς εὐφόρου καὶ μακρᾶς ἑπομένης καὶ πάντ' ἐξαλειφούσης τὰ τοῦ παροξυσμοῦ συμπτώματα. καὶ γὰρ καὶ ταῦτα σαφέστατα παρακμῆς γνωρίσματα, καθάπερ τὰ πρότερα τῆς ἐπιδόσεως. ὅταν δ' ἰσάζῃ πως μάλιστα τὰ κατὰ τοὺς παροξυσμοὺς ἀμφοτέρους, ἀκμάζειν δηλοῖ τὴν νόσον.

[30] Galen recognizes that identifying the precise temporal boundaries of these phases is not always so simple. Sometimes the physician is forced to assume that the moment of transition occurred somewhere between two paroxysms and thus was not directly observable. See Cris. 9.557.2–9.562.6 K.

Here, Galen explicitly identifies the temporal aspects of paroxysms and their abatements as "indicators" (*gnōrismata*) of the fever's most recent *kairos* (in this case, its moment of transition into the stage of "growth," "acme," or "past-prime").[31] He goes on, as we saw in the previous chapter, to offer case histories in which these temporal factors are specified in seasonal hours.[32] For example, in his case history of a woman suffering from what he deems to be a semitertian fever combined with a tertian, Galen meticulously chronicles the pathological events that lead up to the kairotic "acme" of each fever:[33]

> Recently, a little woman had a shivering paroxysm on the even days that was precisely semitertian. And on the odd days, from dawn, she had one much smaller than the one that came before, so that the semitertian seemed to have stronger paroxysms on the alternate days, which indeed is customary in a semitertian. And around the *eighth hour of the day*, another paroxysm came on with powerful shivering, bearing precisely all the indicators of a tertian fever. This one broke during *some hours of the third night*, with sweat and vomiting of bile, and extended into the *second hour of the even [day]*. Indeed, after it had approached a fever-free state, the shivering fever that had an anomalous paroxysm came on again, namely the one which we said arose on even days. Then, withdrawing and attacking again malignantly throughout the whole day, [the fever] only just culminated in acme when the sun went down. It began to abate during *some hours of the fourth night*, and then from dawn again a not very

---

[31] Galen credits Hippocrates with these ideas, citing esp. his advice at *Aph.* 1.12.1–9, which Galen quotes as follows: "As for the paroxysms and constitutions, the diseases themselves and the seasons of the year, as well as the relations of the periods to each other, will make clear whether it is every day or throughout the day or happens over a longer period of time. This is so also with respect to the phenomena, such as in pleuritic cases, whether the saliva appears right away with a short beginning, or whether it appears later with a lengthened one. And urine and stool and sweat, doubtful and certain, short and long—the things that appear elucidate the diseases" (*Cris.* 9.562.11–18 K). Here again, as in earlier chapters, we see Galen build upon the pithy dicta of his predecessor to support his own theories.

[32] Galen says that hourly timekeeping is especially useful for anticipating when two paroxysms are going to overlap and thereby confuse the physician's reading of the symptoms (*Cris.* 9.674.14–9.675.9 K).

[33] *Cris.* 9.675.16–9.677.4 K: ἔναγχος γοῦν τι γύναιον ἐν μὲν ταῖς ἀρτίαις φρικώδη παροξυσμὸν εἶχεν ἀκριβῶς ἡμιτριταϊκόν, ἐν δὲ ταῖς περιτταῖς ἕωθεν μέν τινα τοῦ προγεγονότος πολὺ μικρότερον ὡς ἐπιπαροξύνεσθαι δοκεῖν τῇ ἑτέρᾳ τῶν ἡμερῶν τὸν ἡμιτριταῖον, ὅπερ δὴ καὶ γίνεται συνήθως ἐν αὐτῷ, περὶ δὲ τὴν ὀγδόην ὥραν τῆς ἡμέρας ἕτερος εἰσέβαλλε μετὰ ῥίγους ἰσχυροῦ παροξυσμὸς ἀκριβῶς ἅπαντα τριταίου πυρετοῦ φέρων τὰ γνωρίσματα. θραυόμενος δ' οὗτος ὥρας που νυκτὸς τρίτης ἱδρῶτί τε καὶ χολῆς ἐμέτῳ παρέτεινεν εἰς τὴν τῆς ἀρτίου δευτέραν ὥραν. ἤδη δ' αὐτὸν ἐγγὺς ἀπυρεξίας ἥκοντα διεδέχετο πάλιν ὁ φρικώδης ἐκεῖνος καὶ ἀνώμαλος παροξυσμὸς ὃν ἐν ταῖς ἀρτίαις ἔφαμεν γίνεσθαι, κἄπειτα δι' ὅλης τῆς ἡμέρας κακοήθως ὑποστελλόμενός τε καὶ αὖθις ἐπιτιθέμενος μόλις που τὴν ἀκμὴν ἀπελάμβανεν ἡλίου δύνοντος. ἤρχετο δὲ παρακμάζειν ὥρας που ἀκμὴν ἀπελάμβανεν ἡλίου δύνοντος. ἤρχετο δὲ παρακμάζειν ὥρας που τετάρτης τῆς νυκτός, εἶθ' ἕωθεν πάλιν ἐπὶ δαψιλεῖ λειψάνῳ παραύξησις οὐ λίαν ἀνώμαλος ἕως τῆς ὀγδόης ὥρας ἐπετείνετο κἄπειτα δοκοῦντα σαφῶς ἀκμάζειν ἤδη τὸν πυρετὸν τοῦτον ἕτερος ἐξεδέχετο παροξυσμὸς ἀκριβῶς τριταϊκός. τοῦτον ἡμεῖς τὸν πυρετὸν εὐθὺς μὲν κἂν τῇ δευτέρᾳ τῶν ἡμερῶν ἐξ ἐπιπλοκῆς ἡμιτριταίου τε καὶ τριταίου διέγνωμεν γεγονέναι, πολὺ δὲ πιστοτέραν ἡ τρίτη τὴν διάγνωσιν παρέσχεν ἠλπίσαμέν τε τὴν μὲν ἑτέραν περίοδον τὴν τοῦ τριταίου παύεσθαι ταχέως, τὴν δ' ἑτέραν μηκύνειν, καὶ οὕτως ἀπέβη.

anomalous increase to the ample remnants lasted until the *eighth hour*. Then, when this fever seemed clearly to reach its acme, another paroxysm succeeded it, strictly tertian. We diagnosed this fever right away on the second day as being from an interweaving of a tertian and a semitertian, and the third day confirmed the diagnosis much more trustworthily. We expected the next period, the one belonging to the tertian, to stop quickly while the other was protracted, and so it was.

In *On Crises*, as in the other fever literature we have examined,[34] Galen uses hours to *describe* and *classify* febrile diseases with the goal of anticipating their future behaviors. This, as we are about to see, is not the case in *On Hygiene*.

Kairos and Health Maintenance

On Hygiene is a six-volume work, dated to approximately 180 CE,[35] that investigates the condition of "health" (*hygieia*) and how an individual might maintain it throughout his or her life.[36] Galen's stated motivation for writing this treatise is to refute the "majority" of doctors and trainers, who write about personal hygiene "as if they were talking about a single man."[37] Galen charges these writers with assuming that all human bodies are the same, and that therefore a single health regimen is best for everyone. Galen challenges this position by demonstrating, over the course of the text, that health is actually relative and contingent, rather than universal and fixed.[38] Health is relative, because human bodies differ from one another with respect to their age, build, and what we might call their biochemical makeup or constitution, which for Galen meant their particular mixture of the four humors and qualities (phlegm, yellow bile, black bile, and blood; cold, hot, dry, and moist). Health will manifest differently, for example, in an athletic and phlegmatic 20-year-old than in a frail and bilious 60-year-old. To further elucidate this concept, Galen offers the following metaphor: "For just as

---

[34] See Chapters 3 and 4.
[35] On the dating of this treatise, see Koch et al. 1923, 5.4.2.7 and Wilkins 2015, 415.
[36] On the concept of regimen or *diaita* in Galen's writings, see Capriglione 2000, and within *On Hygiene* specifically, see Wöhrle 1990; Grimaudo 2008; Wilkins 2015. On diet and dietetics in Greco-Roman antiquity, see, e.g., Edelstein 1967a; W. D. Smith 1980; Lain-Entralgo 1987; Craik 1995; Bartoš 2009. For an overview of the concept of health and the cult delivered to the goddess Hygieia in antiquity, see King 2005.
[37] *San. tu.* 6.306.5–6 K = 135.13–14 Koch: οὕτω δὲ γράφουσιν οἱ πλείους αὐτῶν ὑπὲρ τῆς ὑγιεινῆς ἀγωγῆς, ὡς ὑπὲρ ἑνὸς ἀνθρώπου διαλεγόμενοι. His intended audience seems to be educated elites who are not themselves medical specialists, but have enough leisure and training in logical methods to be able to apply Galen's dietetic principles to their own lives (*San. tu.* 6.449.5–6.450.4 K = 197.2–17 Koch). For further discussion of Galen's audience in this text, see Wilkins 2015, 416.
[38] See *San. tu.* 6.164.1–6.165.13 K = 73.1–21 Koch, where Galen provides an overview of the structure and argument of *On Hygiene*.

the whiteness in snow does not differ from the whiteness in milk in being white, but rather in being more or less white, so in the same way the health in Achilles, for instance, does not differ from the health in Thersites in being health—that is the same. But they (i.e., the two healths) differ in another way, in the degree of more or less."[39] Achilles, in this example, has "more" of health than Thersites to the extent that his body can perform better; he is stronger, faster, and more dexterous with a blade. But even though Thersites's version of health may pale in comparison to Achilles's, this does not mean that Thersites is unhealthy. His body is simply different from that of Achilles, and therefore has different capacities and characteristics when healthy.

For Galen, then, health is not monomorphic; it takes a different form in each body. However, it is clear that people often fail to achieve their personal version of best health, whether compelled by external forces, such as overwork or foul weather, or by internal forces, such as laziness or gluttony.[40] Thus, Galen insists that a person's health is not only relative, but also contingent upon his or her environment, mental state, and behavioral patterns. At the beginning of book 2, he puts it this way:[41]

> Just as the diversity of bodies themselves has been shown to be very great, in the same way, the forms of the lives we live are also very great. Thus, it is impossible to recommend the best treatment of the body for every given life, but it is possible to recommend what is best for each particular life, since the absolute best is not achievable in all lives. For many men have a life involved in the circumstances of their business and must necessarily be harmed by the things they do; it is impossible for them to avoid this. Some fall into such lives through poverty, others through slavery.... But the man who is completely free, both by fortune and by choice, him it is possible to instruct as to how he may be most healthy and least sick, and age in the best manner.

Galen points out that, in order for a person to maintain maximum health, he must devote virtually all of his time and energy to the care of his own body. Since few

---

[39] San. tu. 6.16.15–6.17.5 K = 9.30–5 Koch: ὥσπερ γὰρ ἡ ἐν τῇ χιόνι λευκότης τῆς ἐν τῷ γάλακτι λευκότητος, ᾗ μὲν λευκόν ἐστιν, οὐ διαφέρει, τῷ μᾶλλον δὲ καὶ ἧττον διαφέρει, τὸν αὐτὸν δὴ τρόπον ἡ ἐν τῷ Ἀχιλλεῖ, φέρε εἰπεῖν, ὑγεία τῆς ἐν τῷ Θερσίτῃ ὑγείας, καθ' ὅσον μὲν ὑγεία, ταὐτόν ἐστιν, ἑτέρῳ δέ τινι διάφορος· καὶ τοῦτο τὸ ἕτερον οὐδὲν ἄλλο ἐστὶν ἢ τὸ μᾶλλόν τε καὶ ἧττον. Cf. 6.22.3–6 K = 11.33–6 Koch; 6.164.1–6.165.13 K = 73.1–21 Koch; 6.372.6–14 K = 164.15–23 Koch.

[40] See, e.g., San. tu. 6.9.14–6.10.10 K = 6.31–7.9 Koch; 6.404.2–12 K = 177.29–178.6 Koch.

[41] San. tu. 6.82.1–6.83.3 K = 38.9–26 Koch: ὥσπερ αὐτῶν τῶν σωμάτων ἐδείχθη παμπόλλη τις οὖσα διαφορά, κατὰ τὸν αὐτὸν τρόπον καὶ τῶν βίων, οὓς βιοῦμεν, εἴδη πάμπολλά ἐστιν. οὔκουν ἐγχωρεῖ τὴν ἀρίστην τοῦ σώματος ἐπιμέλειαν ἐν ἅπαντι τῷ προχειρισθέντι βίῳ συστήσασθαι ἀλλὰ τὴν μὲν ὡς ἐν ἑκάστῳ βελτίστην οἷόν τε, τὴν δ' ἁπλῶς ἀρίστην οὐκ ἐγχωρεῖ κατὰ πάντας τοὺς βίους ποιήσασθαι. πολλοῖς γὰρ τῶν ἀνθρώπων μετὰ περιστάσεως πραγμάτων ὁ βίος ἐστί. καὶ βλάπτεσθαι μὲν ἀναγκαῖόν ἐστιν αὐτοῖς ἐξ ὧν πράττουσιν, ἀποστῆναι δ' ἀδύνατον. ἔνιοι μὲν γὰρ ὑπὸ πτωχείας εἰς τοὺς τοιούτους ἐμπίπτουσι βίους, ἔνιοι δ' ὑπὸ δουλείας ... ὅστις δὲ ἀκριβῶς ἐλεύθερος ὑπάρχει καὶ τύχῃ καὶ προαιρέσει, δυνατὸν ὑποθέσθαι τῷδε, ὡς ἂν ὑγιαίνοι τε μάλιστα καὶ ἥκιστα νοσήσειε καὶ γηράσειεν ἄριστα.

people, in reality, have the leisure and inclination to do this, it is important that a physician, when recommending a day-to-day personal care regimen, take a patient's lifestyle-constraints into consideration, along with other circumstantial factors.

The concept of *kairos* plays an important role in Galen's argument because he keeps returning to the following question: given a particular patient's body and circumstances, what is the "right moment" for him or her to perform a certain activity, like exercising, eating dinner, or taking a bath? As he himself puts it, "It is necessary for the one who pursues the hygienic art to understand the properties of all the hygienic materials. The deft use of these things results from that. And the deft use of these things, in turn, arises if we should discover the right time and amount of each."[42] Time and again, Galen emphasizes the contingency of these individual *kairoi*. In book 1, for example, Galen addresses the appropriate time (*kairos*) to bathe or rub down a healthy baby. He recommends waiting until after the infant has awoken from a prolonged sleep, because at that time there is likely to be little or no food in the baby's stomach which could be disturbed by the rubbing or bathing motions. Galen then cautions against doing "what some nurses do," namely "designating one particular time of day" for bathing or rubbing, or simply performing these tasks "whenever they have leisure."[43] Instead, Galen says, "the right moment designated by us occurs sometimes at one time of day or night, sometimes at another."[44] Galen draws a distinction between *kairos* and *khronos* here that is similar to the one we saw above, in the opening line of the Hippocratic *Precepts*. Galen agrees that, when the bathing or rubbing takes place, it will necessarily do so at a *khronos*, a particular interval of time within the day or night. But in order for that action to be most effective, and least harmful to the child, it must also take place at a *kairos*, or opportune moment, that is dependent on additional factors, such as the amount of food in the baby's belly and, as an index of this, how recently the baby has slept. Note that here too, as in the passages from *On Periods* and *On Treatment by Venesection* considered in the previous chapter, we see Galen pushing back against a group of practitioners (in this case, nurses rather than physicians) whom he accuses of being overly quantitative and programmatic in their timing. In this context, Galen urges such nurses to take their temporal cues from their infant charges rather than from the clock.

Elsewhere in *On Hygiene*, however, Galen suggests that hourly timekeeping can, in fact, prove useful to the healer interested in zeroing in on a particular

---

[42] *San. tu.* 6.80.8–11 K = 37.6–9 Koch: τῶν δ' ὑλῶν ἁπασῶν τῶν ὑγιεινῶν ἐπίστασθαι ἀναγκαῖον τὰς δυνάμεις τῷ τὴν ὑγιεινὴν τέχνην μετιόντι. καὶ γὰρ καὶ ἡ ἐπιδέξιος αὐτῶν χρῆσις ἐντεῦθεν ὥρμηται. γίνεται δ' ἡ ἐπιδέξιος αὐτῶν χρῆσις, ἐὰν τόν γε καιρὸν ἑκάστου καὶ τὸ μέτρον εὕρωμεν.

[43] *San. tu.* 6.49.5–6 K = 23.24–6 Koch: οὐδ', ὥσπερ νῦν ποιοῦσι ἔνιαι μὲν τῶν τροφῶν, ἕνα τινὰ χρόνον ἀφορίσασαι τῆς ἡμέρας, ἔνιαι δ', ὅταν αὐταὶ σχολάσωσι, [τοῦ] τηνικαῦτα προνοούμεναι.

[44] *San. tu.* 6.49.7–9 K = 23.27–9 Koch: ὁ γὰρ ὑφ' ἡμῶν ἀφοριζόμενος καιρὸς ἄλλοτε εἰς ἄλλον ἐμπίπτει χρόνον ἤτοι τῆς ἡμέρας ἢ τῆς νυκτός.

kairos. Hourly timekeeping can help the healer, on the one hand, to identify one of the factors, or antecedent causes, upon which a *kairos* is contingent, and on the other, to define the specific *khronos* (hour or interval) that turns out to be kairotic in a given case. Let us consider some examples of the former before turning to the latter. According to Galen, one of the additional factors that can influence a *kairos* is the length of daylight that a patient experiences, as measured in equinoctial hours. We recall that Greeks and Romans under the Empire typically marked off their days according to seasonal hours, which were derived by dividing the period from sunrise to sunset (or sunset to sunrise) into twelve equal parts whose absolute lengths varied throughout the year. Hours of consistent length, like the ones we use today, were employed primarily by astronomers and, for the most part, were not integrated into other cultural contexts. There was, however, one exception. By the mid-fourth century BCE, Greek scientists had discovered that one could map the world by dividing it up into latitudes, which they called *klimata*.[45] A city's latitude was expressed as a ratio of the lengths of its longest and shortest days, lengths which differ more dramatically as one moves north or south of the equator due to the curvature of the earth. To give these lengths in seasonal hours would be meaningless, as every day in every location is, by definition, twelve seasonal hours long; therefore, equinoctial hours had to be used instead. By the Imperial period, knowledge of *klimata* and the absolute daylight lengths associated with them had spread beyond the bounds of technical astronomy, and we see latitudes (with their ratios expressed in equinoctial hours) popping up not only in writers like Ptolemy and Cleomedes, but also in Strabo and Pliny.[46] As we noted in Chapter 1, we even have examples of portable sundials that users could adjust to different latitudes.[47]

On Hygiene makes it clear that Galen had been exposed to these ideas, whether through his studies of basic astronomy or simply through cultural osmosis, and that he found them applicable to his medical practice.[48] In this treatise, Galen points out that, because the length of daylight varies seasonally and geographically, inhabitants of different regions will have different quantities of time available for personal care. Any "hygienist" worth his salt should therefore take special note of the location in which a given patient lives and thus of the amount of time that he or she will realistically have for bathing, exercising, etc. between the end of

---

[45] Our earliest references to the term are found in Hypsicles, *Anaphoricus* (Krause and de Falco 1966, 36) and Hipparchus, *Commentary on Aratus*, where it is embedded within a quotation from Eudoxus (1.2.22 Manitius p. 23 = F 68 Lasserre [1966]; 1.3.9–10 Manitius p. 28 = F 67 Lasserre). For discussion, see Neugebauer 1975, 34–44; Nicolet 1991, 59; Shcheglov 2003, 2. Some of the foundational works on ancient *klimata* include Honigmann 1929; Diller 1934; Dicks 1955.

[46] See, e.g., Strabo *Geogr.* 2.5.14, 36, 38–42; Plin. *NH* 2.77.

[47] Such sundials look rather like makeup compacts. They can open and shut on a hinge and carry inside them a series of discs engraved with different networks of hour- and date-lines. Each disc corresponded to a specific latitude and could be swapped out in exchange for other discs as one moved from place to place. For catalogues of portable sundials, see Price 1969; Wright 2000; Talbert 2017.

[48] On the likely sources of Galen's astronomical knowledge, see Cooper 2011b, 62–3.

the workday and bedtime. As an example, Galen asks us to imagine a servant or attendant (*hypēretēs*) of the Emperor, in excellent physical condition, who is released from duty at the end of each day.[49] Galen then asserts that, if one is to determine how much time the attendant has available to complete his evening healthcare regimen, one must define the phrase "end of day" much more precisely:[50]

> If I should say that he first departs for the care of his body when the sun sets, if I do not then add what sort of day I am talking about—whether one around the summer or winter solstice, or at one of the equinoxes, or during one of the other times in between those mentioned—it will be impossible to provide beneficial instructions [for his personal care]. In the city of the Romans, the longest days and nights are a little longer than fifteen equinoctial hours, just as, again, the shortest are fewer than nine. In the great city of Alexandria, however, the longest [days and nights] are fourteen hours, and the shortest are ten. If an attendant

---

[49] The exact meaning of *hypēretēs* here is ambiguous. Is Galen describing an enslaved person or a member of the Emperor's entourage, who may himself have belonged to the elite classes? If the former, then this passage offers some insight into how the lives of enslaved individuals could be affected, and additionally constrained, by sociotemporal frameworks. Galen's example here would then suggest that, in certain locations and during certain times of the year, time limitations compelled enslaved individuals to curtail or even to eliminate activities which they were nominally allowed to perform, such as bathing in a public bath-house or exercising in a gymnasium. (Although sometimes the bathing opportunities of enslaved persons were also restricted by law. E.g., a first-century BCE inscription from a bath-house in Puteoli [*AE* 1971.88 II lines 3–4] forbids a contractor's enslaved workers from bathing there until the first hour of the night. I am grateful to R. Talbert for bringing this inscription to my attention.) However, P. Singer notes (2023, 49–50) that, in this text specifically, Galen tends to use words like *hypēretēs* and *doulos* not as civic categories but as metaphors for individuals of any status whose time and agency is in some way constrained. However we interpret Galen's word choice here, this passage should encourage scholars to be alert to the ways in which Greco-Roman timekeeping tools and practices might have served to deepen or reinforce inequities between enslaved and free persons. A non-medical example can be found in an inscription from Puteoli, dating to the first century BCE, concerning the disposal of bodies following public executions (*AE* 1971. 88 II lines 22–3): "Should instructions be given [to remove] a hanged man, he [i.e., the contractor] is to see to their fulfilment and the removal within the hour. In the case of a male or female slave (*servom servamve*), if instructions be given before the tenth hour of the day, removal is to be effected that day; if after the tenth, on the next day before the second hour." The underlying message of this inscription is clear: because the body of a slave is valued less than that of a free person, it is less important that the slave's body receive prompt treatment and burial. Thus, the differences in how enslaved and free persons were valued within Roman society is reflected and reinforced by the different time frames given here for tending to their bodies after execution. On this inscription, see Talbert 2020, 539.

[50] *San. tu.* 6.405.6–6.406.2 K = 178.15–28 Koch: ἐὰν γοῦν εἴπω χωρίζεσθαι τηνικαῦτα πρῶτον εἰς τὴν ἐπιμέλειαν τοῦ σώματος, ἡνίκα ὁ ἥλιος δύῃ, μὴ προσθείς, ὁποίας ἡμέρας λέγω, πότερον τῆς περὶ τὰς θερινὰς τροπὰς ἢ χειμερινὰς ἢ κατά τινα τῶν ἰσημερινῶν ἢ χρόνον ἑκάτερον ἐν τῷ μεταξὺ τῶν εἰρημένων καιρῶν, ἀδύνατον ἔσται συμφερούσας ποιήσασθαι ὑποθήκας. κατὰ γοῦν τὴν Ῥωμαίων πόλιν αἱ μέγισται μὲν ἡμέραι καὶ νύκτες βραχὺ μείζους ὡρῶν ἰσημερινῶν πεντεκαίδεκα γίνονται, καθάπερ γε πάλιν αἱ ἐλάχισται μικρὸν ἀποδέουσι τῶν ἐννέα, κατὰ δὲ τὴν μεγάλην Ἀλεξάνδρειαν τεσσάρων καὶ δέκα μὲν ὡρῶν αἱ μέγισται, δέκα δὲ αἱ σμικρόταται. ὁ μὲν οὖν ἐν ταῖς σμικροτάταις [μὲν] ἡμέραις [μεγίσταις δὲ νυξὶν] ἀφιστάμενος τῆς ὑπηρεσίας ἡλίου δυομένου καὶ τρίψασθαι κατὰ σχολὴν καὶ λούσασθαι δύναται καὶ κοιμηθῆναι συμμέτρως, ὁ δ' ἐν ταῖς μεγίσταις οὐδ' ἓν τούτων οἷός τ' ἐστὶ πρᾶξαι μετρίως. Boudon-Millot cites this passage as an example of how difficult it was for an Imperial-period doctor "fixer des repères temporels stables et fiables dont dépendront l'exactitude et la juste observance de ses prescriptions" (2008, 77).

departs at sunset during the shortest days and longest nights, then he can be rubbed down and bathed at leisure and go to bed at a suitable time, whereas the [servant] who is gone during the longest days is not able to do any of these things suitably.

Galen sees the length of daylight (as measured in equinoctial hours, according to both the time of year and geographical location) as an external factor that can influence one's *kairoi* for bathing, massage, and other activities. Galen encourages his fellow physicians to recognize that, while it might ultimately be healthier for a patient to pay a leisurely visit to the bath-house every evening of the year, there will be places and seasons in which this is simply not possible.[51] In response to such constraints, the effective physician will adjust his own assessment of the "right time" for a given patient to perform this activity.

This passage also highlights one way in which the growing awareness of clocks and hours (both seasonal and equinoctial) influenced the development of medical theory under the Empire. The Hippocratic authors were revolutionary, at least among our extant sources, in taking account of environmental and seasonal conditions when treating patients.[52] But the world in which many of them lived and wrote had not yet been divided into *klimata*, and geographical locations were not yet being translated regularly into daylight lengths.[53] Once again, as in the fever case histories we examined in Chapter 5, Galen builds upon an originally Hippocratic premise so as to take into account the ideas and technologies of his own time.

Let us now direct our attention to the other way in which Galen connects hourly timekeeping to the idea of *kairos*, namely, to define that *kairos* within a given situation. In *On Hygiene*, Galen devotes a good deal of space to the question of the temporal relationship between eating and bathing: which activity should take place first, and what interval of time should separate them.[54] After setting

---

[51] As discussed in Chapter 2, inscriptions from Crete (*SEG* 26: 1044) and Portugal (*CIL*² 05181, 19–23) suggest that bath-houses could have specific opening and closing times, as well as designated periods for male or female bathing. See Ducrey and van Effenterre 1973. Hadrian also decreed separate bathing times for the sick and the healthy at Rome (*Hist. Aug. Had.* 22.7). Such rules would also have restricted the times at which a person could choose to bathe. On Roman bathing habits more geneally, see Yegül 2010.

[52] The Hippocratic *Regimen in Health*, e.g., recommends different sorts of daily regimens according to the season and the patient's body type and humoral balance.

[53] Strohmaier observes, "Hippocrate croyait encore que la terre était un disque plat et que le soleil se mouvait au-dessus d'elle en une courbe relativement basse. Galien sait que la terre est seulement 'un point' au milieu des sphères célestes.... Pour Galien, comme pour Ptolémée et pour Aristote, la terre est une sphère et les grandes différences météorologiques dépendent du fait que les zones ou les ceintures déterminées par une certain latitude géographique sont plus ou moins inclinées par rapport aux rayons du soleil" (1997, 211).

[54] See, e.g., *San. tu.* 6.88.2–6.89.2 K = 40.33–41.12 Koch; 6.369.3–13 K = 162.32–163.7 Koch.

forth some general principles, such as the dangers of exercising on a full stomach, Galen offers up his own practice as an illustrative example:[55]

> For since, as I said, for some people it is better to eat before bathing, I ought to speak about the hour at which they should do this, and about the quantity and quality of what should be administered to them. Thus, I will not hesitate to say what I myself am accustomed to do on a day in which I decide to bathe later on account of either examining patients or attending to some civic matter. Let us consider a day, on which this happens, of thirteen equinoctial hours, and let us anticipate that the care of the body will take place around the tenth hour. According to this assumption, it seems to me that around the fourth hour the simplest meal should be taken.... But let the amount of food be only so much that it can be digested in the stomach by the tenth hour. For, if [patients] should wish to exercise, in this way they would exercise least harmfully.

Note the care that Galen takes here to specify the length of daylight available to him on this particular day (i.e., thirteen equinoctial hours). On a day of such a length, Galen tells us, he is often in the habit of exercising around the tenth hour (presumably seasonal here).[56] From these two temporal "givens," the length of daylight and his anticipated hour of exercise, Galen then calculates the *khronos* that is the *kairos* for his mid-day meal: i.e., around the fourth seasonal hour. Galen selects this hour as the right moment to lunch because it leaves a six-hour window of time between his meal and his training session, which will allow him plenty of time to digest before subjecting his body to vigorous motion.

This passage showcases how Galen uses hours both as units for articulating and as factors in calculating *kairoi*—in other words, hours can be *kairoi* themselves but can also be antecedent causes of *kairoi*. The length of daylight, in equinoctial hours, is a prior, external variable that helps Galen to determine the right time,

---

[55] *San. tu.* 6.411.13–6.412.15 K = 181.13–29 Koch: ἐπεὶ δέ, ὡς ἔφην, ἐνίοις ἐσθίειν ἄμεινόν ἐστι πρὸ τοῦ λουτροῦ, περί τε τῆς ὥρας, ἐν ᾗ τοῦτο πρακτέον αὐτοῖς ἐστι, καὶ περὶ τῆς ποσότητός τε καὶ ποιότητος, ὧν χρὴ προσφέρεσθαι, λεκτέον ἐστίν. ὅπερ οὖν εἴωθα ποιεῖν αὐτὸς ἐγὼ καθ' ἣν ἂν ἡμέραν ὀψιαίτερον ἡγῶμαι λούσασθαι δι' ἀρρώστων ἐπισκέψεις ἤ τινα πολιτικὴν πρᾶξιν, εἰπεῖν οὐκ ὀκνήσω. ὑποκείσθω γοῦν ἡμέρα, καθ' ἣν τοῦτο γίνεται, τριῶν καὶ δέκα τῶν ἰσημερινῶν ὡρῶν, ἐλπιζέσθω δὲ περὶ δεκάτην ὥραν ἡ τοῦ σώματος ἐπιμέλεια γενήσεσθαι. κατὰ ταύτην τὴν ὑπόθεσιν ἔδοξέ μοι περὶ τετάρτην ὥραν προσφέρεσθαι τροφὴν ἁπλουστάτην, ἥτις ἐστὶν ἄρτος μόνος...ἔστω δ' αὐτῶν πλῆθος ἑκάστου τοσοῦτον, ὅσον ἄχρι τῆς δεκάτης ὥρας ἐν τῇ γαστρὶ πεφθῆναι δύναται· καὶ γὰρ εἰ γυμνάζεσθαι βούλοιντο, μάλιστα ἂν οὕτως ἀβλαβῶς γυμνάσαιντο.

[56] I infer that Galen switches from equinoctial to seasonal hours here because of the conventions common in his day and in his other writings. While, for the reasons stated above, day-lengths must be given in equinoctial hours, very few Greco-Roman sundials or water clocks were designed to indicate equinoctial hours (e.g., the earliest extant stone sundial from Greece, a fourth-century BCE dial from Oropos; see Schaldach 2006, 116–21, 196–8). The vast majority tracked seasonal hours, suggesting that these were favored for coordinating the activities of day-to-day life. Our literary sources seem to confirm this hypothesis, in that they frequently describe activities like exercising and eating dinner as taking place at a particular hour "of the day" (i.e., at a particular seasonal hour).

defined as a specific (again, presumably seasonal) hour, at which he will go to the gymnasium. The time of this gym appointment, in turn, becomes a factor, or antecedent cause, in determining the kairotic seasonal hour for his meal. For Galen, then, the process of designing a person's daily health regimen involves, first, establishing a set of assumptions (based on body type, environmental conditions, personal preferences, etc.), and then reasoning from those assumptions in order to create a coherent system. In other words, Galen uses logic to move from axioms to recommended actions. Now, at this point in our investigations, whenever we see logical reasoning emphasized in a Galenic text, we should be on the alert for references to experience, as well, since logic and experience are the foundational principles of Galen's scientific method. And sure enough, Galen's system for constructing hygienic regimens involves verification and refinement through experience, and hours play a role in this part of the process, as well.

To see how Galen interweaves logic and experience to solve a hygienic problem, we will turn to his discussion of massage. In this discussion, Galen first stresses the importance of fixing one's variables: "Up to this point, only two things have been defined: the condition of the patient's body and his age. But it has not been additionally determined in which part of the world he was raised, nor the time at which he intends to exercise currently, nor what season of the year it is or what hour of the day. And yet the measure of massage varies according to all of these things."[57] Galen then describes the kind of massage that would be appropriate for an ideal young man (i.e., one in unrealistically perfect health) who trains at the gym under ideal conditions. He lays out those conditions in the following manner, beginning with the ideal geographical location:[58]

> Of our very own country... the most central part is the most temperate, such as is the case in the fatherland of Hippocrates. For this region is temperate both in winter and in summer, and still more so in the spring and autumn. Thus, let us assign such a country to our exemplary patient's body, and let us assume additionally that the season of the year in this case is the very middle of spring [i.e., the equinox]. And within that day on which he is going to be first trained by

---

[57] *San. tu.* 6.125.12–16 K = 56.12–16 Koch: μόνα γὰρ ἐπ' αὐτοῦ δύο διώρισται, τό τε τῆς κατασκευῆς τοῦ σώματος καὶ τὸ τῆς ἡλικίας· οὔτε δὲ ἐν ᾧτινι τέθραπται μέρει τῆς γῆς, οὔτε ἐν ᾧ μέλλει γυμνάσασθαι νῦν, οὐδὲ καθ' ἥντινα τοῦ ἔτους ὥραν ἢ καὶ τῆς ἡμέρας, προσδιώρισται, καίτοι παρὰ ταῦτα πάντα τὸ μέτρον τῆς τρίψεως ὑπαλλάττεται.

[58] *San. tu.* 6.127.2–13 K = 56.32–57.7 Koch: καὶ αὐτῆς δὲ τῆς ἡμετέρας χώρας ... οἱόνπερ ὑπάρχει τὸ κατὰ τὴν Ἱπποκράτους πατρίδα· καὶ γὰρ χειμῶνος αὕτη καὶ θέρους ἐστὶν εὔκρατος, ἔτι δὲ δὴ μᾶλλον ἦρός τε καὶ φθινοπώρου. τοιαύτην οὖν τινα χώραν ὑποθέμενοι τῷ προκειμένῳ σώματι, τὴν ὥραν τοῦ ἔτους αὐτῷ προσυποθώμεθα τὸ μεσαίτατον τοῦ ἦρος. ἔστω δὲ καὶ τῆς ἡμέρας ἐκείνης, ἐν ᾗ μέλλει πρὸς ἡμῶν γυμνάζεσθαι τὸ πρῶτον, ὡς οἷόν τε τὸ μεσαίτατον, ἵνα κατὰ μηδένα τρόπον ὑπὸ τοῦ περιέχοντος ἐξαλλαχθῇ πως ἡ φυσικὴ δύναμις τῆς κράσεως αὐτοῦ. διὰ δὲ τὸν αὐτὸν λογισμὸν οὐδὲ τὸν οἶκον, ἐν ᾧ γυμνάζεσθαι μέλλει, θερμότερον ἢ ψυχρότερον εἶναι προσήκει κατά γε τὴν ἡμέραν ἐκείνην τοῦ κοινοῦ τῆς πόλεως ὅλης ἀέρος.

us, let it be as close to noon as possible in order that the physical power of the patient's constitution might in no way be changed by the environment. According to the same reasoning, it is best that the building in which he is going to train be neither warmer nor colder than the general air of the whole city on that day.

This is Galen's best-case scenario: a perfect physical specimen comes for his massage after training in a temperate environment that will neither chill his muscles nor overheat him. But, as we have seen, Galen recognizes that such bodies and days come around but rarely. How, then, is a masseur to determine what kind of massage is most appropriate for various clients at different times of year? The key, Galen asserts, is not only to reason logically from physical and circumstantial givens, but also to check and refine one's system by recourse to experience:[59]

> How great the quantity of massage should be cannot be clarified [simply] by logic, but the knowledgeable person who is beginning to massage such people [i.e., teenage boys] must not use a precise conjecture on the first day, but in the following days, gaining some experience of that body's nature, should make the rule ever more precise. And moreover, with regard to training, on the first day it is not possible to be precise about the amount, but in the days after this it becomes entirely possible.

The "precise conjecture" to which Galen refers, about when and how to massage, is ultimately the product of reasoning, but when a masseur or trainer is first getting to know his client, he does not yet have all of the information necessary for establishing his axioms. As a result, it is important, especially in the early days, that he closely observe his client to see what works and what does not, and to parse his client's physical constitution (in particular, the mixture of his humors). Only by pairing logic and empiricism in this way can a practitioner hope to minimize errors and learn how to correct those that occur.[60] Thus in *On Hygiene*, as in the other texts we have examined, Galen's interest in hours is informed by his approach to scientific inquiry.[61]

---

[59] *San. tu.* 6.129.4–11 K = 57.28–58.2 Koch: ὁπόσον δ' ἐστὶ τὸ πλῆθος τῶν ἀνατρίψεων, οὐχ οἷόν τε λόγῳ δηλῶσαι, ἀλλὰ χρὴ τὸν ἐπιστατοῦντα, τρίβωνα τῶν τοιούτων ὑπάρχοντα, κατὰ μὲν τὴν πρώτην ἡμέραν οὐκ ἀκριβεῖ στοχασμῷ χρήσασθαι, κατὰ δὲ τὰς ἑξῆς ἐμπειρίαν ἤδη τινὰ τῆς τοῦ σώματος ἐκείνου φύσεως ἔχοντα τὸν στοχασμὸν ἀεὶ καὶ μᾶλλον ἐξακριβοῦν. καὶ μὲν δὴ καὶ κατὰ τὰ γυμνάσια τῇ μὲν πρώτῃ τῶν ἡμερῶν οὐ δυνατὸν ἀκριβῶσαι τὸ μέτρον, ἐν δὲ ταῖς μετὰ τήνδε καὶ πάνυ δυνατόν. Cf. 6.189.2–9 K = 83.27–34 Koch.

[60] Galen states this most explicitly at *San. tu.* 6.314.11–16 K = 139.4–9 Koch; 6.373.2–7 K = 164.25–30 Koch.

[61] The noun occurs eleven times in this work, and the related verb *apodeiknumi* seventeen times.

Kairos and the Elderly

Up to this point, we have explored how Galen uses hours to determine and describe specific *kairoi* both for diseased patients and for the predominantly healthy: when patients are sick, Galen seeks to attain a relatively high degree of precision as soon as possible; when they are healthy, he allows himself more time and latitude to discover the precise regimen best suited to the patient. To conclude our discussion, let us examine an intermediate case, that of the elderly. In Galen's opinion, gerontology, the topic of *On Hygiene*'s fifth book, is an especially tricky subject.[62] He tells us that there is some debate over whether old age, though a universal life stage, should actually be considered a kind of a disease.[63] He compares old people to patients convalescing from illness, noting that individuals in both groups occupy a liminal position between sickness and health:[64]

> Both of these dispositions [i.e., the elderly and the convalescent] seem not to be in accordance with strict health, but rather they seem to be in the middle between sickness and health.... Therefore, whether one ought to call old age a "disease," or a "diseased disposition," or a "disposition in between health and disease," or "health in accordance with one's state"... still, one must understand the condition of old bodies, because they can slide into disease due to minor causes, in a manner similar to those convalescing from illness.

The bodies of elderly people are in a precarious position. As their bones become more brittle and their internal systems start to break down, it becomes easier and easier for small shifts in their habits and environments to have an outsized negative impact on their health.[65]

---

[62] On the health regimens of the elderly in antiquity, see Orth 1963; Cokayne 2003, 34–56.

[63] Aristotle takes this position at *Gen. An.* 5.784b32–4, as does Seneca at *Ep.* 108.28–9 (in which he also cites Vergil's *Aen.* 6.275). On the ancient debate over how to characterize old age, see Cockayne 2003: 34–56 (on aging in Galen's *On Hygiene*, see specifically pp. 41–4). Galen also mentions a physician, Philippus of Egypt (100–70 CE), who contributed two volumes to this debate: one on the best regimen for maintaining youth and another, titled *Amazing Agelessness*, which argues that one must begin such a regimen in infancy in order for it to work (Keyser and Irby-Massie 2008, 646).

[64] *San. tu.* 6.330.7–6.331.2 K = 142.23–31 Koch: καὶ δοκοῦσιν αἱ διαθέσεις ἀμφοῖν οὐ κατὰ τὴν ἀκριβεστάτην ὑπάρχειν ὑγείαν, ἀλλ' ἤτοι μέσαι τινὲς εἶναι νόσου τε καὶ ὑγείας... εἴτ' οὖν νόσον εἴτε νοσώδη διὰ θέσιν εἴτε μέσην ὑγείας τε καὶ νόσου διάθεσιν εἴτε κατὰ σχέσιν ὑγείαν ὀνομάζειν χρὴ τὸ γῆρας... ἐπίστασθαι δὲ χρὴ τὴν κατάστασιν τοῦ τῶν γερόντων σώματος, ὅτι ἐπὶ σμικροῖς αἰτίοις εἰς νόσον μεθισταμένη ὁμοίως τοῖς ἀναλαμβάνουσιν ἐκ νόσου τὴν ἀρχαίαν ὑγείαν διαιτᾶν χρή.

[65] This idea is also expressed by the Pneumatist Athenaeus in his work *On Healthy Regimen*: "Old age requires a more exact regimen and additional aids. For the psychic and physical capacities which hold us together and preserve us are quenched, their functions are dissolved, and the body wrinkles and becomes malnourished, loose, and dry. When, therefore, the capacity which keeps the body upright, which offers resistance against external things that cause us injury, and which fights in accordance with certain spermatic principles and natural necessities should fall to its feet, and the body is easily affected and easily injured, it [sc. the body] needs a small cause and chance influence for harm" (Orib. *Coll. med.* [*libri incerti*] 39.15–16 Raeder 1933). Tr. Coughlin n.d.

Galen observes that the bodies of younger people, in contrast, can handle even substantial changes to their circumstances without suffering many ill effects.[66] It may be in part for this reason that in the four preceding books of *On Hygiene*—in which Galen discusses the life stages of infancy, childhood, adolescence, and maturity—he never offers a detailed case history that exemplifies how an individual structures (or should structure) his day. The closest Galen comes, in fact, is in the passage we examined above, wherein Galen shares the times at which he is often (though not always) accustomed to take lunch and visit the gymnasium.[67] The absence of extended case histories seems to suit Galen's emphasis, in these earlier books, on the seasonal and environmental contingency of personal care regimens.[68] The regimen of a teenager living in Alexandria must necessarily be different from that of a fully grown man living in Rome, and the regimen of that man must itself be different in summer than it is in winter. Thus, why waste ink on detailed case histories that could only rarely be used as models?

Galen points out, however, that the frequent regimen adjustments appropriate for the young would be downright dangerous if adopted by the old. Galen proposes that an elderly person should try to adhere to a single, constant regimen throughout the year. He then provides not one, but two detailed examples of personal care regimens that proved successful for different geriatric patients. The first is Antiochus, a fellow physician who managed to keep himself in relatively fine fettle past the age of 80. He accomplished this in the following manner:[69]

> [Every day, Antiochus] walked to the agora from his house, a road up to three stades long.... He had a room in his house that was heated by a furnace in the winter but had temperate air in the summer, even apart from the fire. He had massages in it regularly, both in winter and in summer, *during the mornings*, apparently after having earlier produced a stool. At a place in the agora, around

---

[66] Cf. *San. tu.* 6.331.11–6.332.2 K = 143.10–16 Koch: "For those who are habitually healthy, not even powerful causes change their bodies. But for the elderly, the smallest things produce the greatest disruption. This holds true also with regard to the quantity and quality of food. For in these matters, too, if old men deviate a little bit from what is proper, they are harmed in no small degree, although young men are harmed little by the greatest deviations."

[67] *San. tu.* 6.411.13–6.412.15 K = 181.13–29 Koch.

[68] On the concept of the ordered or regimented body in this and other Imperial-period medical texts, see Ker 2023, 141–50. For other examples of elite Roman's daily schedules, see also Ker 2020.

[69] *San. tu.* 6.332.7–6.333.10 K = 143.21–144.3 Koch: ἀλλ' εἰς μὲν τὴν ἀγορὰν ἀπὸ τῆς οἰκίας ἐβάδιζεν, ὁδὸν ἕως τριῶν σταδίων... ἦν δ' αὐτῷ τι κατὰ τὴν οἰκίαν οἴκημα διὰ καμίνου θερμαινόμενον ἔν γε τῷ χειμῶνι, θέρους δὲ εὔκρατον ἔχον ἀέρα καὶ χωρὶς τοῦ πυρός. ἐν τούτῳ πάντως ἀνετρίβετο καὶ χειμῶνος καὶ θέρους ἕωθεν ἀποπατήσας δηλονότι πρότερον. ἐν δὲ τῷ κατὰ τὴν ἀγορὰν χωρίῳ περὶ τρίτην ὥραν ἢ τὸ μακρότερον περὶ τετάρτην ἤσθιεν ἄρτον μετὰ μέλιτος Ἀττικοῦ, πλειστάκις μὲν ἐφθοῦ, σπανιώτερον δ' ὠμοῦ. καὶ μετὰ ταῦτα τὸ μέν τι συγγιγνόμενος ἑτέροις διὰ λόγων, τὸ δέ τι καθ' ἑαυτὸν ἀναγινώσκων εἰς ἑβδόμην ὥραν παρέτεινε, μεθ' ἣν ἐτρίβετό τε κατὰ τὸ δημόσιον βαλανεῖον ἐγυμνάζετό τε τὰ πρέποντα γέροντι γυμνάσια... κἄπειτα λουσάμενος ἤριστα σύμμετρον, πρῶτα μὲν ὅσα λαπάττει τὴν γαστέρα προσφερόμενος... τούτῳ μὲν οὖν τῷ τρόπῳ γηροκομῶν ἑαυτὸν ὁ Ἀντίοχος ἕως ἐσχάτου διετέλεσεν ἀπήρωτος ταῖς αἰσθήσεσι καὶ τοῖς μέλεσιν ἄρτιος ἅπασι. Further discussion of the role of time management in this passage can be found at Singer 2022, 29–30.

the *third hour* or later around the *fourth hour*, he ate bread with Attic honey, usually boiled, rarely raw. After this, he passed the time until the *seventh hour*, sometimes by entering into debate with other people, at other times by reading on his own. Afterward, he got a rub-down at the public bath-house and performed the exercises appropriate for an old man.... Then, after bathing, he ate an appropriate second meal, in which he set before himself basic things....[70] In this way Antiochos tended himself in old age until he died, with his senses unimpaired and his limbs intact.

In this passage, Galen carefully lists in succession each of the activities that Antiochus engaged in over the course of a typical day: his meals, exercise, massages, even his bowel movements. What is more, Galen is clearly concerned with the timing of these activities, whether absolute (e.g., "until the seventh hour"), general (e.g., "during the morning"), or relative (e.g., "after this"). In fact, after our earlier investigations, we might be inclined to compare the way in which Galen carefully tracks Antiochus's activity cycle over time to the way in which he tracks the condition of fever patients over time in texts like *On Crises*. This is probably no coincidence; if Galen viewed elderly people as being somewhat akin to sick patients, it makes sense that he would subject their conditions to more rigorous observation and analysis.[71]

It should be noted that Galen does not advise every aged person to adopt the *same* strict regimen. Each body—in old age, as in any other—still has its own constitution, needs, tastes, and customs. For this reason, it is sensible that Galen includes two distinct case histories back-to-back, so that no reader could get the unintended impression that he himself should adopt the exact same habits. Telephus the grammarian, the subject of the second case history, apparently surpassed 100 years of age by maintaining a schedule very different from that of Antiochus:[72]

However, Telephus the grammarian reached an even greater age than Antiochus, living almost a hundred years. He was in the habit of bathing twice a month in winter and four times a month in summer. In the seasons between these, he bathed three times a month. On the days he didn't bathe, he was anointed *around the third hour* with a brief massage. Then he used to eat gruel boiled in water mixed with raw honey of the best quality, and this alone was enough for him at

---

[70] For other examples of old men eating light meals, see Suet. *Aug.* 75–7, Sen. *Ep.* 83.6, and Plin. *Ep.* 3.10.

[71] Indeed, Galen himself compares the effects of old age to the effects of a ravaging fever: "For the thing that all men properly call old age is the dry and cold constitution of the body from having existed for many years. But sometimes it also arises from febrile disease, and we call that 'old age from disease,' as I have said in my book *On Marasmus*" (*San. tu.* 6.357.3–7 K = 154.19–23 Koch).

[72] *San. tu.* 6.333.10–6.334.4 K = 144.3–12 Koch.

the first meal. He also dined *at the seventh hour or a little sooner*, taking vegetables first and next tasting fish or birds. In the evening, he used to eat only bread, moistened in wine that had been mixed.

Around the third hour, when Antiochus typically ate a light lunch of bread and honey, Telephus would receive massage. Then, at the seventh hour, the usual time for Antiochus to see the masseur, Telephus would be sitting down to dinner. Each body, like each febrile illness, Galen seems to suggest, has its own natural rhythms. The physician is responsible for identifying these and working harmoniously with them.

But what does all of this mean for our understanding of the relationship between hourly timekeeping and the *kairoi* of the elderly? Shortly before embarking upon these two case histories, Galen makes an important observation about how the nature of *kairos* differs in health and in disease: "For in the case of these things [i.e., those that pertain to hygienic regimen], the *kairos* is not sharp, as in diseases. Rather, it is possible to begin from the surest things in each measure, and then, once you have examined the outcome, reduce or increase a bit each day so as to correct what was omitted."[73] We have seen that, for Galen, the concept of *kairos* is fundamental both in combatting disease and in maintaining relative health. But what Galen has learned is that the *kairos* in each of these contexts has a very different margin of error. In healthy patients, who are unlikely to be affected even by substantial disruptions to their lifestyles, the "right moment" to engage in a particular activity can encompass a wide range of absolute times. The constitutions of sick patients, on the other hand, are so imbalanced and volatile that any minor change can cause their conditions to degenerate rapidly. In these cases, the window of opportunity for stabilizing the patient can be very narrow, which is why, as Galen so often insists, physicians must pay close attention to their patients' conditions and have a plan of action ready in case things go awry. Because elderly people fall somewhere in the middle between sick patients and hale youngsters, the aperture of their kairotic window is similarly intermediate, and hence their behavior must be watched and regimented closely.[74]

It is also worthwhile to note the specific contexts within Galen's case studies (those pertaining to Antiochus, Telephus, and Galen's own habits) where he pins a patient's activity to a particular numbered hour of the day. In all three, numbered hours appear most often in connection with the activities of eating and going to the bath-house or gymnasium. In Chapter 2, we discussed how the chronotopes associated with the bath-house and gymnasium seem to have been oriented to the

---

[73] *San. tu.* 6.326.13–6.327.2 K = 157.13–6 Koch: οὐ γάρ ἐστιν ἐπ᾽ αὐτοῖς ὀξὺς ὁ καιρός, ὡς ἐν ταῖς νόσοις, ἀλλ᾽ ἔνεστιν ἄρξασθαι μὲν ἀπὸ τῶν ἀσφαλεστάτων ἐν ἑκάστῳ μέτρῳ, ἐπισκοπούμενον δὲ τὸ ἀποβαῖνον ἤτοι γ᾽ ἀφαιρεῖν ἢ προστιθέναι τι, καθ᾽ ἑκάστην ἡμέραν ἐπανορθούμενον τὸ παροφθέν.

[74] Galen seems to have been somewhat innovative in recognizing the elderly as a distinct medical class that required special treatment. See Cokayne 2003, 39.

clock. Extant archaeological and papyrological evidence suggest that what we might call the chronotope of the "dining room" was oriented in a similar fashion, with mealtimes often specified to the numbered hour.[75] Some Imperial-period mosaic assemblages, for example, not only illustrate the close connection between dinner time and clocks, but even help us to identify the ninth hour of the day as a conventional dinner time. One of these assemblages (Figure J), from fourth-century CE Daphne in Turkey, depicts a man wearing a toga and looking up at a sundial mounted on a column. Above his head is written "The ninth hour has gone by," which seems to caution the viewer against arriving late to dinner.[76]

Figure J Mosaic depicting a man looking at a sundial. Fourth century CE, Daphne, housed in the Hatay Arkeoloji Müzesi. Image © NPL—De Agostini Picture Library/Bridgeman Images.

---

[75] A line from Aristophanes's play *Assemblywomen* suggests that, even before the widespread use of hourly timekeeping, fifth-century Athenians used the length of their own shadows to determine when it was time to go to dinner (Ar. *Ekk.* 651–2).
[76] Hatay Archaeological Museum, Inv. 865. Another assemblage, found in Tarsos and dated to the late second or early third century CE, tells a story, over three mosaic panels, of a man who is excluded

Another, recently excavated from Syrian Antioch and dated preliminarily to the third century CE, also shows men in togas looking up at a sundial that reads an evening hour (judging by the position of the sun rendered above it). The text labels one figure as "one who hurries to dinner" and the other as "Akairos," the concept of "lateness" or "untimeliness" personified.[77] Above the sundial itself, a small theta is depicted which corresponds, in Greek alphabetic numeral notation, to the number nine and seems to reflect the dinner hour. Dinner invitations from the Imperial period, preserved among the extant papyri from Egypt, corroborate the notion that dinner was typically eaten around the ninth hour. In *P. Coll. Youtie* 1.52, for example, a woman named Herais invites the recipient to dine at the Sarapeion on the eleventh of the month, "beginning from the ninth hour."[78]

This interest in using numbered hours to specify mealtimes may have arisen because meals often require a high degree of interpersonal coordination: hosts need to coordinate the arrival of their guests, and even on less festive evenings, the members of an individual family need to coordinate their appearances at table. However, it is also possible that the dining room chronotope derived its clock-awareness from the common notion (which Galen frequently articulates) that a certain interval must be allowed between bathing and exercising, on the one hand, and eating, on the other. Thus, it may also be that the chronotopes of the bath-house, gymnasium, and dining room developed in mutual response to one another and in active dialogue with medical and wellness practices.

Conclusion

This chapter explored how Galen's interest in hourly timekeeping relates to his understanding of *kairos*, the "right time" to perform a given activity. By examining passages drawn first from *On Crises* and then from *On Hygiene*, we were able to see how Galen's interpretation of *kairoi*, and their relationship to hours, changed depending on whether the patient in question was sick, healthy, or of advanced age. If the patient was sick, Galen used hourly timekeeping to track the progress of the disease and thereby to determine the *kairoi* at which the illness would transition from one stage to the next. If the patient was healthy, Galen still used the time and duration of phenomena as indicators of *kairoi*, but he also considered

---

from a dinner party for arriving late, although the exact time of the dinner is not given. Sundials feature prominently in this assemblage. In the first panel, the tardy dinner guest is shown hurling a rock at a raven perched ominously atop a sundial, and in the third panel it seems (though the panel is badly damaged) that the column with the sundial is toppling down, perhaps upon the guest himself. For discussion, see Dunbabin et al. 2019.

[77] Pamir and Sezgin 2016.

[78] ἀπὸ ὥρας θ. Cf. *P. Oxy.* 62 4339, 75 5056, 75 5057. I am grateful to S. Toralles Tovar for bringing these papyri to my attention.

the time of day and length of daylight as adjustable variables that could act as antecedent causes influencing the timing of *kairoi*. If the patient was elderly, Galen considered him or her to fall in the middle of the spectrum between sickness and health, and therefore cautioned against experimenting too much with that patient's *kairoi*. Furthermore, we saw that the process by which Galen recommended discovering individual *kairoi* involved a synthesis of reason and experience which is familiar to us, both from Plato's description of how to determine *kairoi* in rhetoric and from our earlier forays into other Galenic writings. Thus, once again, Galen's interest in hourly timekeeping appears to have been closely linked to his desire to prove himself a peerless practitioner of correct scientific method.

Finally, we learned that Galen differed from his predecessors, such as Hippocrates, in discriminating between different kinds of *kairoi*: those for the sick, the elderly, and the healthy. Among the sick and elderly, he understood the kairotic window of action to be narrow (more so for the sick, less so for the elderly). In the passage above, Galen referred to the notion, expressed by Plato and the Hippocratics, that a *kairos* is "sharp" (*oxus*). Yet, in the context of healthy patients, Galen seemed to disagree with this characterization. In health, Galen actually understood the kairotic window to be relatively wide, since deviations in environment and regimen are less problematic for the young and hearty than for the old and infirm. Thus, here too, Galen did not simply inherit the ideas of his intellectual heroes but refined them in order to bring them into closer alignment with his own lived experiences.

Conclusion

From Antiquity to Modernity

On the eve of World War I, Dr. John O'Conor, Senior Medical Officer of the British Hospital in Buenos Aires, contributed a letter to the "Correspondence" section of the *British Medical Journal*. In that letter, which was printed on January 31, 1914, and given the title of "Two Years' Operating to the Clock," Dr. O'Conor criticized the lack of time-sensitivity among his fellow surgeons and encouraged them to adopt a different view:[1]

> Speaking recently to Mr. Maclaren, the famous cricketer, I was much interested by hearing him state, "There are more matches lost by not playing to the clock than many people imagine." Might I venture to apply this remark to our "game," and say there are many more lives saved by a smart operation than probably many of us realize?

Dr. O'Conor's letter goes on to stress the importance of speed and clock-awareness for improving surgical outcomes. Between 1914 and 1916, as World War I began and modern artillery and chemical weapons started to send soldiers to the operating table in unprecedented numbers, Dr. O'Conor's argument was seized and amplified by several more surgeons writing in to the *BMJ*. Dr. R. P. Rowlands, for example, surgeon to Guy's Hospital in London, blamed the recent adoption of anesthesia for tricking surgeons into thinking that "time did not matter" in the operating room, and argued that, in treating serious internal conditions, every hour counts.[2]

These epistles testify to an important moment of transition in how certain medical practitioners thought about the roles of clocks and timekeeping in their practice. The authors of these letters recognized that the longer a surgical patient's body was open to the elements, the greater the risk of infection and other adverse effects. They saw in the clock a useful tool for holding surgeons temporally accountable, making sure that they worked with speed and minimized the amount of time their patients spent in vulnerable states. Yet, while the authors of these letters agreed on the importance of operating by the clock, they also acknowledged

---

[1] O'Conor 1914, 231. [2] Rowlands 1916.

CONCLUSION: FROM ANTIQUITY TO MODERNITY 181

another side of the debate. Dr. A. Ernest Maylard, a surgeon from Glasgow writing in 1916, had this to say about performing abdominal surgery with speed:[3]

> So impressive is the result that if only boldness and rapidity be adequately coupled with discretion and precaution the surgeon may be ready to endure the taunt of having "one eye on the clock and the other on the patient," a reflection, however, which, when cast, is usually the expression of ignorance and inexperience.

Dr. Maylard suggests that the phrase "having one eye on the clock" was more commonly used to critique surgeons than to praise them. In this context, the phrase seems to imply that the surgeon is so busy timing his procedures (or perhaps checking to see how much time remains before his lunch hour) that he overlooks critical elements of his patients' care. In response to this kind of critique, Dr. Rowlands hastily assures his readers, "While always conscious of the time element, [the surgeon] must *not* keep one eye on the clock, but both eyes and all his faculties focused on his task."[4] Thus, while the surgeons who wrote these letters emphasized the utility of clocks and hourly timekeeping for surgical success, others among their colleagues seem to have focused on the risks involved in time-conscious surgery: most specifically, the danger that a clock can easily shift from being a tool to being a tyrant, one that exerts an outsized influence on surgeons' decisions and actions and may lead them to neglect other important features of the cases at hand.

The early twentieth-century surgeons considered here lived during a time when mechanical clocks were prolific and comparatively precise. Such clocks could mark and measure units of time much shorter than the hour (including the minute and the second), and the new cultures of work and metrical standardization that arose from the nineteenth-century industrial revolutions had created an acute awareness of clock time in Western communities. Therefore, one might be tempted to assume that debates over the proper roles of clocks in medical contexts first arose in this post-industrial environment, as a peculiar feature of modern Western biomedicine. The present book, however, has argued that forerunners of these medical debates can actually be found over 2,000 years ago, within the works of Galen of Pergamon and his rivals.

Like Drs. O'Conor, Rowlands, and Maylard, Galen and his contemporaries among the Roman elite operated in physical and psychological environments that were structured to a significant degree around clocks and hourly timekeeping. By Galen's day, sizeable sundials and water clocks were available in both public and private spaces, and miniature examples could be found in the equipage of the

---

[3] Maylard 1916.   [4] Rowlands 1916, 550, emphasis added.

lucky few. Members of the Roman elite were accustomed to coordinating certain of their daily activities with the help of these devices, and the popularity of astrology meant that many of them were in the habit of investing with great significance knowledge of the day and hour at which particular events occurred. Sundials and water clocks were recognized as useful tools that facilitated synchronization and standardization among ever growing populations and that allowed mere mortals to identify and gain some predictive power over both terrestrial and celestial cycles.

At the same time, these clocks also bore symbolic weight. In art, architecture, and literature, they could, for example, represent aspects of Roman imperial power, Greek *paideia*, mathematical ingenuity, or the brevity of human existence. We have seen that individual physicians of the Roman period, when faced with decisions about when and how to incorporate these multifunctional, polyvalent devices into their medical theory and practice, considered not only the accessibility of such tools, but also the degree to which clock use aligned with their own ideological commitments and self-presentational goals. In antiquity, as in twentieth- and twenty-first-century modernity, the simple fact that such tools were available did not mean that in every context they were considered *relevant*.

Through Galen's writings—paying particular attention to the ways in which he engaged with the work of other physicians among his contemporaries and predecessors—we have been able to reconstruct some of the diverse perspectives on hourly timekeeping that Greco-Roman physicians held. Galen himself seems to have viewed clocks as tools with great potential. In general, sundials and water clocks allowed physicians to track and model periodic disease behavior with greater precision and (presumed) predictive accuracy. Furthermore, they offered physicians from diverse cultural backgrounds a common, numerical language for expressing temporal information. More specifically, by incorporating hourly timekeeping into his own clinical theories and practices, Galen could provide further support to a series of claims that he liked to make about himself as a physician. For example, by refining Hippocratic critical-day frameworks to account for what I have called critical hours, Galen could simultaneously assert that he was the one true successor of Hippocrates and that Hippocrates, in turn, was the only one among medicine's eligible "founding fathers" to have anticipated the developments of Roman-period medicine. Furthermore, Galen's calculations involving seasonal and equinoctial hours may have helped him to persuade his readers that he had followed Hippocrates's injunction to become familiar with mathematics and astronomy. In *On Hygiene*, for instance, we saw how Galen applied his knowledge of how absolute daylight length varies with latitude in order to more precisely calculate the amount of time that an imperial "servant" would have available for exercise. We also saw, in *On Critical Days*, how Galen used Hipparchus's length of the year, expressed down to fractions of equinoctial hours, to derive the length of a "medical week," a unit that he then used to mathematically demystify a Hippocratic sequence of critical days. This engagement with mathematics also assisted Galen in his efforts to promote medicine as a formal and rigorous *tekhnē* on par with the highly esteemed mathematical sciences.

Galen's use of clocks, both as metaphors and as heuristic devices, also allowed him to present himself as a meticulous practitioner of correct scientific method, that signature blend of logical reasoning and empirical verification. Galen appreciated that one must activate both of these faculties, logic and empiricism, if one is to either construct or use a clock. In building sundials and water clocks, as Galen emphasizes in *Affections and Errors*, one must first apply logical reasoning in order to progress from a set of trigonometric principles and a schematic *analemma* to the creation of a functioning, three-dimensional clock. To ensure the clock tells time accurately, one can then test it against a variety of visual rubrics, asking, for example, whether the clock's lines are proportional, whether its readings are consistent with the sun's movements, and whether it keeps time in synchrony with other clocks. Sundials and water clocks, therefore, can serve as potent metaphors for an apodeictically inflected approach to science. Likewise, clocks *qua* tools can promote a new kind of heuristic thinking by encouraging doctors to ask themselves: "At what hour did such-and-such a medical event occur? How long did it last? Did it recur, and if so, after what interval? And what patterns are created by these temporal data points?" Asking such questions can help a physician to be more precise about his empirical observations. Then, once these "time-stamped" observations have been aggregated and examined for patterns, the physician can begin to develop rational models for predicting future occurrences. Thus, clocks and hours prove to be not only products, but also facilitators of Galen's scientific methodology.

We have encountered several instances, particularly in Galen's fever literature and in his critiques of Methodist principles, where Galen chastises his predecessors and contemporaries for being insufficiently exact in their daily timekeeping. However, texts like *On Periods* and *On Treatment by Venesection* also reveal that Galen did not promote such temporal exactitude across the board. In those texts, we saw him rail against physicians whom he accused of paying too much attention to their clocks and mathematical models, with the result that they consistently failed to address the idiosyncrasies of each patient's condition—an accusation similar to that reported by Dr. Maylard in his letter on surgical timekeeping. Galen, like our early twentieth-century abdominal surgeons, seems to have presented himself as operating in an agonistic environment in which some physicians viewed clock time as essentially irrelevant to their medical practice, while others—inspired, perhaps, by a fascination with mathematical knowledge and the desire to make of medicine a more exact science—relied on clock time to a significant degree. Debates on this topic tended to be provoked by certain questions such as "When is the best time to let a patient's blood?" or "How can one anticipate the arrival and outcome of a paroxysm in irregular periodic fevers?" How individual physicians responded to these questions had to do not only with their views on mathematics, but also with factors like how they defined health and disease, how wide they felt the relevant kairotic windows might be, and the extent to which they viewed each patient as a unique individual rather than as a representative of a general class.

While certain elements of these debates, such as the approved use of bloodletting, have by now gone out of fashion, others should strike the twenty-first-century reader as uncannily similar to current debates within medical communities over, for example, the relationship between quantitative and qualitative research methods[5] or the merits of "precision medicine," which promises to draw upon the resources of genome sequencing, Big Data, and patient participation (via, e.g., the voluntary sharing of mobile health data) to develop models and recommendations that are more holistic and precisely tailored to the individual patient's needs.[6] I suspect that Galen and his contemporaries would have been particularly intrigued by the aspects of precision medicine facilitated by growing interest in chronobiology and its various subdisciplines, such as chronotoxicology, chronopharmacology, and chronotherapy.[7] Contemporary medicine has become particularly curious about what ancient Greek physicians would have called *kairoi*: the right or opportune moments (in this case, within the day) at which a physician or patient can administer a treatment most effectively. However, while Galen often turned to sundials, water clocks, and the hour unit to identify medical *kairoi*, modern-day physicians are especially concerned with how external clock time interacts with the "molecular clocks" within our own bodies. It is currently believed that "between 30 to 50 percent of our genes have activity regulated by circadian rhythms, including those that are part of our immune system."[8] In 2017, the Nobel Prize in Physiology or Medicine was awarded to three circadian rhythm scientists (J. C. Hall, M. Robash, and M. W. Young), which attests to the widespread recognition that this field has recently garnered. At the time of writing, university research teams are currently racing to develop tests that could help patients figure out the settings of their own biological clocks, in the hope that, one day, such tests might become "a standard part of an annual checkup."[9]

---

[5] In response to the emphasis within twentieth-century medicine on quantitative research and analysis, many physicians have begun to advocate for the inclusion and valorization of more qualitative research methods. Qualitative research, as Dr. N. Black explains, "seeks to answer the 'what' question, not the 'how often' one. Thus, rather than adopting a simplified, reductionist view of the subject in order to measure and count the occurrence of states or events, qualitative methods take an holistic perspective which preserves the complexities of human behavior.... The benefits of qualitative methods are greatest when the subject of study cannot be controlled and is poorly defined" (Black 1994, 425).

[6] On precision medicine in the modern-day United States, see, e.g., Hodson 2016; Vegter 2018.

[7] See, e.g., Beauchamp and Labrecque 2007; Smolensky and Peppas 2007; Dallmann et al. 2016. The journal *Advanced Drug Delivery Reviews*, for instance, devoted two whole issues to the subject in 2007.

[8] Rieland 2018.   [9] Rieland 2018.

CONCLUSION: FROM ANTIQUITY TO MODERNITY   185

Galen and his rough contemporaries often represented "clock time" and "body time" as being diametrically opposed to one another. We may recall, for example, how the freeloader in Plautus's comedy *Boeotian Women* lamented the fact that his dinner times were now determined by the sundial rather than by his own belly. Yet, advancements in chronobiology suggest that, through greater knowledge of our own biorhythms, it may be possible to create greater harmony between the clocks on our phones, walls, wrists, and computers and the molecular clocks that make us tick.

This book opened with a series of questions. In order to be effective, to what extent must physicians' measurements of time (and other variables, like length, weight, etc.) be standardized? To what extent should such measurements be articulated using numbers instead of more qualitative descriptors? And what levels of precision and accuracy are required in given cases? Nowadays, it can be easy to assume that the answers to these questions ought to be "as much as possible" and "as high as current technology permits." While this way of conceptualizing timekeeping's role within medical and other sciences is often thought to have its origins in Enlightenment ideals and to have been jumpstarted by the new technologies and mental frameworks generated by the nineteenth century's industrial revolutions, we have now seen that, embedded within this common narrative of medical timekeeping are two unsupported assumptions: first, that modern medical practitioners are and have always been more or less univocal in their responses to those questions, always favoring the quantitative over the qualitative, and the precise over the approximate; second, that interest in and debates concerning those questions arose for the first time in and around the modern period. Instead, we have seen that many of these debates have their roots in antiquity and persist, in new forms, in the present day.

Time-indication and timekeeping practices, as well as our embodied experiences of moving through time, are fundamental ingredients of our worldviews. Yet so much is still unknown about the roles of short-term timekeeping in different cultures, time periods, and social contexts. A central goal of this book has been to illuminate some of the many, complex ways in which temporal concepts and technologies both inform and are informed by our access to various resources, our ideas of how the world works, and the ways in which we personally and culturally construct value. We have seen that some debates—for example, over the relative merits of quantitative and qualitative approaches to information, or over the tension between "body time" and "clock time"—have recurred in various protean forms over the millennia. Others, meanwhile, like those over the timing of venesections or the use of Big Data in chronobiological modeling, are more localized and contingent upon certain technologies and mental frameworks. It is clear, however, that further research into the sociocultural history of short time enriches our understanding not only of the past but also of the present and may help us to more gracefully negotiate future temporal crises induced by new social disruptions, like those caused by wars, rapid technological advancements, or a global pandemic.

Bibliography

Adams, F. 1886. *The Genuine Works of Hippocrates*. New York: William Wood & Co.
Allen, D. 1996. "A Schedule of Boundaries." *Greece & Rome* 43: 157–68.
Allen, J. 2001. *Inference from Signs: Ancient Debates about the Nature of Evidence*. Oxford: Oxford University Press.
Alonge, M. 2011. "Greek Hymns from Performance to Stone." In *Sacred Words: Orality, Literacy and Religion*, ed. A. Lardinois, J. Blok, and M. G. M. van der Poel, 217–34. Mnemosyne Supplements, 332. Leiden: Brill.
Álvarez Millàn, C. 1999. "Graeco-Roman Case Histories and their Influence on Medieval Islamic Clinical Accounts." *Social History of Medicine* 12(1): 19–43.
Álvarez Millán, C. 2010. "The Case History in Medieval Islamic Medical Literature: Tajārib and Mujarrabāt as Source." *Medical History* 54(2): 195–214.
Armisen-Marchetti, M. 1989. *Sapientiae Facies: Étude sur les images de Sénèque*. Collection d'études anciennes. Paris: Les Belles Lettres.
Armisen-Marchetti, M. 1995. "Sénèque et l'appropriation Du Temps." *Latomus* 54(3): 545–67.
Arnaldi, M., and K. Schaldach. 1997. "A Roman Cylinder Dial: Witness to a Forgotten Tradition." *Journal for the History of Astronomy* 28(2): 107–17.
Asper, M. 2013. "Making up Progress—in Ancient Greek Science Writing." In *Writing Science: Medical and Mathematical Authorship in Ancient Greece*, ed. M. Asper, 411–30. Berlin: de Gruyter.
Aveni, A. F. 1989. *Empires of Time: Calendars, Clocks and Cultures*. New York: Basic Books.
Bachtin, Michail M. 2014. *Chronotopos: Aus dem Russischen von M. Dewey, mit einem Nachwort v. M.C. Frank und K. Mahlke*. 3rd ed. Frankfurt am Main: Suhrkamp.
Baiocchi, V., M. Barbarella, M. T. D'Alessio, K. Lelo, and S. Troisi. 2016. "The Sundial of Augustus and its Survey: Unresolved Issues and Possible Solutions." *Acta Geodaetica et Geophysica* 51(3): 527–40.
Balalykin, D. A. 2020. *Galen on Apodictics*. Stuttgart: ibidem.
Barnes, J. 1991. "Galen on Logic and Therapy." In *Galen's Method of Healing: Proceedings of the 1982 Galen Symposium*, ed. V. Nutton and R. J. Durling, 50–102. Leiden: Brill.
Barnes, J. 1993. "Galen and the Utility of Logic." *ZWG Beihefte* 32: 33–52.
Barras, V., T. Birchler, A.-F. Morand, and J. Starobinski, eds. 1995. *L'âme et ses passions: Les passions et les erreurs de l'âme. Les facultés de l'âme suivent les tempéraments du corps*. Paris: Les Belles Lettres.
Barton, T. 1994. *Power and Knowledge: Astrology, Physiognomics, and Medicine under the Roman Empire*. Ann Arbor, MI: University of Michigan Press.
Barton, T. 1995. "Augustus and Capricorn: Astrological Polyvalency and Imperial Rhetoric." *Journal of Roman Studies* 85: 33–51.
Bartoš, H. 2009. "Dietetic Therapy and its Limitations in the Hippocratic *On Regimen*." In *Sokratika: Sebapoznanie a Starost' o Seba*, ed. V. Suvák, 8–16. Prešov: Acta Facultatis Philosophicae Universitatis Prešoviensis.
Bartsch, S., and D. Wray. 2009. *Seneca and the Self*. Cambridge: Cambridge University Press.

Bates, D. G. 1981. "Thomas Willis and the Fevers Literature of the Seventeenth Century." *Medical History* 25(suppl. 1): 45–70.
Baumlin, J. S. 1984. "Decorum, Kairos, and the 'New' Rhetoric." *Pre/Text* 5(3–4): 171–83.
Beauchamp, D., and G. Labrecque. 2007. "Chronobiology and Chronotoxicology of Antibiotics and Aminoglycosides." *Advanced Drug Delivery Reviews* 59(9): 896–903.
Beckby, H. 1965. *Anthologia Graeca, Buch I–VI*. Munich: E. Heimeran.
Behr, C. A., ed. 1969. *Aelius Aristides and the Sacred Tales*. Amsterdam: Hakkert.
Ben-Dov, J., and L. Doering, eds. 2017. *The Construction of Time in Antiquity*. Cambridge: Cambridge University Press.
Berrey, M. 2017. *Hellenistic Science at Court*. Science, Technology, and Medicine in Ancient Cultures 5. Berlin: De Gruyter.
Bickel, S., and R. Gautschy. 2014. "Eine ramessidische Sonnenuhr im Tal der Könige." *Zeitschrift für Ägyptische Sprache und Altertumskunde* 141(1): 3–14.
Bilfinger, G. 1886. *Die Zeitmesser der antiken Völker*. Stuttgart: W. Kohlhammer.
Bilfinger, G. 1888. *Die antiken Stundenangaben*. Stuttgart: W. Kohlhammer.
Black, N. 1994. "Why we Need Qualitative Research." *Journal of Epidemiology and Community Health* 48: 425–6.
Bodel, J. 1997. "Monumental Villas and Villa Monuments." *Journal of Roman Archaeology* 10: 5–35.
Bonnin, J. 2010a. "Les horologia romana en Hispanie, mobilier, histoire et realités archéologiques." *Archivo Español de Arqueología* 83: 183–98.
Bonnin, J. 2010b. "Timekeepers in Britain, 43–780 AD." Tr. T. Wood. *British Sundial Society Bulletin* 22(3): 34–7.
Bonnin, J. 2012a. "Horologia Romana: Cadrans et instruments à eau." *Les Dossiers d'Archéologie* 354: 18–25.
Bonnin, J. 2012b. "Les horloges au quotidien dans l'Antiquité romaine." *Les Dossiers d'Archéologie* 354: 70–5.
Bonnin, J. 2013. "Horologia et memento mori: Les hommes, la mort et le temps dans l'Antiquité gréco-romaine." *Latomus* 72(2): 468–91.
Bonnin, J. 2015. *La mesure du temps dans l'Antiquité*. Paris: Les Belles Lettres.
Bonnin, J., and D. Savoie. 2013. "Report on the Greek Dial from Delos Stored in the Louvre (Ma 4823)." *British Sundial Society Bulletin* 25(1): 20–2.
Bonomi, S. 1984. "Tomba Romana del medico a Este." *Aquileia Nostra* 55, cols 77–107.
Borchardt, Ludwig. 1910. "Altägyptische Sonnenuhren." *Zeitschrift für Ägyptische Sprache und Altertumskunde* 48: 9–17.
Borg, B. 2004a. "Glamorous Intellectuals: Portraits of Pepaideumenoi in the Second and Third Centuries AD." In *Paideia: The World of the Second Sophistic*, ed. B. Borg, 157–78. Millennium Studies 2. Berlin: De Gruyter.
Borg, B., ed. 2004b. *Paideia: The World of the Second Sophistic*. Berlin: De Gruyter.
Bos, G., and Y. T. Langermann. 2015. *The Alexandrian Summaries of Galen's On Critical Days: Editions and Translations of the Two Versions of the Jawami, with an Introduction and Notes*. Leiden: Brill.
Bostock, J., and H. T. Riley, tr. 1855. *Pliny the Elder. The Natural History*. London: Taylor & Francis.
Boudon-Millot, V., ed. 2000. *Galien. Œuvres. Tome II. Exhortation à l'étude de la médecine. Art médical*. Paris: Les Belles Lettres.
Boudon-Millot, V. 2008. "Galien de Pergame, témoin de son temps: l'acculturation de la médecine grecque à la société romaine du IIe siècle de notre ère." *Semitica et Classica* 1: 71–80.

Boudon-Millot, V. 2009. "Galen's Bios and Methodos: From Ways of Life to Paths of Knowledge." In *Galen and the World of Knowledge*, ed. T. Whitmarsh, C. Gill, and J. Wilkins, 175–89. Cambridge: Cambridge University Press.

Bowen, A. C., and B. R. Goldstein. 1988. "Meton of Athens and Astronomy in the Late Fifth Century BC." In *A Scientific Humanist: Studies in Memory of Abraham Sachs*, ed. E. Leichty and M. deJ Ellis, 39–81. Occasional Publications of the Samuel Noah Kramer Fund 9. Philadelphia, PA: University Museum.

Bowen, A. C., and F. Rochberg, eds. 2020. *Hellenistic Astronomy: The Science in its Contexts*. Leiden: Brill.

Brain, P. 1986. *Galen on Bloodletting: A Study of the Origins, Development and Validity of his Opinions, with a Translation of the Three Works*. Cambridge: Cambridge University Press.

Brauneiser, M. 1944. *Tagzeiten und Landschaft im Epos der Griechen und Römer*. Würzburg: Triltsch.

Brown, D., J. Fermor, and C. Walker. 1999. "The Water Clock in Mesopotamia." *Archiv für Orientforschung* 46–7: 130–48.

Brumbaugh, R. S. 1975. *Ancient Greek Gadgets and Machines*. Westport, CT: Greenwood Press.

Buchner, E. 1976a. "Römische Medaillons als Sonnenuhren." *Chiron* 6: 329–46.

Buchner, E. 1976b. "Solarium Augusti und Ara Pacis." *Römische Mitteilungen* 83: 319–65.

Buchner, E. 1982. *Die Sonnenuhr des Augustus*. Mainz: Philipp von Zabern.

Burke, P. F. 1996. "Malaria in the Greco-Roman World: A Historical and Epidemiological Survey." In *Aufstieg und Niedergang der Römischen Welt II*, ed. W. Haase and H. Temporini, 37.3: 2252–81. Berlin: de Gruyter.

Camp, J. M. 1986. *The Athenian Agora: Excavations in the Heart of Classical Athens*. New York: Thames & Hudson.

Campbell-Kelly, M., ed. 2003. *The History of Mathematical Tables: From Sumer to Spreadsheets*. Oxford: Oxford University Press.

Capriglione, J. C. 2000. "La diaita secondo Galeno." *Cuadernos de filología clásica: Estudios griegos e indoeuropeos* 10: 155–72.

Carman, C. C., A. Thorndike, and J. Evans. 2012. "On the Pin-and-Slot Device of the Antikythera Mechanism, with a New Application to the Superior Planets." *Journal for the History of Astronomy* 43(1): 93–116.

Catamo, M., N. Lanciano, K. Locher, M. Lombardero, and M. Valdés. 2000. "Fifteen Further Greco-Roman Sundials from the Mediterranean Area and Sudan." *Journal for the History of Astronomy* 31(3): 203–21.

Chiaradonna, R. 2014. "Galen on What is Persuasive (Pithanon) and What Approximates to Truth." *Bulletin of the Institute of Classical Studies* (supplement): 61–88.

Chiaradonna, R. 2019. "Galen and Middle Platonists on Dialectic and Knowledge." In *Dialectic After Plato and Aristotle*, ed. T. Bénatouïl and K. Ierodiakonou, 320–49. Cambridge: Cambridge University Press.

Clarke, M. L. 1963. "The Architects of Greece and Rome." *Architectural History* 6: 9–22.

Cokayne, K. 2003. *Experiencing Old Age in Ancient Rome*. London: Routledge.

Cooper, G. M. 2004. "Numbers, Prognosis, and Healing: Galen on Medical Theory." *Journal of the Washington Academy of Sciences* 90(2): 45–60.

Cooper, G. M. 2011a. "Astronomy, Medicine, and Galen: The Beginnings of Empirical Science." In *The Traditional Mediterranean: Essays from the Ancient to the Early Modern Era*, ed. J. Che and N. C. J. Pappas, 161–72. Athens: Athens Institute for Education and Research.

Cooper, G. M., ed. 2011b. *Galen, De Diebus Decretoriis, from Greek Into Arabic: A Critical Edition, with Translation and Commentary, of Ḥunayn Ibn Isḥāq, Kitāb Ayyām Al-Buḥrān*. Tr. G. M Cooper. Farnham: Ashgate.
Cotterell, B., F. P. Dickson, and J. Kamminga. 1986. "Ancient Egyptian Water-Clocks: A Reappraisal." *Journal of Archaeological Science* 13(1): 31–50.
Coughlin, S., ed. N.d. "Athenaeus of Attalia: The Complete Fragments with Translation and Commentary." Unpublished draft.
Craik, E. 1995. "Diet, Diaita and Dietetics." In *The Greek World*, ed. A. Powell, 387–402. London: Routledge.
Craik, E. 2018. "The 'Hippocratic Question' and the Nature of the Hippocratic Corpus." In *Cambridge Companion to Hippocrates*, ed. P. Pormann, 25–37. Cambridge: Cambridge University Press.
Cramer, F. H. 1954. *Astrology in Roman Law and Politics*. Philadelphia, PA: American Philosophical Society.
Cumont, F. 1942. *Recherches sur le symbolisme funéraire des Romains*. Paris: Geuthner.
Cuomo, S. 2007. "Measures for an Emperor: Volusius Maecianus' Monetary Pamphlet for Marcus Aurelius." In *Ordering Knowledge in the Roman Empire*, ed. J. König and T. Whitmarsh, 206–28. Cambridge: Cambridge University Press.
Curtis, T. 2014. "Genre and Galen's Philosophical Discourses." In *Philosophical Themes in Galen*, ed. R. Hansberger, P. Adamson, and J. Wilberdings, 39–58. Bulletin of the Institute of Classical Studies Supplement 114. London: Institute of Classical Studies, School of Advanced Study, University of London.
Dallmann, R., A. Okyar, and F. Lévi. 2016. "Dosing-Time Makes the Poison: Circadian Regulation and Pharmacotherapy." *Trends in Molecular Medicine* 22(5): 430–45.
Darbo-Peschanski, C. 2000. *Constructions du temps dans le monde grec ancien*. Paris: CNRS Éditions.
Daumas, François. 1995. *Valeurs phonétiques des signes hiéroglyphiques d'époque gréco-romaine*. 4 vols, vol. 4. Montpellier: Université de Montpellier.
Davis, H. H. 1956. "The Horologium and Symbolism." *Classical Weekly* 49(6): 69–71.
De Lacy, P. H. 1979. "Galen's Concept of Continuity." *Greek, Roman and Byzantine Studies* 20(4): 355–69.
Deichgräber, K. 1971. *Die Epidemien und das Corpus Hippocraticum: Voruntersuchungen zu einer Geschichte der koischen Ärzteschule*. First published 1933 by the Preußischen Akademie der Wissenschaften. Berlin: de Gruyter.
Dewald, C. 2006. "Paying Attention: History as the Development of a Secular Narrative." In *Rethinking Revolutions through Ancient Greece*, ed. S. Goldhill and R. Osborne, 164–82. Cambridge: Cambridge University Press.
Dewald, C. 2007. "The Construction of Meaning in the First Three Historians." In *A Companion to Greek and Roman Historiography*, ed. J. Marincola. Oxford: Blackwell Publishing.
Dicks, D. R. 1955. "The ΚΛΙΜΑΤΑ in the Greek Geography." *Classical Quarterly* 5: 248–55.
Diels, H. 1904. *Laterculi Alexandrini aus einem Papyrus ptolemäischer Zeit*. Berlin: G. Reimer.
Diller, A. 1934. "Geographical Latitudes in Eratosthenes, Hipparchus and Posidonius." *Klio* 27: 258–69.
Dillon, J. M. 1977. *The Middle Platonists: A Study of Platonism, 80 B.C. to A.D. 220*. London: Duckworth.

Dohrn-van Rossum, G. 1996. *History of the Hour: Clocks and Modern Temporal Orders*. Chicago: University of Chicago Press.

Donderer, M. 1996. *Die Architekten der späten römischen Republik und der Kaiserzeit: Epigraphische Zeugnisse*. Erlangen: Universitätsbund Erlangen-Nürnberg.

Donini, P. 1988. "Tipologia degli errori e loro correzione secondo Galeno." In *Le opere psicologiche di Galeno: Atti del terzo colloquio galenico internazionale. Pavia 10–12 settembre 1986*, ed. P. Manuli and M. Vegetti, 10–12. Naples: Bibliopolis.

Downie, J. 2013. *At the Limits of Art: A Literary Study of Aelius Aristides'* Hieroi Logoi. Oxford: Oxford University Press.

Drachmann, A. G. 1963. *The Mechanical Technology of Greek and Roman Antiquity*. Copenhagen: Munksgaard.

Ducrey, P., and H. van Effenterre. 1973. "Un règlement d'époque romaine sur les bains d'Arcades." *Kretika Chronika* 25: 281–90.

Dunbabin, K. 1986. "Sic Erimus Cuncti... The Skeleton in Graeco-Roman Art." *Jahrbuch des deutschen archäologischen Instituts* 101: 185–255.

Dunbabin, K., I. A. Adıbelli, M. Çavuş, and D. Alper. 2019. "The Man Who Came Late to Dinner: A Sundial, a Raven, and a Missed Dinner Party on a Mosaic at Tarsus." *Journal of Roman Archaeology* 32: 329–58.

Edelstein, L. 1967a. "The Dietetics of Antiquity." In *Ancient Medicine: Selected Papers of Ludwig Edelstein*, ed. C. L. Temkin, 303–16. Baltimore, MD: Johns Hopkins University Press.

Edelstein, L. 1967b. *The Idea of Progress in Classical Antiquity*. Baltimore, MD: Johns Hopkins University Press.

Ehrlich, S. 2012. "'Horae' in Roman Funerary Inscriptions." Master's thesis, University of Western Ontario.

Eijk, P. van der. 2012. "Exegesis, Explanation and Epistemology in Galen's Commentaries on Epidemics, Books One and Two." In *Epidemics in Context: Greek Commentaries on Hippocrates in the Arabic Tradition*, ed. P. E. Pormann, 25–48. Berlin: de Gruyter.

Eijk, P. van der. 2014. "An Episode in the Historiography of Malaria in the Ancient World." In *Medicine and Healing in the Ancient Mediterranean*, ed. D. Michaelides, 112–17. Oxford: Oxbow Books.

Elsner, J. 1996. *Art and Text in Roman Culture*. Cambridge: Cambridge University Press.

Elsner, J. 2007. *Roman Eyes: Visuality and Subjectivity in Art and Text*. Princeton: Princeton University Press.

Enos, R. L. 1995. *Roman Rhetoric: Revolution and the Greek Influence*. Prospect Heights, IL: Waveland Press.

Eskin, C. R. 2002. "Hippocrates, Kairos, and Writing in the Sciences." In *Rhetoric and Kairos: Essays in History, Theory, and Praxis*, ed. P. Sipiora and J. S. Baumlin, 97–113. Albany, NY: State University of New York Press.

Evans, J. 1999. "The Material Culture of Greek Astronomy." *Journal for the History of Astronomy* 30(3): 238–307.

Evans, J. 2004. "The Astrologer's Apparatus: A Picture of Professional Practice in Greco-Roman Egypt." *Journal for the History of Astronomy* 35: 1–44.

Evans, J. 2005. "Gnōmonikē Technē." In *The New Astronomy: Opening the Electromagnetic Window and Expanding our View of Planet Earth*, ed. W. Orchiston, 273–92. Dordrecht: Springer.

Ewald, B. C. 1999. *Der Philosoph als Leitbild ikonographische Untersuchungen an römischen Sarkophagreliefs*. Mainz: P. von Zabern.

Feeney, D. C. 2007. *Caesar's Calendar: Ancient Time and the Beginnings of History*. Berkeley, CA: University of California Press.

Feke, J. 2018. *Ptolemy's Philosophy: Mathematics as a Way of Life*. Princeton: Princeton University Press.
Fermor, J., and J. M. Steele. 2000. "The Design of Babylonian Waterclocks: Astronomical and Experimental Evidence." *Centaurus* 42(3): 210–22.
Finley, M. I. 1965. "Technical Innovation and Economic Progress in the Ancient World." *Economic History Review* 18(1): 29–45.
Finzenhagen, U. 1939. *Die geographische Terminologie des Griechischen*. Berlin: Friedrich-Wilhelms-Universitat.
Fıratlı, N., M. Akok, and N. Olcay. 1971. *Izmit Şehri ve Eski Eserleri Rehberi*. Istanbul: Millî Eğitim Basımevi.
Flemming, R. 2007a. "Galen's Imperial Order of Knowledge." In *Ordering Knowledge in the Roman Empire*, ed. J. König and T. Whitmarsh, 241–77. Cambridge: Cambridge University Press.
Flemming, R. 2007b. "Women, Writing and Medicine in the Classical World." *Classical Quarterly* 57(1): 257–79.
Flemming, R. 2013. "Gendering Medical Provision in the Cities of the Roman West." In *Women and the Roman City in the Latin West*, ed. E. Hemelrijk and G. Woolf, 271–93. Leiden: Brill.
Frankfort, H. 1933. *The Cenotaph of Seti I at Abydos*. London: Egypt Exploration Society.
Freeth, T., A. Jones, J. M. Steele, and Y. Bitsakis. 2008. "Calendars with Olympiad Display and Eclipse Prediction on the Antikythera Mechanism." *Nature* 454(7204): 614–17.
Friedrich, H.-V., ed. 1968. *Thessalos von Tralles*. Beiträge zur klassischen Philologie 28. Meisenheim am Glan: Hain.
Frischer, B., and J. Fillwalk. 2013. "A Digital Simulation of the Northern Campus Martius in the Age of Augustus: Preliminary Results of New Studies of the Relationship of the Obelisk, Meridian, and Ara Pacis of Augustus." Delivered at the Vatican Pontifical Academy of Archaeology, Rome.
Frischer, B., J. Pollini, N. Cipolla, G. Capriotti, J. Murray, M. Swetnam-Burland, K. Galinsky, et al. 2017. "New Light on the Relationship of the Montecitorio Obelisk and the Ara Pacis of Augustus." *Studies in Digital Heritage* 1(1): 18–119.
Furtwängler, A. 1900. *Die antiken Gemmen: Geschichte der Steinschneidekunst im klassischen Altertum*. Leipzig: Gesecke & Devrient.
Gaiser, K. 1980. *Das Philosophenmosaik in Neapel: Eine Darstellung der platonischen Akademie*. Heidelberg: C. Winter.
Gamble, J. 2016. "Life in Circadia: The Ticking of the Bodyclock Can Help us Fight Cancer, Safeguard our Hearts, Time our Meals, and Enhance our Intelligence." *Aeon Essays*, June 2.
Garofalo, I. 2003. "Note sui giorni critici in Galeno." In *Rationnel et irrationnel dans la médecine ancienne et médiévale: aspects historiques, scientifiques et culturels*, ed. N. Palmieri, 45–58. Saint-Etienne: Publications de l'Université de Saint-Etienne.
Gellar, M. 2010. *Ancient Babylonian Medicine: Theory and Practice*. Oxford: Wiley-Blackwell.
Gibbs, S. L. 1976. *Greek and Roman Sundials*. New Haven: Yale University Press.
Gleason, M. 2007. *Making Men: Sophists and Self-Presentation in Ancient Rome*. Princeton: Princeton University Press.
Glennie, P., and N. Thrift. 1996. "Reworking EP Thompson's 'Time, Work-Discipline and Industrial Capitalism.'" *Time and Society* 5(3): 275–99.
Goldin, O. 1996. *Explaining an Eclipse: Aristotle's Posterior Analytics 2.1–10*. Ann Arbor, MI: University of Michigan Press.

Gourevitch, D. 1999. "The Paths of Knowledge: Medicine in the Roman World." In *Western Medical Thought from Antiquity to the Middle Ages*, ed. M. D. Grmek and B. Fantini, tr. A. Shugaar, 104-38. Cambridge, MA: Harvard University Press.

Gow, A. S. F., ed. 1965. *Machon: The Fragments*. Cambridge: Cambridge University Press.

Grabar, A. 1967. *L'arte paleocristiana (200-395)*. Paris: Rizzoli.

Graßhoff, G., E. Rinner, K. Schaldach, B. Fritsch, and L. Taub. 2015. "Ancient Sundials." Online database. TOPOI. 2015. https://doi.org/10.17171/1-1.

Gratwick, A. S. 1979. "Sundials, Parasites, and Girls from Boeotia." *Classical Quarterly* NS 29 (02): 308-23.

Greenbaum, D. G. 2020a. "Divination and Decumbiture: Katarchic Astrology and Greek Medicine." In *Divination and Knowledge in Greco-Roman Antiquity*, ed. C. Addey. London and New York: Routledge.

Greenbaum, D. G. 2020b. "Hellenistic Astronomy in Medicine." In *Hellenistic Astronomy: The Science in Its Contexts*, ed. A. C. Bowen and F. Rochberg, 350-80. Leiden: Brill.

Greenbaum, D. G. 2020c. "The Hellenistic Horoscope." In *Hellenistic Astronomy: The Science in its Contexts*, ed. A. C. Bowen and F. Rochberg, 443-71. Leiden: Brill.

Griffith, F. L., and W. M. F. Petrie. 1889. *Two Hieroglyphic Papyri from Tanis*. London: Egypt Exploration Society.

Grimaudo, S. 2008. *Difendere la salute: Igiene e disciplina del sogetto nel* De sanitate tuenda *di Galeno*. Naples: Bibliopolis.

Grmek, M. D. 1989. *Diseases in the Ancient Greek World*. Tr. L. Muellner and M. Muellner. Baltimore, MD: Johns Hopkins University Press.

Hankinson, R. J. 1991a. "Galen on the Foundations of Science." In *Galeno: Obra, Pensamiento e Influencia*, ed. J. A. López Férez, 15-29. Madrid: Universidad Nacional de Educación a Distancia.

Hankinson, R. J. 1991b. *Galen* On the Therapeutic Method *Books I and II*. Oxford: Oxford University Press.

Hankinson, R. J. 1994a. "Galen's Concept of Scientific Progress." In *Aufstieg und Niedergang der römischen Welt II*, ed. W. Haase and H. Temporini, 37.2: 1775-89. Berlin: de Gruyter.

Hankinson, R. J. 1994b. "Usage and Abusage: Galen on Language." In *Language*, ed. S. Everson, 166-87. Companions to Ancient Thought 3. Cambridge: Cambridge University Press.

Hankinson, R. J. 1998. *Galen: On Antecedent Causes*. Cambridge Classical Texts and Commentaries 35. Cambridge: Cambridge University Press.

Hankinson, R. J. 2008a. "Epistemology." In *The Cambridge Companion to Galen*, ed. R. J. Hankinson, 157-83. Cambridge: Cambridge University Press.

Hankinson, R. J. 2008b. "The Man and his Work." In *The Cambridge Companion to Galen*, ed. R. J. Hankinson, 1-33. Cambridge: Cambridge University Press.

Hankinson, R. J. 2009. "Galen on the Limitations of Knowledge." In *Galen and the World of Knowledge*, ed. C. Gill, T. Whitmarsh, and J. Wilkins, 206-42. Cambridge: Cambridge University Press.

Hankinson, R. J. 2022. "Discovery, Method, and Justification: Galen and the Determination of Therapy." In *Galen's Epistemology: Experience, Reason, and Method in Ancient Medicine*, ed. R. J. Hankinson and M. Havrda, 79-115. Cambridge: Cambridge University Press.

Hankinson, R. J., and M. Havrda, eds. 2022. *Galen's Epistemology: Experience, Reason, and Method in Ancient Medicine*. Cambridge: Cambridge University Press.

Hannah, R. 2008. "Timekeeping." In *The Oxford Handbook of Engineering and Technology in the Classical World*, ed. J. P. Oleson, 740–58. Oxford: Oxford University Press.

Hannah, R. 2009. *Time in Antiquity*. London: Routledge.

Hannah, R. 2011. "The Horologium of Augustus as a Sundial." *Journal of Roman Archaeology* 24: 87–95.

Hannah, R. 2020. "The Sundial and the Calendar." In *Hellenistic Astronomy: The Science in Its Contexts*, ed. A. C. Bowen and F. Rochberg, 323–39. Leiden: Brill.

Harkins, P. W., tr. 1963. *Galen on the Passions and Errors of the Soul*. Columbus, OH: Ohio State University Press.

Harper, K. 2017. *The Fate of Rome: Climate, Disease, and the End of an Empire*. Princeton: Princeton University Press.

Harris, W. V., and B. Holmes, eds. 2008. *Aelius Aristides between Greece, Rome, and the Gods*. Leiden: Brill.

Haselberger, L. 2011. "A Debate on the Horologium of Augustus: Controversy and Clarifications." *Journal of Roman Archaeology* 24: 47–73.

Havrda, M. 2022. "From Problems to Demonstrations: Two Case Studies of Galen's Method." In *Galen's Epistemology: Experience, Reason, and Method in Ancient Medicine*, ed. R. J. Hankinson and M. Havrda, 116–35. Cambridge: Cambridge University Press.

Heilen, S. 2018. "Galen's Computation of Medical Weeks: Textual Emendations, Interpretation History, Rhetorical and Mathematical Examinations." *SCIAMVS* 19: 201–79.

Heilen, S. 2020. "Short Time in Greco-Roman Astrology." In *Down to the Hour: Short Time in the Ancient Mediterranean and Near East*, ed. K. J. Miller and S. L. Symons, 239–70. Time, Astronomy, and Calendars. Leiden: Brill.

Heseltine, M., tr. 1987. *Petronius. Satyricon*. Loeb Classical Library 15. Cambridge, MA: Harvard University Press.

Heslin, P. 2007. "Augustus, Domitian and the So-Called Horologium Augusti." *Journal of Roman Studies* 97(1): 1–20.

Himmelmann-Wildschütz, N. 1973. *Typologische Untersuchungen an römischen Sarkophagreliefs des 3. und 4. Jahrhunderts n. Chr.* Mainz: Zabern.

Hintikka, J. 1980. "Aristotelian Axiomatics and Geometrical Axiomatics." In *Theory Change, Ancient Axiomatics, and Galileo's Methodology. Proceedings of the 1978 Pisa Conference on the History and Philosophy of Science*, ed. J. Hintikka, D. Gruender, and E. Agazzi, 1: 133–44. Dordrecht: Springer.

Hodson, R., ed. 2016. *Precision Medicine*. Nature 537(S49), https://doi.org/10.1038/537S49a.

Hoffmann, F. 2016. "Ägyptische astronomische Texte." In *Translating Writings of Early Scholars in the Ancient Near East, Egypt, Rome, and Greece—Methodological Aspects with Examples*, ed. A. Imhausen and T. Pommerening, 335–78. Beiträge zur Altertumskunde 344. Berlin: De Gruyter.

Holmes, B. 2012. "Sympathy between Hippocrates and Galen: The Case of Galen's Commentary on Hippocrates' 'Epidemics', Book Two." In *Epidemics in Context: Greek Commentaries on Hippocrates in the Arabic Tradition*, ed. P. E. Pormann, 49–70. Berlin: de Gruyter.

Honigmann, E. 1929. *Die sieben Klimata und die Poleis epísēmoi: Eine Untersuchung zur Geschichte der Geographie und Astrologie im Altertum und Mittelalter*. Heidelberg: C. Winter.

Hörle, J. 1929. *Catos Hausbücher: Analyse seiner Schrift* De Agricultura *nebst Wiederherstellung seines Kelterhauses und Guthofes*. Paderborn: F. Schöningh.
Horstmanshoff, H. F. and J., M. Stol, ed. 2004. *Magic and Rationality in Ancient Near Eastern and Graeco-Roman Medicine*. Leiden: Brill.
Hübner, W. 2020. "The Professional Ἀστρολόγος." In *Hellenistic Astronomy: The Science and its Contexts*, ed. A. C. Bowen and F. Rochberg, 297–320. Leiden: Brill.
HUJI. 2017. "The Day Unit in Antiquity and the Middle Ages [Conference Website]." <http://ias.huji.ac.il/dayunit>.
Israelowich, I. 2012. *Society, Medicine and Religion in the Sacred Tales of Aelius Aristides*. Leiden: Brill.
Jenzen, I. A., and R. Glasemann. 1989. *Uhrzeiten: Die Geschichte der Uhr und ihres Gebrauchs*. Frankfurt am Main: Historisches Museum Frankfurt.
Johnson, W. A., ed. 2017. *The Oxford Handbook of the Second Sophistic*. Oxford: Oxford University Press.
Johnston, I., and G. H. R. Horsley, tr. 2011. *Galen. Method of Medicine*. Loeb Classical Library 516-18. Cambridge, MA: Harvard University Press.
Jones, A. 1991. "The Adaptation of Babylonian Methods in Greek Numerical Astronomy." *Isis* 82(3): 440–53.
Jones, A., ed. 1999. *Astronomical Papyri from Oxyrhynchus (P. Oxy. 4133-4300a). Edited with Translation and Commentaries*. 2 vols. Philadelphia, PA: American Philosophical Society.
Jones, A. 2003. "The Stoics and the Astronomical Sciences." In *The Cambridge Companion to the Stoics*, ed. B. Inwood, 328–34. Cambridge: Cambridge University Press.
Jones, A. 2009. "Mathematics, Science, and Medicine in the Papyri." In *Oxford Handbook of Papyrology*, ed. R. Bagnall, 338–57. Oxford: Oxford University Press.
Jones, A. 2012. "The Antikythera Mechanism and the Public Face of Greek Science." *Proceedings of Science* 38: 1–22.
Jones, A, ed. 2016. *Time and Cosmos in Greco-Roman Antiquity*. Princeton and New York: Princeton University Press and the Institute for the Study of the Ancient World at New York University.
Jones, A. 2017. *A Portable Cosmos: Revealing the Antikythera Mechanism, Scientific Wonder of the Ancient World*. Oxford: Oxford University Press.
Jones, A. 2020. "Greco-Roman Sundials: Precision and Displacement." In *Down to the Hour: Short Time in the Ancient Mediterranean and Near East*, ed. K. J. Miller and S. L. Symons, 125–57. Time, Astronomy, and Calendars. Leiden: Brill.
Jones, W. H. S. 1909. *Malaria and Greek History*. Manchester: Manchester University Press.
Jones, W. H. S., ed. 1931. *Hippocrates Volume IV: Nature of Man. Regimen in Health. Humours. Aphorisms. Regimen 1-3. Dreams*. Loeb Classical Library 150. Cambridge, MA: Harvard University Press.
Jones, William H.S. 1907. *Malaria: A Neglected Factor in the History of Greece and Rome*. Cambridge: Macmillan & Bowes.
Jouanna, J., tr. 1999. *Hippocrates*. Baltimore, MD: Johns Hopkins University Press.
Jouanna, J. 2012. *Greek Medicine from Hippocrates to Galen: Selected Papers*, ed. P. van der Eijk, tr. N. Allies. Studies in Ancient Medicine 40. Leiden: Brill.
Jouanna, J, ed. 2013. *Hippocrate, Pronostic*. Paris: Les Belles Lettres.
Ker, J. 2009. "Drinking from the Water-Clock: Time and Speech in Imperial Rome." *Arethusa* 42: 279–302.
Ker, J. 2012. *The Deaths of Seneca*. Oxford: Oxford University Press.

Ker, J. 2022. "Diurnal Selves in Ancient Rome." In *Down to the Hour: Short Time in the Ancient Mediterranean and Near East*, ed. K. J. Miller and S. L. Symons, 184–213. Leiden: Brill.

Ker, J. 2023. *The Ordered Day*. Baltimore. MD: Johns Hopkins University Press.

Keyser, P. T. 2013. "The Name and Nature of Science: Authorship in Social and Evolutionary Context." In *Writing Science: Medical and Mathematical Authorship in Ancient Greece*, ed. M. Asper, 17–61. Berlin: de Gruyter.

Keyser, P. T., and G. L. Irby-Massie. 2008. *Encyclopedia of Ancient Natural Scientists: The Greek Tradition and its Many Heirs*. New York: Routledge.

Kieffer, J. S. 1964. *Galen's Institutio Logica: English Translation, Introduction, and Commentary*. Baltimore, MD: Johns Hopkins University Press.

Kim, S. 2017. "Toward a Phenomenology of Time in Ancient Greek Art." In *The Construction of Time in Antiquity*, ed. J. Ben-Dov and L. Doering, 142–172. Cambridge: Cambridge University Press.

King, H., ed. 2005. *Health in Antiquity*. London and New York: Routledge.

King, H. 2019. *Hippocrates Now: The "Father of Medicine" in the Internet Age*. Bloomsbury Studies in Classical Reception. London: Bloomsbury.

Kinneavy, J. L., and C. R. Eskin. 2000. "Kairos in Aristotle's Rhetoric." *Written Communication* 17(3): 432–44.

Knorr, W. 1982. "Techniques of Fractions in Ancient Egypt and Greece." *Historia Mathematica* 9(2): 133–71.

Koch, G., and H. Sichtermann. 1982. *Römische Sarkophage*. Munich: Beck.

Koch, K., G. Helmreich, K. Kalbfleisch, O. Hartlich, and W. John, eds. 1923. *Galeni De sanitate tuenda, De alimentorum facultatibus. De bonis malisque sucis, De victu attenuante. De ptisana*. Corpus medicorum Graecorum 5.4.2. Leipzig: Teubner.

Kondoleon, C. 1999. "Timing Spectacles: Roman Domestic Art and Performance." *Studies in the History of Art* 56: 320–41.

Kosmin, P. J. 2018. *Time and its Adversaries in the Seleucid Empire*. Cambridge, MA: Belknap Press of Harvard University Press.

Krause, M., and V. de Falco, eds. 1966. *Hypsikles: Die Aufgangszeiten der Gestirne*. Göttingen: Vandenhoeck & Ruprecht.

Kucharski, P. 1963. "Sur la notion Pythagoricienne du καιρός." *Revue philosophique de la France et de l'étranger* 153: 141–69.

Kuriyama, S. 1999. *The Expressiveness of the Body*. New York: Zone Books.

Lackeit, C. D. E. 1916. *Aion: Zeit und Ewigkeit in Sprache und Religion der Griechen*. Königsberg: Hartung.

Laín-Entralgo, P. 1987. "El sentido de la díaita en la Grecia Clásica, II." In *Athlon: Satura Grammatica in Honorem Francisci R.z. Adrados*, ed. A. Bernabé, 2: 485–97. Gredos: Editorial Gredos.

Landels, J. G. 1979. "Water-Clocks and Time Measurement in Classical Antiquity." *Endeavour* 3(1): 32–37.

Landes, D. S. 1983. *Revolution in Time: Clocks and the Making of the Modern World*. Cambridge, MA: Belknap Press of Harvard University Press.

Lang, J. 2012. *Mit Wissen geschmückt? Zur bildlichen Rezeption griechischer Dichter und Denker in der römischen Lebenswelt*. Wiesbaden: Reichert.

Langermann, Y. T. 2012. "Critical Notes on a Study of Galen's On Critical Days in Arabic, or a Study in Need of Critical Repairs." *Aestimatio* 9: 220–40.

Langholf, V. 1973. ""Ώρα—Stunde zwei Belege aus dem Anfang des 4. Jh. v. Chr." *Hermes* 101: 382–4.

Langholf, V. 1990. *Medical Theories in Hippocrates: Early Texts and the "Epidemics."* Berlin: de Gruyter.
Lasserre, F., ed. 1966. *Die Fragmente des Eudoxos von Knidos.* Texte und Kommentare 4. Berlin: de Gruyter.
Lehoux, D. 2005. "The Parapegma Fragments from Miletus." *Zeitschrift für Papyrologie und Epigraphik* 152: 125–40.
Lehoux, D. 2007. *Astronomy, Weather, and Calendars in the Ancient World: Parapegmata and Related Texts in Classical and Near-Eastern Societies.* Cambridge: Cambridge University Press.
Leith, D. 2008. "The Diatritus and Therapy in Graeco-Roman Medicine." *Classical Quarterly* 58(2): 581–600.
Leitz, Christian. 2014. *Die Gaumonographien in Edfu und ihre Papyrusvarianten: ein überregionaler Kanon im spätzeitlichen Ägypten: Soubassementstudien III.* 2 vols. Studien zur spätägyptischen Religion 9. Wiesbaden: Harrassowitz.
Levi, D. 1924. "Il kairos attraverso la letteratura greca." *Reconditi della Reale Accademia Nazionale dei Lincei classe di scienzia morali, RV* 32: 260–81.
Lewis, M. 2000. "Theoretical Hydraulics, Automata, and Water Clocks." In *Handbook of Ancient Water Technology*, ed. Ö. Wikander, 343–69. Leiden: Brill.
Lieven, A. von, and A. Schomberg. 2020. "The Ancient Egyptian Water Clock between Religious Significance and Scientific Functionality." In *Down to the Hour: Short Time in the Ancient Mediterranean and Near East*, ed. K. J. Miller and S. L. Symons, 52–89. Time, Astronomy, and Calendars. Leiden: Brill.
Liewert, A. 2015. *Die meteorologische Medizin des Corpus Hippocraticum.* Berlin: de Gruyter.
Lin, J.-L., and H.-S. Yan. 2016. *Decoding the Mechanisms of the Antikythera Astronomical Device.* Berlin: Springer.
Lloyd, G. E. R., ed. 1978. *Hippocratic Writings.* Tr. J. Chadwick, W. N. Mann, I. M. Lonie, and E. T. Withington. Penguin Classics. London: Penguin Books.
Lloyd, G. E. R. 1979. *Magic, Reason, and Experience: Studies in the Origin and Development of Greek Science.* Cambridge: Cambridge University Press.
Lloyd, G. E. R. 1987. *The Revolutions of Wisdom: Studies in the Claims and Practice of Ancient Greek Science.* Berkeley, CA: University of California Press.
Lloyd, G. E. R. 1988. "Scholarship, Authority and Argument in Galen's *Quod Animi Mores.*" In *Le opere psicologiche di Galeno: Atti del terzo colloquio Galenico internazionale, Pavia, 10–12 Settembre 1986*, ed. P. Manuli and M. Vegetti, 11–42. Naples: Bibliopolis.
Lloyd, G. E. R. 1996. "Theories and Practices of Demonstration in Galen." In *Rationality in Greek Thought*, ed. M. Frede and G. Striker, 255–77. Oxford: Oxford University Press.
Lloyd, G. E. R. 2006. "Mathematics as a Model of Method in Galen." In *Principles and Practices in Ancient Greek and Chinese Science*, ed. G. E. R. Lloyd. Burlington, VT: Variorum.
Lloyd, G. E. R. 2008. "Galen and his Contemporaries." In *The Cambridge Companion to Galen*, ed. R. J. Hankinson, 34–48. Cambridge: Cambridge University Press.
Lloyd, G. E. R. 2009. "Galen's Un-Hippocratic Case Histories." In *Galen and the World of Knowledge*, ed. T. Whitmarsh, C. Gill, and J. Wilkins, 115–31. Cambridge: Cambridge University Press.
Long, A. A. 2002. *Epictetus: A Stoic and Socratic Guide to Life.* Oxford: Oxford University Press.
Longhi, V. 2020. *Krisis ou la décision génératrice.* Cahiers de philologie 36. Villeneuve d'Ascq: Presses Universitaires du Septentrion.

Longrigg, J. 1993. *Greek Rational Medicine: Philosophy and Medicine from Alcmaeon to the Alexandrians.* London: Routledge.
Longrigg, J. 2013. *Greek Medicine: From the Heroic to the Hellenistic Age. A Source Book.* London: Routledge.
Lonie, I. M. 1981. "Fever Pathology in the Sixteenth Century: Tradition and Innovation." *Medical History* 25(suppl. 1): 19–44.
Lotito, G. 2001. *Suum esse: Forme dell'interiorità senecana.* Bologna: Pàtron.
Lowrie, M. 2009. *Writing, Performance, and Authority in Augustan Rome.* Oxford: Oxford University Press.
Manetti, D. 2003. "Galeno, la lingua di Ippocrate e il tempo." In *Galien et la philosophie: Huit exposés suivis de discussions,* ed. J. Barnes and J. Jouanna, 171–228. Geneva: Fondation Hardt.
Manetti, D. 2009. "Galen and Hippocratic Medicine: Language and Practice." In *Galen and the World of Knowledge,* ed. C. Gill, T. Whitmarsh, and J. Wilkins, 157–74. Cambridge: Cambridge University Press.
Manetti, D., and A. Roselli. 1994. "Galeno commentatore di Ippocrate." In *Aufstieg und Niedergang der Römischen Welt II,* ed. W. Haase and H. Temporini, 37.2: 1529–1637. Berlin: De Gruyter.
Mansfeld, J. 1991. "The Idea of the Will in Chrysippus, Posidonius, and Galen." In *Proceedings of the Boston Area Colloquium in Ancient Philosophy* 7: 107–45. Leiden: Brill.
Manuli, P. E. 1983. "Lo stile del commento. Galeno e la tradizione ippocratica." In *Formes de pensée dans la collection Hippocratique,* ed. F. Lasserre and P. Mudry, 471–8. Geneva: Droz.
Marrou, H.-I. 1938. *Mousikos anēr: Etude sur les scènes de la vie intellectuelle figurant sur les monuments funéraires romains.* Grenoble: Didier & Richard.
Mattern, S. P. 2008. *Galen and the Rhetoric of Healing.* Baltimore, MD: Johns Hopkins University Press.
Mattern, S. P. 2013. *The Prince of Medicine: Galen in the Roman Empire.* Oxford: Oxford University Press.
Maylard, A. E. 1916. "Time in Surgery." *British Medical Journal,* April 29.
McKirahan, R. D. 1992. *Principles and Proofs: Aristotle's Theory of Demonstrative Science.* Princeton: Princeton University Press.
Miller, C. R. 1992. "Kairos in the Rhetoric of Science." In *A Rhetoric of Doing: Essays on Written Discourse in Honor of James L. Kinneavy,* ed. S. P. Witte, N. Nakadate, and R. D. Cherry, 310–27. Carbondale, IL: Southern Illinois University Press.
Miller, K. J. 2018. "From Critical Days to Critical Hours: Galenic Refinements of Hippocratic Models." *TAPA* 148(1): 111–38.
Miller, K. J. 2020. "Hourly Timekeeping and the Problem of Irregular Fevers." In *Down to the Hour: Short Time in the Ancient Mediterranean and Near East,* ed. K. J. Miller and S. L. Symons, 271–92. Time, Astronomy, and Calendars. Leiden: Brill.
Miller, K. J., and S. L. Symons, eds. 2020. *Down to the Hour: Short Time in the Ancient Mediterranean and Near East.* Time, Astronomy, and Calendars. Leiden: Brill.
Mills, A. A. 1996. "Altitude Sundials for Seasonal and Equal Hours." *Annals of Science* 53 (1): 75–84.
Morison, B. 2008a. "Language." In *The Cambridge Companion to Galen,* ed. R. J. Hankinson, 116–56. Cambridge: Cambridge University Press.
Morison, B. 2008b. "Logic." In *The Cambridge Companion to Galen,* ed. R. J. Hankinson, 66–115. Cambridge: Cambridge University Press.
Müller, W. 1989. *Architekten in der Welt der Antike.* Leipzig: Koehler & Amelang.

Mumford, L. 1934. *Technics and Civilization*. New York: Harcourt, Brace, & Co.
Murray, J. 2021. "Exemplary Biography: Reading Valerius Maximus Writing the Life of Cicero." *Mnemosyne* 76(2): 1–20.
Neugebauer, O. 1975. *A History of Ancient Mathematical Astronomy*. 3 vols. New York: Springer.
Neugebauer, O. 1983. "Astronomical Fragments in Galen's Treatise on Seven-Month Children." In *Astronomy and History: Selected Essays*, 298–300 (first published 1949 in RSO 24: 92–4). New York: Springer.
Neugebauer, O., and H. B. Van Hoesen. 1957. *Greek Horoscopes*. Philadelphia, PA: American Philosophical Society.
Nicolet, C. 1991. *Space, Geography, and Politics in the Early Roman Empire*. Ann Arbor, MI: University of Michigan Press.
Nijf, O. van. 2001. "Local Heroes: Athletics, Festivals and Elite Self-Fashioning in the Roman East." In *Being Greek under Rome: Cultural Identity, the Second Sophistic and the Development of Empire*, ed. S. Goldhill, 309–34. Cambridge: Cambridge University Press.
Nilsson, M. P. 1920. *Primitive Time-Reckoning: A Study in the Origins and First Development of the Art of Counting Time among the Primitive and Early Culture Peoples*. Lund: C. W. K. Gleerup.
Nutton, V. 1992. "Healers in the Medical Market Place: Towards a Social History of Graeco-Roman Medicine." In *Medicine in Society*, ed. A. Wear, 1–58. Cambridge: Cambridge University Press.
Nutton, V. 2008. "The Fortunes of Galen." In *The Cambridge Companion to Galen*, ed. R. J. Hankinson, 355–90. Cambridge: Cambridge University Press.
Nutton, V. 2013. *Ancient Medicine*. 2nd ed. New York: Routledge.
Nutton, V. 2020. *Galen: A Thinking Doctor in Imperial Rome*. Routledge Ancient Biographies. New York: Routledge.
O'Conor, J. 1914. "Two Years' Operating to the Clock." *British Medical Journal*, January 31.
Oestmann, G., H. D. Rutkin, and K. von Stuckrad, eds. 2005. *Horoscopes and Public Spheres: Essays on the History of Astrology*. Religion and Society 42. Berlin: de Gruyter.
Orlandos, A. K., and J. N. Tavlos. 1986. *Lexicon Archaiôn Architektonikôn Horôn*. Athens: Athens Archaeological Association.
Orth, H. 1963. "Diaita γερόντων, die Geriatrie der griechischen Antike." *Centaurus* 8: 19–47.
Pamir, H., and N. Sezgin. 2016. "The Sundial and Convivium Scene on the Mosaic from the Rescue Excavation in a Late Antique House of Antioch." *Adalya* 19: 251–80.
Pampana, E. 1963. *A Textbook of Malaria Eradication*. Oxford: Oxford University Press.
Parrish, D. 1993. "The Mosaic of Aion and the Seasons from Haïdra (Tunisia): An Interpretation of its Meaning and Importance." *Antiquité Tardive* 3: 167–91.
Patterson, R. 2007. "Diagrams, Dialectic, and Mathematical Foundations in Plato." *Apeiron* 40(1): 1–34.
Pearcy, L. T. 1985. "Galen's Pergamum." *Archaeology* 38(6): 33–39.
Pearcy, L. T. 1992. "Diagnosis as Narrative in Ancient Literature." *American Journal of Philology* 113(4): 595–616.
Pérez Cañizares, P. 2002. "Duration of Diseases and Duration of Therapy in Internal Affections." In *Le normal et le pathologique dans la Collection Hippocratique*, ed. A. Thivel and A. Zucker, 551–62. Nice: Publication de la Faculté des Lettres, Arts et Sciences Humaines de Nice-Sophia Antipolis.
Pérez Cañizares, P. 2005. "Special Features in Internal Affections: Comparison to Other Nosological Treatises." In *Hippocrates in Context*, ed. P. van der Eijk, 363–70. Leiden: Brill.

Pfeiffer, R. 1924. *Callimachi fragmenta nuper reperta*. Bonn: Marcus et Weber.
Phillips, J. H. 1983. "The Hippocratic Physician and Astronomy." In *Formes de pensée dans la Collection Hippocratique*, ed. F. Lasserre and P. Mudry, 427–34. Geneva: Droz.
Pomata, G. 2014. "The Medical Case Narrative: Distant Reading of an Epistemic Genre." *Literature and Medicine* 32(1): 1–23.
Potter, P., tr. 2010. *Hippocrates Vol. IX. Coan Prenotions. Anatomical and Minor Clinical Writings*. Loeb Classical Library 509. Cambridge, MA: Harvard University Press.
Poulakos, J., and S. Whitson. 2002. "Kairos in Gorgias' Rhetorical Compositions." In *Rhetoric and Kairos: Essays in History, Theory, and Praxis*, ed. P. Sipiora and J. S. Baumlin, 89–96. Albany, NY: State University of New York Press.
Price, D. J. de Solla. 1969. "Portable Sundials in Antiquity." *Centaurus* 14(1): 242–66.
Rackham, H., tr. 2005. *Cicero. De Natura Deorum. Academica*. Loeb Classical Library 268. Cambridge, MA: Harvard University Press.
Raeder, J. 1933. *Oribasii Collectionum Medicarum Reliquiae*, vol. 4. Corpus Medicorum Graecorum 6.2.2. Leipzig: Teubner.
Ramsay, G. G., ed. 1928. *Juvenal and Persius with an English Translation*. London: William Heinemann.
Reckford, K. J. 1997. "Horatius: The Man and the Hour." *American Journal of Philology* 118 (4): 583–612.
Rehm, A. 1913. "Horologium." In *Paulys Real-Encyclopädie der Classischen Altertumswissenschaft*, ed. A. Pauly and G. Wissowa, 2416–32. Stuttgart: J. B. Metzler.
Remijsen, S. 2007. "The Postal Service and the Hour as a Unit of Time in Antiquity." *Historia: Zeitschrift für alte Geschichte* 56(2): 127–40.
Remijsen, S. 2021. "Living by the Clock: The Introduction of Clock Time in the Greek World." *Klio* 103(1): 1–29.
Remijsen, S. 2023. "Women on Time: Gendered Temporalities in Greco-Roman Egypt." In *The Public Lives of Ancient Women (500 BCE–650 CE)*, ed. L. Dirven, M. Icks, and S. Remijsen, 158–172. Leiden: Brill.
Rieland, R. 2018. "A New Blood Test Can Determine your Biological Clock." *Smithsonian Magazine*, October 4.
Riggsby, A. M. 2009. "For Whom the Clock Drips." *Arethusa* 42(3): 271–78.
Ripat, P. 2011. "Expelling Misconceptions: Astrologers at Rome." *Classical Philology* 106(2): 115–54.
Robert, C. 1897. *Einzelmythen. Die antiken Sarkophag-Reliefs 3/3*. Berlin: Grote.
Rocca, J. 2006. ""Plato will Tell you": Galen's Use of the *Phaedrus* in *De Placitis Hippocratis et Platonis* IX." In *Reading Plato in Antiquity*, ed. H. Tarrant and D. Baltzly, 49–59. London: Bristol Classical Press.
Rochberg, F. 2020a. "Hellenistic Babylonian Astral Divination and Nativities." In *Hellenistic Astronomy: The Science and its Contexts*, ed. A. C. Bowen and F. Rochberg, 472–89. Leiden: Brill.
Rochberg, F. 2020b. "The Babylonian Contribution to Greco-Roman Astronomy." In *Hellenistic Astronomy: The Science in its Contexts*, ed. F. Rochberg and A. C. Bowen, 147–59. Leiden: Brill.
Rochberg-Halton, F. 1989. "Babylonian Seasonal Hours." *Centaurus* 32(2): 146–70.
Rodriguez-Almeida, E. 1978. "Il Campo Marzio Settentrionale: Solarium e Pomerium." *Rendiconti della pontificia accademia di archeologia* 51-2: 195–212.
Roselli, A. 1990. "Some Remarks about the Account of Symptoms in *Diseases II* and *Internal Affections*." In *La maladie et les maladies dans la Collection Hippocratique*, ed. P. Potter, G. Maloney, and J. Desautels, 159–70. Québec: Sphinx.

Rostagni, A. 2002. "A New Chapter in the History of Rhetoric and Sophistry." In *Rhetoric and Kairos: Essays in History Theory and Practice*, ed. P. Sipiora and J. S. Baumlin, tr. P. Sipiora, 23–45. Albany, NY: State University of New York Press.
Rowlands, R. P. 1916. "Time in Surgery." *British Medical Journal*, April 15.
Rumor, M. 2021. "Babylonian Astro-Medicine, Quadruplicities and Pliny the Elder." *Zeitschrift für Assyrologie* 111(1): 47–76.
Rüpke, J. 2011. *The Roman Calendar from Numa to Constantine: Time, History, and the Fasti*. Chichester, UK: Wiley-Blackwell.
Sachs, A. J., and H. Hunger, eds. 1988. *Astronomical Diaries and Related Texts from Babylonia*. 4 vols. Vienna: Österreichishe Akademie der Wissenschaften.
Sallares, R. 2002. *Malaria and Rome: A History of Malaria in Ancient Italy*. Oxford: Oxford University Press.
Samama, E. 2003. *Les médecins dans le monde grec: Sources épigraphiques sur la naissance d'un corps médical*. Hautes études du monde greco-romain 31. Geneva: Droz.
Samuel, A. E. 1972. *Greek and Roman Chronology: Calendars and Years in Classical Antiquity*. Munich: Beck.
Sattler, B. M. 2020. "Cosmology and Ideal Society: The Division of the Day into Hours in Plato's Laws." In *Down to the Hour: Short Time in the Ancient Mediterranean and Near East*, ed. K. J. Miller and S. L. Symons, 158–83. Time, Astronomy, and Calendars. Leiden: Brill.
Sauter, M. J. 2007. "Clockwatchers and Stargazers: Time Discipline in Early Modern Berlin." *American Historical Review* 112(3): 685–709.
Schaldach, K. 1998. *Römische Sonnenuhren: eine Einführung in die antike Gnomonik*. Frankfurt: Deutsch.
Schaldach, K. 2006. *Die antiken Sonnenuhren Griechenlands*. 2 vols. Frankfurt: Deutsch.
Schaldach, K., and O. Feustel. 2013. "The Globe Dial of Prosymna." *BBS Bulletin* 25: 6–12.
Scheidel, W. 2001. *Death on the Nile: Disease and the Demography of Roman Egypt*. Leiden: Brill.
Schlange-Schöningen, H. 2003. *Die römische Gesellschaft bei Galen: Biographie und Sozialgeschichte*. Berlin: de Gruyter.
Schmidt, M. C. P. 1906. *Kulturhistorische Beiträge zur Kenntnis des Griechischen und römischen Altertums: Hft. Die Entstehung der antiken Wasseruhr*. 2 vols. Leipzig: Dürr'sche Buchhandlung.
Schöne, H. 1933. "Galenos' Schrift über die Siebenmonatskinder." *Quellen und Studien zur Geschichte der Naturwissenschaften und Medizin* 3(4): 127–30.
Schütz, M. 1990. "Zur Sonnenuhr des Augustus auf dem Marsfeld: Eine Auseinandersetzung mit E. Buchners Rekonstruktion und seiner Deutung der Ausgrabungsergebnisse, aus der Sicht eines Physikers." *Gymnasium* 97: 432–57.
Scurlock, J., ed. 2014. *Sourcebook for Ancient Mesopotamian Medicine*. Atlanta, GA: SBL Press.
Shcheglov, D. 2003. "Hipparchus' Table of Climata and Ptolemy's Geography." *Orbis Terrarum* 9: 159–92.
Simpson, C. J. 1992. "Unexpected References to the 'Horologium Augusti' at Ovid 'Ars Amatoria' 1, 68 and 3,388." *Athenaeum* 80: 478.
Singer, P. N. 2013. *Galen: Psychological Writings: Avoiding Distress, Character Traits, the Diagnosis and Treatment of the Affections and Errors Peculiar to Each Person's Soul, the Capacities of the Soul Depend on the Mixtures of the Body*. Cambridge: Cambridge University Press.
Singer, P. N. 2022. *Time for the Ancients: Measurement, Theory, Experience*. Berlin: de Gruyter.

Singer, P. N. 2023. *Galen: Writings on Health. Thrasybulus and Health (De Sanitate Tuenda)*. Cambridge: Cambridge University Press.
Sipiora, P. 2002. "Introduction: The Ancient Concept of Kairos." In *Rhetoric and Kairos: Essays in History, Theory, and Praxis*, ed. P. Sipiora and J. S. Baumlin, 1–22. Albany, NY: State University of New York Press.
Sipiora, P., and J. S. Baumlin, eds. 2002. *Rhetoric and Kairos: Essays in History, Theory, and Praxis*. Albany, NY: State University of New York Press.
Slater, W. J. 1971. "Pindar's House." *Greek, Roman and Byzantine Studies* 12(2): 141.
Slater, W. J. 1972. "Simonides' House." *Phoenix* 26(3): 232–40.
Smith, J. E. 2002. "Time and Qualitative Time." In *Rhetoric and Kairos: Essays in History, Theory, and Praxis*, ed. P. Sipiora and J. S. Baumlin, 46–57. Albany, NY: State University of New York Press.
Smith, M. M. 1996. "Time, Slavery and Plantation Capitalism in the Ante-Bellum American South." *Past and Present* February (150): 142–68.
Smith, R. 2009. "Aristotle's Theory of Demonstration." In *The Cambridge Companion to Aristotle*, ed. J. Barnes, 51–65. Cambridge: Cambridge University Press.
Smith, R. R. R. 1999. "Late Antique Portraits in a Public Context: Honorific Statuary at Aphrodisias in Caria, A.D. 300-600." *Journal of Roman Studies* 89: 155–89.
Smith, W. D. 1979. *The Hippocratic Tradition*. Ithaca, NY: Cornell University Press.
Smith, W. D. 1980. "The Development of Classical Dietetic Theory." In *Hippocratica: actes du Colloque Hippocratique de Paris, 4–9 Septembre 1978*, ed. M. D. Grmeks, 1: 439–48. Paris: CNRS.
Smith, W. D. 1981. "Implicit Fever Theory in Epidemics 5 and 7." In *Theories of Fever from Antiquity to the Enlightenment*, ed. W. F. Bynum and V. Nutton, 1–18. London: Wellcome Institute for the History of Medicine.
Smith, W. D., tr. 1994. *Hippocrates Volume VII: Epidemics 2, 4–7*. Loeb Classical Library 477. Cambridge, MA: Harvard University Press.
Smolensky, M. H., and N. A. Peppas. 2007. "Chronobiology, Drug Delivery, and Chronotherapeutics." *Advanced Drug Delivery Reviews* 59(9): 828–51.
Sontheimer, W. 2011. "Tageszeiten." *Paulus Real-Encyclopädie der Classischen Altertumswissenschaft. Neue Bearbaitung. Zweite Reihe [RZ]. Vierter Band: Stoa-Tauris*, 2011–23. Stuttgart: J. B. Metzler.
Sorabji, R. 2006. *Time, Creation, and the Continuum: Theories in Antiquity and the Early Middle Ages*. Chicago, IL: University of Chicago Press.
Staden, H. von. 1975. "Experiment and Experience in Hellenistic Medicine." *Bulletin of the Institute of Classical Studies* 22: 178–99.
Staden, H. von. 1989. *Herophilus: The Art of Medicine in Early Alexandria: Edition, Translation and Essays*. Cambridge: Cambridge University Press.
Staden, H. von. 1997. "Galen and the Second Sophistic." In *Aristotle and After*, ed. R. Sorabji, 33–54. Bulletin of the Institute of Classical Studies, Suppl. 68. London: Institute of Classical Studies, School of Advanced Study, University of London.
Staden, H. von. 2002. "'A Woman does Not Become Ambidextrous': Galen and the Culture of Scientific Commentary." In *The Classical Commentary: Histories, Practices, Theory*, ed. R. K. Gibson and C. S. Kraus, 109–40. Leiden: Brill.
Steele, J. M. 2016. "Near East Relations: Mesopotamia and Egypt." In *Time and Cosmos in Greco-Roman Antiquity*, ed. A. Jones, 45–62. New York and Princeton: Institute for the Study of the Ancient World and Princeton University Press.
Steele, J. M. 2020. "Short Time in Mesopotamia." In *Down to the Hour: Short Time in the Ancient Mediterranean and Near East*, ed. K. J. Miller and S. L. Symons, 90–124. Time, Astronomy, and Calendars. Leiden: Brill.

Stern, S. 2012. *Calendars in Antiquity: Empires, States, and Societies.* Oxford: Oxford University Press.
Sticker, G. 1928. "Fieber und Entzündung bei den Hippokratikern." *Archiv für Geschichte der Medizin* 20: 150–74.
Sticker, G. 1929. "Fieber und Entzündung bei den Hippokratikern." *Archiv für Geschichte der Medizin* 22: 313–43 and 361–81.
Sticker, G. 1930. "Fieber und Entzündung bei den Hippokratikern." *Archiv für Geschichte der Medizin* 23: 140–67.
Storey, C., tr. 2011. *Fragments of Old Comedy. Volume I. Alcaeus to Diocles.* Loeb Classical Library 513. Cambridge, MA: Harvard University Press.
Strohmaier, G. 1997. "La question de l'influence du climat dans la pensée arabe et le nouveau commentaire de Galien sur le traité hippocratique des *Airs, eaux et lieux*." In *Perspectives arabes et médiévales sur la tradition scientifique et philosophique grecque. Actes du colloque e la Société internationale d'histoire des sciences et de la phliosophie arabes et islamiques. Paris, 31 mars–3 avril 1993,* ed. A. Elamrani-Jamal, A. Hasnawi, and M. Aouad, 209–16. Leuven: Peeters.
Strouhal, E., B. Vachala, and H. Vymazalová, eds. 2014. *The Medicine of the Ancient Egyptians I: Surgery, Gynecology, Obstetrics, Pediatrics.* Cairo: American University in Cairo Press.
Strouhal, E., B. Vachala, and H. Vymazalová, eds. 2021. *The Medicine of the Ancient Egyptians 2: Internal Medicine.* Cairo: American University in Cairo Press.
Stutzinger, D. 2001. *Eine römische Wasserauslaufuhr.* Berlin: Kulturstiftung der Länder.
Stylianos, A. 1968. *Guide to the Archaeological Museum of Heraclion.* 5th ed. Archaeological Guides 19. Athens: General Direction of Antiquities and Restoration.
Swain, S. 2008. "Social Stress and Political Pressure: *On Melancholy* in Context." In *Rufus of Ephesus: On Melancholy,* ed. P. E. Pormann, 113–38. Tübingen: Mohr Siebeck.
Symons, S. L. 1998. "Shadow Clocks and Sloping Sundials of the Egyptian New Kingdom and Late Period: Usage, Development, and Structure." *British Sundial Society Bulletin* 98: 30–36.
Symons, S. L. 2016. "Challenges of Interpreting Egyptian Astronomical Texts." In *Translating Writings of Early Scholars in the Ancient Near East, Egypt, Greece and Rome: Methodological Aspects with Examples,* ed. A. Imhausen and T. Pommerening, 379–401. Beiträge Zur Altertumskunde 344. Berlin: de Gruyter.
Symons, S. L. 2020. "Sun and Stars: Astronomical Timekeeping in Ancient Egypt." In *Down to the Hour: Short Time in the Ancient Mediterranean and Near East,* ed. K. J. Miller and S. L. Symons, 14–51. Time, Astronomy, and Calendars. Leiden: Brill.
Symons, S. L., R. Cockcroft, J. Bettencourt, and C. Koykka. 2013. "Ancient Egyptian Astronomy." Online database. 2013. <http://aea.physics.mcmaster.ca>.
Talbert, R. A. 2017. *Roman Portable Sundials: The Empire in your Hand.* Oxford: Oxford University Press.
Talbert, R. A. 2020. "Roman Concern to Know the Hour in Broader Historical Context." In *Homo omnium horarum: Symbolae ad anniversarium septuagesimum Professoris Alexandri Podosinov Dedicatae,* ed. A. Belousov and C. J. Ilyushechkina, 534–55. Moscow: Academia Pozharskiana.
Taylor, B. N., and A. Thompson, eds. 2006. *The International System of Units (SI).* NIST Special Publication 330. Gaithersburg: National Institute of Standards and Technology.
Tecusan, M., ed. 2007. *The Fragments of the Methodists. Methodism Outside Soranus. Volume One: Text and Translation.* Leiden: Brill.
Temkin, O. 1973. *Galenism: Rise and Decline of a Medical Philosophy.* Ithaca, NY: Cornell University Press.

Thalheim, T. 1921. "Klepsydra 2." In *Real-Encyclopädie der klassischen Altertumswissenschaft*, ed. A. Pauly and G. Wissowa, 1: 807–9. Stuttgart: J. B. Metzler.
Thom, J. C. 2001. "Cleanthes, Chrysippus and the Pythagorean Golden Verses." In *Acta Classica: Proceedings of the Classical Association of South Africa*, 44: 197–219. Pretoria: Classical Association of South Africa.
Thomas, E. 2007. *Monumentality and the Roman Empire: Architecture in the Antonine Age*. Oxford: Oxford University Press.
Thompson, E. P. 1967. "Time, Work-Discipline, and Industrial Capitalism." *Past and Present* 38: 56–97.
Thomson, A. 1889. *Suetonius: The Lives of the Twelve Caesars*. Philadelphia, PA: Gebbie & Co.
Thomssen, H. 1994. "Die Medizin des Rufus von Ephesos." In *Aufstieg und Niedergang der römischen Welt II*, ed. W. Haase and H. Temporini, 37.2: 1254–92. Berlin: de Gruyter.
Tieleman, T. 1996. *Galen and Chrysippus on the Soul: Argument and Refutation in the De Placitis Books II–III*. Leiden: Brill.
Tieleman, T. 2008. "Methodology." In *The Cambridge Companion to Galen*, ed. R. J. Hankinson, 49–65. Cambridge: Cambridge University Press.
Tonchéva, G. 1969. "La sculpture dans la Ville d'Odessos du Veme au Ier siècle avant notre ère." *Bulletin du Musée National à Varna* 5(20): 3–47.
Toomer, G. J. 1985. "Galen on the Astronomers and Astrologers." *Archive for the History of Exact Sciences* 32(3): 193–206.
Toomer, G. J. 1988. "Hipparchus and Babylonian Astronomy." In *A Scientific Humanist: Studies in Memory of Abraham Sachs*, ed. E. Leichty, M. deJ. Ellis, and P. Gerardi, 353–62. Occasional Publications of the Samuel Noah Kramer Fund 9. Philadelphia, PA: Samuel Noah Kramer Fund, University Museum.
Totelin, L. M. V., and R. Flemming, eds. 2020. *Medicine and Markets in the Graeco-Roman World and Beyond: Essays on Ancient Medicine in Honour of Vivian Nutton*. Swansea: Classical Press of Wales.
Traversari, G. 1991. "Il 'Pelecinum': Un particolare tipo di orologia solare raffigurato su alcuni rilievi di sarcofagi di età romana." In *Archeologia e astronomia (colloquio internazionale, Venezia 3–6 maggio 1989)*, ed. M. Fano Santi, 66–73. Rome: G. Bretschneider.
Tucci, P. L. 2008. "Galen's Storeroom, Rome's Libraries, and the Fire of AD 192." *Journal of Roman Archaeology* 21: 133–49.
Turcan, R. 1968. "Note sur les sarcophages 'au Prométhée'." *Latomus* 27(3): 630–34.
Turcan, R. 1999. *Messages d'outre-tombe: L'iconographie des sarcophages romains*. Paris: De Boccard.
Turner, A. J. 1989. "Sun-Dials: History and Classification." *History of Science* 27(3): 303–18.
Ullmann, M. 1978. *Rufus of Ephesus. Krankenjournale*. Wiesbaden: Harrassowitz.
Vadan, P. 2018. "Crisis Management and Political Risk in the Hellenistic Age." Doctoral dissertation, University of Chicago.
Van Brummelen, G. 2009. *The Mathematics of the Heavens and the Earth: The Early History of Trigonometry*. Princeton: Princeton University Press.
Van Brummelen, G. 2013. *Heavenly Mathematics: The Forgotten Art of Spherical Trigonometry*. Princeton: Princeton University Press.
Vegter, M. W. 2018. "Towards Precision Medicine: A New Biomedical Cosmology." *Medicine, Health Care and Philosophy* 21: 443–56.
Veyne, P. 1985. "Les saluts aux dieux, le voyage de cette vie et la 'réception' en iconographie." *Revue Archéologique*, NS 1: 47–61.
Wagman, R. 1995. *Inni di Epidauro*. Pisa: Giardini.

Walzer, R. 1935. "Galens Schrift Über die Siebenmonatskinder." *Rivista degli studi orientali* 15: 323–57.
Webb, R. 2016. *Ekphrasis, Imagination and Persuasion in Ancient Rhetorical Theory and Practice*. London: Routledge.
Webster, C. 2015. "Heuristic Medicine: The Methodists and Metalepsis." *Isis* 106(3): 657–68.
Webster, C. 2023. *Tools and the Organism: Technology and the Body in Ancient Greek and Roman Medicine*. Chicago, IL: University of Chicago Press.
Wee, J. Z. 2015. "Case History as Minority Report in the Hippocratic Epidemics." In *Homo Patiens: Approaches to the Patient in the Ancient World*, ed. G. Petridou and C. Thumiger, 138–65. Leiden: Brill.
Wegner, M. 1966. *Die Musensarkophage*. Antiken Sarkophagreliefs. Berlin: Mann.
Wilkins, J. 2007. "Galen and Athenaeus in the Hellenistic Library." In *Ordering Knowledge in the Roman Empire*, ed. J. König and T. Whitmarsh, 69–87. Cambridge: Cambridge University Press.
Wilkins, J. 2015. "Treatment of the Man: Galen's Preventive Medicine in the *De Sanitate Tuenda*." In *Homo Patiens: Approaches to the Patient in the Ancient World*, ed. G. Petridou and C. Thumiger, 413–31. Studies in Ancient Medicine 45. Leiden: Brill.
Wilkinson, R. H. 1999. *Symbol and Magic in Egyptian Art*. London: Thames and Hudson.
Wilson, M. 1987. "Seneca's Epistles to Lucilius: A Reevaluation." *Ramus* 16 (1–2): 102–21.
Wittern, R. 1978. "Zur Krankenheitserkennung in der Knidischen Schrift *De Internis Affectionibus*." In *Medizinische Diagnostik in Geschichte und Gegenwart. Festschrift für Heinz Goerke zum 60. Geburtstag*, ed. C. Habrich, F. Marguth, and J. Hennigwolf, 101–19. Munich: Fritsch.
Wittern, R. 1989. "Die Wechselfieber bei Galen." *History and Philosophy of the Life Sciences* 11: 3–22.
Wöhrle, G. 1990. *Studien zur Theorie der antiken Gesundheitslehre*. Stuttgart: F. Steiner.
Wolfsdorf, D. 2008. *Trials of Reason: Plato and the Crafting of Philosophy*. Oxford: Oxford University Press.
Wolkenhauer, A. 2011. *Sonne und Mond, Kalender und Uhr: Studien zur Darstellung und poetischen Reflexion der Zeitordnung in der römischen Literatur*. Untersuchungen zur antiken Literatur und Geschichte 103. Berlin: de Gruyter.
Wolkenhauer, A. 2018. "Dividio Diei: The Hour in Latin Letters 100 BCE-500 CE." Delivered at the Israel Institute for Advanced Study, Jerusalem.
Wolkenhauer, A. 2020. "Time, Punctuality, and Chronotopes: Concepts and Attitudes Concerning Short Time in Ancient Rome." In *Down to the Hour: Short Time in the Ancient Mediterranean and Near East*, ed. K. J. Miller and S. L. Symons, 214–38. Time, Astronomy, and Calendars. Leiden: Brill.
Wright, M. T. 2000. "Greek and Roman Portable Sundials: An Ancient Essay in Approximation." *Archive for History of Exact Sciences* 55 (2): 177–87.
Yegül, F. K. 2010. *Bathing in the Roman World*. Cambridge: Cambridge University Press.
Yeo, I.-S. 2005. "Hippocrates in the Context of Galen: Galen's Commentary on the Classification of Fevers in *Epidemics* VI." In *Medicine and Philosophy in Classical Antiquity. Doctors and Philosophers on Nature, Soul, Health and Disease*, ed. P. J. van der Eijk, 433–43. Studies in Ancient Medicine 31. Cambridge: Cambridge University Press.
Zaccaria Ruggiu, A. 2006. *Le forme del tempo. Aion, Chronos, Kairos*. Padova: Il Poligrafo.
Zanker, P. 1995. *The Mask of Socrates: The Image of the Intellectual in Antiquity*. Sather Classical Lectures 59. Berkeley, C.A.: University of California Press.

Zerubavel, E. 1979. *Patterns of Time in Hospital Life: A Sociological Perspective*. Chicago, IL: University of Chicago Press.

Zerubavel, E. 1980. "The Benedictine Ethic and the Modern Spirit of Scheduling: On Schedules and Social Organization." *Sociological Inquiry* 50 (2): 157-69.

Zhelezcheva, T., and J. S. Baumlin. 2002. "A Bibliography on Kairos and Related Concepts." In *Rhetoric and Kairos: Essays in History, Theory, and Praxis*, ed. P. Sipiora and J. S. Baumlin, 237-45. Albany, N.Y.: State University of New York Press.

Zulueta, J. de. 1973. "Malaria and Mediterranean History." *Parassitologia* 15: 1-15.

Zuntz, G. 1989. *Aion, Gott des Romerreichs*. Heidelberg: C. Winter.

Zuntz, G. 1991. *Aiōn im Römerreich: die archäologischen Zeugnisse*. Heidelberg: Winter.

Zuntz, G. 1992. *Aiōn in der Literatur der Kaiserzeit*. Vienna: Verlag der Österreichischen Akademie der Wissenschaften.

Index

Note: Tables and figures are indicated by an italic "*t*" and "*f*", respectively, following the page number.

For the benefit of digital users, indexed terms that span two pages (e.g., 52–53) may, on occasion, appear on only one of those pages.

Achaemenid era 39
Adams, F.
  The Genuine Works of Hippocrates 43–4
Aeficianus 62
Aelius Aristides *see* Aristides
Aeneas Tacticus 17–18
Aeschrion 62
Aesculapius (Roman god of healing) 52–3
Affections (Hippocratic text) 43–4
Affections and Errors (Galen) 63, 76–9, 84
  see also Galen of Pergamon (129–216 CE);
    Galen's clock-making paradigm
  clock-making paradigms/examples 63–6, 77, 80–1, 87
    paradeigmata 68–70, 79–80, 89–90, 92, 111
    as a process/model versus finished product 104–5, 108, 111–12, 115
    role of sundials and water clocks 9–10, 63–4, 87, 89–90, 95, 100, 183
    social currency of sundials *see* sundials
  critique by Galen of sectarian philosophers 9–10, 64, 68–70, 74, 76–86, 92
  dating of 64–5
  distinguishing between affections and errors 65
  end of 78, 83–4
  mistakes of judgment 65–6
  progress narrative 82–3
  Stoicism in 100
  types of affections 65
Afterlife *see* death and the Afterlife, clocks as symbols of
afternoons 38–9
  see also time of day
  autumn 152–3
  late 38, 153
Akades, Crete 55–6
Albinus 62
Alexandria, Egypt 30–1, 45–6
  Mousaion of 48
  training of Galen in 62

Aljustrel, Portugal 55–6
Álvarez Millàn, C. 120–2, 124n.42, 125n.49
Amenemhet (nobleman) 28
Ammon at Karnak, temple of 28
Amphiareion of Oropus 45
amulet-makers 5–6
"Anaximander Mosaic" 98–9
ancient Egypt *see* Egypt (ancient)
ancient Mesopotamia *see* Mesopotamia (ancient)
Antiochus 174–8
Antiphanes 23
Antonius the Epicurean
  Galen's criticism of 64–5
Antyllus 152–3
Aphorisms (Hippocratic text) 119–20
Arabic texts 124–6, 129
Ara Pacis monument 90
archaeological remains, evidence from 6, 9, 15–16, 55
Archaic period 21–2
Archimedes 82–3
architectural writing, clocks in 105–8
  buildings reflecting moral dispositions of owners 106–7
  ekphrasis 105–6
  façades, building 105–6
Aristides 52–4
  use of sundials and water clocks 56–7
  on bath-houses 54–5
  Sacred Tales 52–7
Aristophanes 17–18
Aristotle 66–7, 69, 81, 160–1
  Nicomachean Ethics 159–60
  Politics 36
arkhitektonia ("architecture"/"engineering") 61–2, 69–72, 78–9, 83, 85
Armisen-Marchetti, M. 101
Arnaldi, M. 15
Artemidorus 107–8, 107n.77
Asclepius (Greek healing god) 61–2
Asklepieion, Epidauros 45
Asper, M. 82–3

Assyro-Babylonian physicians 8–9, 18–19
  approaches to predictive healing 2
  astral medicine 39–40
  and Hippocratic evidence 40
  hourly timekeeping, emergence in medical contexts 34–40
  incorporation of quantitative timekeeping by astronomers 38–9
astral medicine, Assyro-Babylonian 39–40
astrology 4–5, 145
  astrological view of medicine 58–9
  astrological week 57–8
  catarchic 57–9, 133n.92
  and Galen 59–60
  horoscopic 57–9, 108–11
  in Roman Empire
    Imperial Rome 57–8
    popularity 34–5, 59–60, 108–9, 181–2
  and sundials 58–9, 108–9
  theory 131
astronomy 4–5, 63
  Babylonian astronomy 48
  and fixed-length time-units 38–9
  geometrical 93
  and hours 126–33
  and mathematics 126–7
  period of the moon 126–31
  periods of planets 131–3
  table-making 29–30, 132
Athenaeus of Attalia 152–3
Augustus (Rome's first emperor) 87–91
Aulus Gellius 22
autumn
  afternoons 152–3
  perceived decrease of blood in 151–2

Babylonians
  see also Assyro-Babylonian and Pharaonic Egyptian physicians
  archives 29–30
  astronomers 48
  attention to timing by physicians 38
  hours 20–1
  interaction between medicine and omen-interpretation 37
  precedents, medical timekeeping 35–40
  seasonal hours 29–30
  sundials 25–6, 29
  System B lunar theory 29–30
  water clocks 28–31
bath-houses
  Aristides on 54–5
  chronotope of 55, 176–8

  epigraphic and archaeological evidence from 55
  opening hours 55–6
  proliferation of clocks in 55–7
  when to visit 55, 169
bathing 41–2
  see also bath-houses
  Aristides on 54–5
  and eating 169–70
  separation of female and male bathers 55–6
  timing in relation to meals 55
bearded intellectual, on elite sarcophagi 98–9
Ben-Dov, J. 8–9
Berrey, M. 47–8
bēru (fixed-length time-units) 20–1, 38–9
bile see black bile; yellow bile
biomedicine, twenty-first century 11, 156
black bile 124, 137–8, 151, 164–5
  septic 138
blood
  and kairos 164–5
  perceived to be in excess 149–51, 154
  in spring 151–2
  type 1
  wet and hot qualities 151–2, 164–5
bloodletting 10, 115–16, 125, 135–6, 184
  and fevers 146–54
  optimum time of day for 148–52, 154–5, 183–4
Bonnin, J. 8–9, 23–4, 26, 100n.43
Bonomi, S. 15
Borchardt, L. 28–9
Borg, B. 96, 100nn.42,43
Borvo-Apollo (healing god) 30
Boudon-Millot, V. 76–7
Buchner, E. 89

Caelius Aurelianus 144–5
calendars 8, 39, 57–9
  astrological 58n.62
  civil 89–90
  modern 21–2
  Roman 8n.22, 89–90
  stone fragments of 58–9
Campus Martius assembly ground, Rome
  obelisk 62, 87–92
case histories
  see also critical days; febrile illnesses/fevers; Galen of Pergamon (129–216 ce); Hippocratic authors; Hippocratics (Greek physicians)
  Arabic 124–6
  and critical-day doctrines 125–6
  early Hippocratic 125–6

fever 10, 54, 120–6
  forms 125–6
catarchic astrology 57–9, 133n.92
  see also astrology
Cato the Elder 33–5, 40–1
  On Agriculture 32
celestial bodies 1–2, 17–18, 29, 39–40, 57–8
  planets 131–3
celestial cycles
  and the planets 131–2
celestial geometry 93–6
celestial phenomena, indicating time by 17–18
Celsus 52
China, shadow- and water-based clocks in 7–8
chronobiology 1–2, 11, 184–5
chronocrator (time-ruler) 57–9
chronotopes 34–5, 45, 61
  of bath-house and gymnasium 55, 176–8
  concept 33–4
  of dining room 33–4, 176–8
  of military camps 63
  of performance venues 62
  regimented 63
  of villa 91
chronotype 1
Cicero 106–8
  On the Nature of the Gods 93–4
clarity, concept of 9–10
Classical period 21–2, 34–6
  end of 132
  intellectual achievements 81
  late 2, 9, 18–19, 117, 131–3, 136
Cleomedes 166–7
clock-construction 9–10, 111–12
  see also *Affections and Errors* (Galen); clocks; Galen of Pergamon (129–216 ce); Galen's clock-making paradigm
  clock-making paradigms/examples 63–6, 77, 80–1, 87
  exposure of Galen to processes of 61–2
  lifestyle 111–12
  paradeigmata, in *Affections and Errors* 68–70, 79–80, 89–90, 92, 111
  philosophers compared with clock-makers 84
clocks
  accuracy, level of 7, 26
  in architectural writing 105–8
  in bath-houses and gymnasia 55
  and celestial geometry 93–6
  "clock time" vs. "body time" 23, 185
  design 9–10
  and educated elite 96–100
  evidence for, within Greco-Roman medicine
    see evidence for clocks and hourly timekeeping (Greco-Roman medicine)
  faces 50–2
  Hellenistic Greek 26
  and horoscopic astrology 108–11
  influence on medical theory under Empire 169
  making see clock-construction
  as metaphors 92, 104, 183
  molecular 1
  monumental 50–2
  moral link between clock and individual (Galen) 108
  multivalent 21–2
  origin, development and impact in the West 7
  and philosophers 100–5
  semiology of 87
  shadow-based 7–8
  social history 7–9
  as symbols see symbols, clocks as
  telling the time without 17–19
  use for patient diagnosis 2, 46, 112, 115, 154
  use for patient prognosis 2, 34–5, 46, 112, 115, 154
  water see water clocks
Cnidos (Greek island) 43–4
Column of Antoninus Pius 91
Commodus, Roman Emperor 4
computers, digital timekeeping on 3
concord, concept of 9–10
Cooper, G. 127–8
Cos (Greek island) 43–4
  see also Hippocrates of Cos
Coughlin, S. 152–3
council chambers, proliferation of clocks in 56–7
Craik, E. 43–4
critical days
  see also *On Critical Days* (Galen)
  configurations of 119–20
  construction by Hippocratic authors 119–20, adjustment by Galen 10, 116, 126
  continuity with Galen's temporal framework 125–6
  dominant organizing principle in case histories of *Epidemics* 1 and 3 121
  and hourly timekeeping 116
  patterns 127–8
  primary cycle of one week 121, 128
  secondary cycle of four days 127–8
  theories / schemes / frameworks 10, 116, 119, 121, 133–4, 182–3
  and waxing and waning of the moon 127–8
Cumont, F. 100n.43

daylight
  length of 19, 80n.95, 89, 131–2, 166–71, 178–9, 182–3

daylight (cont.)
  number of hours 21
  period of 131–2, 152
  waxing and waning 89
days
  see also critical days
  divisions of 21
  as microcosms of human life 103–4
  number of in year 128–9
  between paroxysms 136–7
  time of day see time of day
  in week 128–9
daytime 19, 21, 152–3
  midday 154
death and the Afterlife, clocks as symbols of 108–11
decans (asterisms) 19–20
decumbiture (astrology type) 145
deities 5–6
Dewald, C. 77n.81
Diagnostic and Prognostic Series (DPS), The 37–8
diaita (regimen) 40–1
diatritos (therapeutic intervals system), Methodists 36, 144–6
dietary regimes 125
Diodorus Siculus 36
Diphilos 23
diseases
  disease process 161
  febrile see febrile illnesses/fevers
  liver 43–4
  metaphysical interpretations 59–60
  microcosmic life of 161
  non-metaphysical theories 59–60
  perceived influence of zodiac on 39–40
  persistent affection 144
  prediction of outcomes 118–19
  symptoms and affections 135–6, 143
  temporal markers/patterns 53–4, 139, 162
  temporal trajectories 2, 34–8, 40, 122
  documentary papyri, evidence from 6, 58–9, 96–8, 156, 176–8
    astronomical 133n.94
    medical 35–7
    rolls 96, 100
Diseases 2 (Hippocratic text) 43–4
Dohrn-van Rossum, G.
  History of the Hour 7–8
Domitian 91
"double hours" 20–1, 29
  see also hours
dreams, sundial symbolism in 107–8
drugs see medications

egg-timers 17–18
Egypt (ancient)
  see also Assyro-Babylonian and Pharaonic Egyptian physicians
  astrology in 57
  evidence for clocks and hourly timekeeping 15–16
  hour attestations 19–20
  hourly timekeeping, emergence in medical contexts 35–40
  New Kingdom Egypt (c. 1543–1078 bce) 35–6
  Pharaonic periods 19–20, 25, 35–6
  shadow- and water-based clocks in 7–8
  sundials 25, 30–1
  therapeutics 35–6
  water clocks 28–31, 48
Einstein Center, Berlin 8–9
ekphrasis (detailed description of work of art) 105–6
elderly people see older people, and *kairos*
Empiricists 66–7, 116–17
Ephemerides 133
Epictetus 92, 100, 104, 111–12, 152
  see also Seneca; Stoics/Stoic philosophers
  Discourse 103
Epicureans 64, 66–7, 84–5
Epidemics 17–18, 43, 118–22, 124–5, 133
  see also Hippocratic authors
  case histories in 120–1
  dating and authorship 119
  Epidemics 1 119–21, 123, 125–6, 133
  Epidemics 2 44, 119, 129–31, 153
  Epidemics 3 119, 121, 123, 125–6, 133
  Epidemics 4 43–4, 119, 122
  Epidemics 5 119, 122
  Epidemics 6 44, 119
  Epidemics 7 119, 119n.19, 122
  later books 122
  temporal structures 124–6
epigraphic evidence and inscriptions 6, 31, 45, 55–6, 55n.55, 58–9, 90n.5, 91n.12, 168n.50, 169n.52
  inscriptional *parapēgmata* 132n.86
equinoctial hours 19, 21, 129–30, 166–70
  see also seasonal hours
  of astronomy 39
  Babylonian 29–30
  day length 169–71, 171n.57
  fractions of 182–3
Erasistratus 30–1, 148
Esagil-kīn-apli 37
"Este Dial" (portable sundial) 15, 16f, 48–9
Euclid 81
Eudemus 62–3

Euryphon 43–4
evenings 53, 102–3, 125, 152–3, 167–9, 176–8
  see also time of day
  timing of symptoms in 38–9
evidence for clocks and hourly timekeeping
  (Greco-Roman medicine)
  see also medical contexts, emergence of hourly
    timekeeping in
  archaeological 9, 15–16, 55
  epigraphic 55, 58–9
  Hellenistic 45–8
  Hippocratic 40–5
  sundials 9
  types 6
exactitude
  excessive 146–54
  insufficient 136–46
  quantitative 154
  temporal, when desirable 10, 135–55
exercise 41–2, 147, 170, 174–5, 182–3
  gym attendance 168n.50, 170–1, 174, 176–8
  walking 32–3, 41–2, 55n.55

Facundus Novus 88–9, 92
Fates 103, 109
febrile illnesses/fevers
  Aristides's personal experiences, account
    of 53–4
  continuous fevers 138–9
  "crisis," concept of 118–20, 127–9
  "dodecan" fevers 147
  Galen's works on 10, 54, 115–16, 118–21
  growth or abatement 162–3
  importance of timekeeping in 54, 115–16,
    118–19, 135, 157–8, 161–2, 176
  intermittent fevers 135–9, 142
  irregular fevers 137–8, 146
  and *kairos* 157–8, 161–4
  no limit to types 146
  paroxysms *see* paroxysms (fevers)
  and phlebotomy 146–54
  quartan (seventy-two-hour cycle) 138–9
  quotidian (twenty-four-hour cycle) 138–9
  septic fevers 137–8
  stoking of fire 137
  and symptoms 136–46
  temporality in Galen's case histories 54, 120–6
  tertian (forty-eight-hour cycle) 138–9, 142
fixed-length time-units 20–1, 38–9
Franklin, Benjamin 90
friezes 10, 105–6
funerary iconography, clocks and sundials in 50,
  102–3, 107n.78, 109–10
funerary *stēlai* 108–10

Galen of Pergamon (129–216 CE) 61–86
  see also Galen's clock-making paradigm
  and astrology 59–60
  birth/early life 4, 61–2
  case histories, temporality in 6–7, 120–6
  chronotopic biography 61–3
  corpus of 52, 56–7
  criticism of Antonius the Epicurean 64–5
  death of 63
  early career as a physician 62
  education 62
  elite readership 87, 89–90
  Hippocratic doctrines, reinterpreting 42–4,
    55, 82, 116–21, 118n.14, 123–4, 126n.51,
    129–31
    critical-day structures 10, 123, 125–8,
      133–4, 182–3
  Hippocratism of 116–18
  hourly timekeeping, incorporation into
    medical practice 115–18
    fever case histories 10, 54, 115–16,
      118–21, 124
  on lack of standardization 2–3, 135, 145–6
  medical theories 134
  physician to members of imperial family 62–3
  relationship to philosophical
    traditions 66n.28, 68–9
  revisionism of 117–18, 126
  in Rome 62–3
  scientific method *see* scientific method of
    Galen
  self-presentation as true successor of
    Hippocrates 10, 116, 182–3
  texts 4, 52, 182–3
    Affections and Errors see *Affections and
      Errors* (Galen)
    The Best Doctor is also a Philosopher 66–7,
      126–7
    Character Traits 64–5
    On Crises see *On Crises* (Galen)
    On Critical Days see *On Critical Days*
      (Galen)
    On the Distinct Types of Fevers 120–1,
      135–7, 142–3
    On the Eighth-Month Child 129–31
    On Hygiene see *On Hygiene* (Galen)
    On the Method of Healing see *On the
      Method of Healing* (Galen)
    On the Method of Medicine 10
    On Prognosis 120–1
    On the Seventh-Month Child 129–30
    *Against Those Who Have Written on
      (or On Periods)* 135–6, 136n.3, 146–9,
      166, 183–4

Galen of Pergamon (129–216 CE) (*cont.*)
   On Treatment by Venesection 135–6,
     149–51, 154, 166, 183–4
   On Types 135–7
Galen's clock-making paradigm 61–86
   see also Galen of Pergamon (129–216 ce)
   in *Affections and Errors* see *Affections and
     Errors* (Galen)
   and Galen's chronotopic biography 61–3
   Galen's scientific method see scientific method
     of Galen
Gamble, J. 1
gemstones 10, 97f
   Types A–E 96–8
geometry 93–6, 126–7
gerontology 173–8
Gibbs, S. L. 19n.13, 23–4
gnomon (sundial shadow-caster) 15, 23–5, 50–2,
   54, 62, 93
Greco-Roman antiquity
   see also Classical period; clocks; Imperial
     period (Roman); Roman elite, clocks as
     symbols among; Roman period/Roman
     Empire; sundials; telling the time; time
   available temporal frameworks 9
   clocks, telling time without 17–19
   Hellenistic period see Hellenistic period
   hour-units, emergence of 19–23
   lack of formal institutions for medical
     education/accreditation 5
   late Classical period 2
   medical marketplaces 5–6
   sundials in 1–2, 23–6
   water clocks 3–4, 27–30
gymnasium
   chronotope of 55, 176–8
   clocks in 55

Hall, J. C. 184
Hankinson, R. J. 69, 82–3
Hannah, R. 8–9
Havrda, M. 69
healers, in Greco-Roman era
   see also Aesculapius (Roman god of healing),
     Asclepius (Greek god of healing)
   among the Roman elite 6
   medical marketplace 5–6
   metaphysically informed 59–60
health
   see also health regimens; *kairos* (opportune
     moment for medical intervention)
   contingent factors 165–6
   forms in body 165
   maintaining of 164–72
   manifestation of 164–5

Heilen, S. 8–9
Hellenistic period 135, 148
   astrology in 57
   chronotopes 33–4
   early 2, 18–19, 117, 131–3
   emergence of hourly timekeeping 45–8
   Empiricists 116–17
   evidence for hours in medicine 9
   inflow clocks 29
   intellectual achievements 81
   medical timekeeping 34–5
   middle 2
   outflow clocks 29
   sloping sundials, use of 25–6
   tables 132
Heraclianus 62
Herodotus
   Histories 21
Heron of Alexandria 82–3
Herophilus 30–1, 46–8
Hesiod 81
   Hōrai 21–2
   Works and Days 131–2
Heslin, P. 89–90
himation (Greek garment) 96, 98–9
Hipparchus 29–30, 128–30
Hippocrates of Cos 21, 81–2, 128
   see also Hippocratic authors; Hippocratic
     Corpus; Hippocratics (Greek physicians)
   as "Father of Medicine" 4–5, 116–17, 126
   "genuine" texts by 117
   Hippocratic evidence, emergence of hourly
     timekeeping 45–8
   Hippocratic *hōra*, Galenic refinements 115–34
   and *kairos* 159–61
   as a legendary persona 117
   medical supremacy 126
   vindication by Galen 130–1
Hippocratic authors 38, 53, 117–18, 130–1
   see also *Epidemics*
   on fevers/crisis 118–20, 136–7
   inconsistencies, ambiguities and generalities in
     texts 117–18
   and *kairos* 160, 169
   texts 40–3, 116–17
     Affections 43–4
     Aphorisms 158
     Diseases 2 43–4
     Epidemics see *Epidemics*
     Human Nature 151–2
     On Internal Affections 43–4
     On Nourishment 130
     On Seven-Month Children 129–30
     Precepts 158, 160
     Prognostic 136n.3

On Regimen 3 40–3
   writings compared with Galen's temporal narratives 116–18, 123
Hippocratic Corpus 40–1, 43–4, 55, 116–17, 130–1
   additions to 119
Hippocratics (Greek physicians) 2, 39–41, 77
   critical-day schemes 10, 116, 119–21
   development of Hippocratism 116–17
   Hippocratic exegesis 116–17
   and *kairos* 44, 179
   medical timekeeping 34–5
   neo-Hippocratics 42–3, 68
hōra/hōrai, concept of 21–2, 40–1
   as "hour" or "season" 41–3
horoscope, meaning as "hour-watcher" 108–9
horoscopic astrology 57–9, 108
   see also astrology
   and clocks 108–11
hourly timekeeping
   Assyro-Babylonian physicians 37–40
   Pharaonic Egyptian physicians 35–6
   in *On Critical Days* 131, 134
   emergence of in medical context *see* medical contexts, emergence of hourly timekeeping in
   evidence for *see* evidence for clocks and hourly timekeeping (Greco-Roman medicine)
   exposure of Galen to 61–2
   gender divisions, reinforcing 55–6
   Hippocratic texts 42–3
   and *kairos* 115–16, 156–79
   "striking" of the hour 52
   tool availability 33–4
hours
   see also hourly timekeeping
   as antecedent causes of *kairoi* 170–1
   and astronomy *see* astronomy
   Babylonian 20–1
   critical 123–6
   Egyptian 29
   emergence of hourly timekeeping in medical contexts 32–60
   fractions of 102, 129
   Greco-Roman conception of 19
   Hippocratic *hōra*, Galenic refinements see *hōra/hōrai*, concept of
   hour-units, emergence of 19–23
   modern conception of 19
   numbered *see* numbered hours
   research sources 21
   seasonal *see* seasonal hours
   twelve-hour day 21
   twenty-four hour day 21
   in writings of Seneca and Epictetus 103–4
human body
   see also diseases
   activities and experiences, time indicated by 42
   architectural writing, in 105–8
   body time 23, 185
   clocks as symbols of 100–11
   death and the Afterlife 108–11
   diversity of 164–5
   of elderly person 173–4
   humans as "early birds" or "night owls" 1
   lifespan 100–8, 110–11
   medications, processing rate 1
   perceived influence of zodiac on course of diseases 39–40
   personal desires of patient 42
   sequence of patient's physical activities 42
   signs perceptible on or within 42
   timekeeping based on signals of 3
   of younger person 174
humors 124–5, 137n.8, 150–2
   see also black bile; blood; phlegm; yellow bile
   clogging of 137
   in equilibrium 151
   in excess 151
   mixture of 164–5, 172
   qualities of 137–8
   seasons, linked to 151–2
hygiene
   see also health; health regimens; *On Hygiene* (Galen); *kairos* (opportune moment for medical intervention)
   personal 164–5
hypēretēs (servant/attendant of Emperor) 167–8
hyper-indicated time 17

Imperial period (Roman)
   astrology 57–8
   case studies 15–16
   Egyptian and Assyro-Babylonian precedents 36
   link with temporal regulation 91–2
   motifs, incorporating a sundial 98–9
   semi-circular sundials, use of 25
   symbols of 95
   timekeeping and Imperial program 87–92
   urban centers in 2–3, 22, 105–6
inclusive counting technique 127
Industrial Revolution 7
inscriptions *see* epigraphic evidence and inscriptions
intellectual and civic communities, Galen on 70–2, 75–81, 85–6, 92

intellectualism, Greek 96, 100, 111
intellectual role models, of Galen 156-8, 179
International Bureau of Weights and Measures 2-3
iPhones 3

Jones, A. 8-9, 23-4, 93, 108n.79
Julian the Methodist 62
Julius Caesar 89-90
Jupiter 57-8
Juvenal, Augustan-era poetry of 49-52
    caricature 52
    Satire 6, portrayal of a superstitious woman in 57-9

Kahun gynaecological papyrus 35
kairos (opportune moment for medical intervention)
    see also time of day
    antecedent causes 157, 166-7, 170-1, 178-9
    assessment of circumstances 160-1
    and astrology 57-8
    calculating *kairoi* 170-1, 178-9
    concept 156-7, 160-1, 166, 178-9
    and the elderly 157-8, 173-8
    and febrile disease 157-8, 161-4
    Galen on 156-79
        in *On Crises* see *On Crises* (Galen)
        in *On Hygiene* see *On Hygiene* (Galen)
        and predecessors 157-61
    general *kairoi* 161
    and health maintenance 164-72
    and Hippocratics 44, 179
    and hourly timekeeping 115-16, 166-7, 169-70
    identifying 42-3, 166
    and *khronos* 158-60
        in *On Hygiene* 166-7
    massage 171-2, 175
    medicine seen as dependent on 159-60
    perceived by the senses 159
    specific *kairoi* 161, 178-9
    timing of a phase-change 162
Kattan Gribetz, S. 8-9
khronos 158-60, 166-7
Kircher, Athanasius 89
klepsydra see water clocks
klimata (latitudes) 166-7
knowledge
    accurate 65-6
    aggregation of 82
    astronomical 10, 126-8
    communities of 76
    conjectural and insufficiently defensible versus error-free 68-9

experiential 82-3
    of functions of clocks and hours 8-9
    gradual accumulation over time 70
    mathematical 183-4
    medical 37, 143, 148, 154-5
    passing down 70
    privileged 78
    quantitative versus qualitative 1, 6-7, 154, 184-5
    scientific 68-9, 79-80
    shared pool of 78
    sociocultural 111
    specialist 38-9
    technical 8
    testable 84
    theoretical 159
    types of 150
Kosmin, P. 18n.10

lamps 18n.7
Langholf, V. 121
Lang, J. 96-8
Leith, D. 144
liver disease 43-4
Lloyd, G. E. R. 120-1
logic 66-7, 71, 82-3
Lonie, I. M. 137n.8
Lucius Verus 91

Machon (New Comic playwright), fragment of 45-6
magical thinking 5-6
malaria 136
Marcellinus 46
Marcus Aurelius, emperor 62-3, 91
Marinus 116-17
Marrou, H.-I. 100n.43
Mars 57-8
Martial 49-52, 57, 91
massage 171-2, 175
mathematics
    see also geometry; logic
    and astronomy 126-7
    clocks as symbols of mathematical ingenuity 93-6, 111
    formal problems 71
    Greco-Roman "hour" perceived as a mathematical object 19-20
    intellectual circles 63
    mathematical knowledge 183-4
    models 3-4, 39-40
    numbered hours *see* numbered hours
    theories 147-8
Mattern, S. 4

Mausoleum 90
Maylard, A. E. 180-2
mealtimes 58-9, 170-1, 174-6, 178
  and bathing 55
  numbered hours 176-8
medical contexts, emergence of hourly
    timekeeping in 9, 32-60
  see also evidence for clocks and hourly
    timekeeping (Greco-Roman medicine)
  digestive conditions 32-3
  Egyptian and Assyro-Babylonian
    precedents 35-40
  Hellenistic evidence 45-8
  Hippocratic evidence 40-5
  numbered hours 33, 40-1, 43-4, 52-3
  Roman period 48-60
medications
  processing rate 1
medicine
  and astronomical principles 126-7
  rational and irrational practices 5-6
  as an inexact science 3-4
  as a scientific discipline 82-3
medieval period 7-8
Menander 23
Mercury 57-8
Mesopotamia (ancient)
  astrology in 57
  clocks and hourly timekeeping 8-9, 15-16,
    18-19
  hour attestations 19-21
    seasonal hours 21
  shadow- and water-based clocks in 7-8
  sundials 21
  short time in 20-1
  water clocks 23
metaphors
  clocks and hours as 92, 183
  health 164-5
  for ideal lifestyle (Galen) 111-12
  Stoics' use of 101-2, 104
  sundials as 183
  water 101
Methodists 135-6, 154-5
  diatritos (therapeutic intervals system) 36,
    144-6
  Galen's contempt for 66-7, 116-17, 143,
    145-6
molecular clocks 1
Montecitorio obelisk 88*f*
month 28, 47, 66-7, 127-8
  average 128, 130-1
  defining 129
  length of 130-1

seven-month infants 129-30
sidereal 128-9
synodic 128-30
moon
  period of 126-31
  in quartile with a malefic planet 131
  waxing and waning 127-8
mornings 20-1, 32-3, 35-6, 38-9, 42, 174-5
  see also time of day
  early 41-2
  spring-like 152-3
  walking in 42
mosaics 27*f*, 98-9
  Imperial-period assemblage 176-8
  Trier 98n.35
motifs, incorporating a sundial 98-9, 109
Mumford, L. 7
Muses 98-9, 99*f*
Myrina, Greece
  terracotta statuette of grieving slave 50, 51*f*

navigation 159-60
neo-Hippocratics 42-3, 68
night-time
  hours 19, 21, 152-3
  length of 19
Nilsson, M. P. 17
NINDA (fixed-length time-units) 20-1, 38-9
Norse mythology 57-8
numbered hours 17, 19-21, 143, 149, 154
  see also hourly timekeeping; hours;
    mathematics
  in Aristides' *Sacred Tales* 53-4, 56-7
  in Hippocratic Corpus 43
  and *kairos* 156, 176-8
  mealtimes 176-8
  medical contexts, emergence of hourly
    timekeeping in 33, 40-1, 43-4, 48-9, 52-3
  seasonal 47
  in Ullman's cases 125
  and work of Galen 22, 61, 63, 112
    case histories 122, 125
numerical measurement 3-4, 10
  see also mathematics; numbered hours
Numesianus 62, 116-17

Obelisk of Augustus 10, 91
  in Campus Martius assembly ground,
    Rome 62, 87-92
  Montecitorio 88*f*
  as solar meridian 89-90
O'Conor, J. 180-2
older people, and *kairos* 157-8, 173-8
  health regimens 173-6

On Crises (Galen) 120–1, 135–7, 139
  and *kairos* 156–7, 164, 175
On Critical Days (Galen) 115–16, 118–20, 182–3
  book 1 123, 127–8
  book 2 127–8
  book 3 126–8, 131
  on crisis 119–20, 127
  fever treatises 120–1, 123
  hourly timekeeping 131, 134
  lengths of solar and lunar periods 130
  revisionist technique 117–18, 126
  temporal structures 124–6
On Hygiene (Galen)
  daylight length 182–3
  eating and bathing relationship 169–70
  elderly, care for 173–4
  and hourly timekeeping 115–16, 166–7, 172
  and *kairos* 10–11, 156–79
On Internal Affections (Hippocratic text) 43–4
On Regimen 3 (Hippocratic text) 40–3
On the African War (unknown author) 22
On the Diagnosis and Treatment of the Affections and Errors of the Individual Human Soul (Galen) see *Affections and Errors* (Galen)
On the Method of Healing (Galen) 82, 126–7, 135–6, 143, 145–6
On the Spanish War (unknown author) 22
oracle-mongers 5–6
Oribasius 152–3

Pacuvius 102
paideia (Greek-inflected education) 96, 100, 111
Palazzina Capodaglio, Este (Italy) 15
papyrus/papyri 22, 35, 46–7, 58–9, 76–7
  evidence from 6, 22, 35, 58–9, 133
  scrolls, depictions of 96–8, 99*f*
parapēgma/ parapēgmata (calendar type) 108n.79, 132, 132nn.86,89
paroxysms (fevers) 119–20, 123–4, 128, 136nn.2,3, 137, 148–9, 154, 161–2
  cycles of 138–40, 141*t*, 142, 146–7, 163–4
  days between 136–7
  duration 142, 147, 162
  fixed and moving types 138, 140
  overlapping of paroxysmal periods 140
  and remission periods 138, 142
  shivering 163–4
  strong 163–4
Pausanias
  Description of Greece 105–6
pelecinum dials 26, 27*f*
Pelops 62

pepaideumenoi (educated Roman elite) 96–100, 116
Pergamon
  see also Galen of Pergamon (129–216 ce)
  healing sanctuary in 61–2
  sundials and water clocks in 61–2
Peripatetics 64, 66–7, 84–5
personal phenomena, indicating time by 17–18
Petronius 102
  Satyricon 49, 102 n.54
philosophers
  on clocks as symbols of human lifespan 100–8
  compared with clock-makers 84
  critique by Galen of 9–10, 64, 68–70, 74, 76–86, 92
Philoxenos 45–6
phlebotomy see bloodletting
phlegm 43–4, 137–8, 151–2, 164–5
  see also Humors
planets 57–8
  days of 59
  malefic 131
  periods of 131–3
Plato 81, 158–61, 178–9
  Academy 98–9
  Laws 21
  Timaeus 106–7
Platonists 66–7
Plautus (Roman comic poet)
  The Boeotian Women 22–3, 185
Pliny 87–90, 92, 166–7
  Natural History 22
Pneumatists 116–17, 152–4
Pontifex Maximus 89–90
portable sundials 3, 49, 62–3, 166–7
  "Este Dial," Italy 15, 16*f*, 48–9
portrait busts 96
prayers 5–6
Proclus 82–3
progress and perpetuity 79–83
propositional logic 66–7
Ptolemaic postal system 22
Ptolemy 64, 166–7
punctuality 56–7
Pythagoras 158–9

Quintus 116–17

Rationalists 66–7, 116–17
rationality/rationalism
  see also Rationalists
  and empiricism 66–7, 87
  rational and irrational practices in medicine 5–6

regimen 40–3, 179
  see also bathing, diaita
  dietary 125
  of the elderly 173–6
  and maintenance of health 164–8, 170–1
  personal care 174
Remijsen, S. 8–9
rhetoric 42–3, 50, 105–6
  forms 159
  and kairos 158–61, 178–9
rites 5–6
rituals 19–20
ritual specialists 5–6
Robash, M. 184
Rodríguez-Almeida, E. 89
Roman elite, clocks as symbols among 87–112, 181–2
  see also Roman period/Roman Empire
  clock-construction 9–10, 111–12
  educated elite 96–100
  funerary iconography 50
  male elite patients, experience of 52–3
  pepaideumenoi, self-representation of elites as 96–100, 116
  symbols of death and the afterlife 108–11
  symbols of mathematical ingenuity 93–6, 111
  symbols of the human body 105–8
  symbols of the human lifespan 100–8, 110–11
  timekeeping and Imperial program 87–92
Roman period/Roman Empire
  see also Greco-Roman antiquity
  astrology in see astrology
  early Rome 9
  elite, clocks as symbols among see Roman elite, clocks as symbols among
  emergence of hourly timekeeping 48–60
  healers in see healers, in Greco-Roman era
  Imperial period see Imperial period (Roman)
  increase in standardization of timekeeping devices 2
  recognition of need for standardization 2–3
  schools of medical thought 4–5
  slaves in see slavery
  sundials and water clocks in see sundials; water clocks
Rome
  see also Campus Martius assembly ground, Rome
  Palazzo Massimo 109
  "typical" day in 91
  visit of Galen to 62–3
root-cutters 5–6
Rowlands, R. P. 180–2

Rufus of Ephesus 52, 124–5
Rumor, M. 39

sarcophagi (coffins) 10, 107n.78, 109–10
  of elite 98–9
  motifs on 109
  muse sarcophagus 99f
  Portonaccio Sarcophagus, lid of 109
  Prometheus sarcophagus 109, 110f
  Type I classification and scenes 98n.39
  Type II classification and scenes 98n.40
  Type III classification and scenes 99n.41
Saturn 57–8
Satyrus 62
Schaldach, K. 15, 23–4
Schomberg, A. 8–9, 28–9
Schütz, M. 89–90
science
  Galen's scientific method see scientific method of Galen
  immortality of scientific reasoning process 110–11
  knowledge, scientific 68–9, 79–80
  language 143
  medicine as a scientific discipline 82–3
  progress, scientific 9–10
scientific method of Galen 9–10, 66–71, 87, 111
  see also Affections and Errors (Galen); clock-construction; Galen of Pergamon (129-216 CE); Galen's clock-making paradigm; symbols, clocks as
  and critique of sectarian philosophers 9–10, 64, 68–70, 74, 76–86, 110–11
  and kairos 159
  logical reasoning combined with observational experience 67–8, 111–12, 170–1, 183
  use of symbols 112
Scurlock, J. 37–8
seasonal hours 19–22, 57–8
  see also equinoctial hours
  Galen on 75
  in On Hygiene 166–7
  numbered 47
seasons 40–1, 152, 160n.19, 163n.32, 169, 175–6
  see also autumn; spring; summer; winter
  humors, linked to 151–2
  solar and lunar 128
seconds
  defining 2–3
Seleucid Empire 18n.10
Seneca 92, 100, 102, 111–12, 152
  see also Epictetus; Stoics/Stoic philosophers
  Epistles to Lucilium 101, 103
Septimius Severus 148

seven-day week 57–9, 127
short time, indicating 17–18, 20–1
sign almanacs 133
Siminius Stephanus, T. 98–9
Skeptics 66–7
slavery
  in the American South 54n.51
  domestic slaves 49–50
  Galen on 165, 168n.50
  public slaves 49–52
  responsibility for reading/announcing time of day 50
  in Roman period 48–50
  statuette of grieving slave with sundial (Necropolis, Myrina) 50, 51*f*
sleep 41–2
Smith, R. R. R. 100n.42
Smith, W. D. 116–17
Smyrna, training of Galen in 62
social phenomena, indicating time by 17–18
Soranus 52, 144
soul
  affections of 64–5
  Aristotle on 69
  forms 158–9
  Galen on 81
  location within the body 106–7
  Plato on 106–7
  rational and irrational parts 65
spring 171–2
  see also autumn; seasons; summer; winter
  blood in 151–2
  Greek sources 21–2
  treatment provided in 150–2
  warm and hot qualities 152–4
standardization
  Galen on lack of 2–3
  increased in the Roman Empire 2–3
  of medicalized time 1–3
  in names of plants and units of measurement 2–3
  recognition of need for 2–3
Steele, J. 8–9
Stern, S. 8–9
Stoics/Stoic philosophers 64, 66–7, 92,
  see also Epictetus; Seneca
  in *Affections and Errors* (Galen) 84–5, 100
  ancient and more recent 147–8
  engagement with clocks and hours 100
  symbolic language in writings of 100–2, 104, 111–12
  sympathetic concept 118n.13
stone inscriptions, evidence from 6
Strabo 166–7

Stratonicus 62
Strouhal, E. 35–6
Suetonius 90
summer
  see also autumn; seasonal hours; seasons; spring; winter
  associated diseases 150–2
  Greek sources 21–2
  health regimens 174–6
  solstice 24*f*, 78n.85, 89, 168–9
  and time of day 152–3
sundials 1–4, 23–6, 61–2
  see also water clocks
  in *Affections and Errors* (Galen) 9–10
  archaeological evidence 6, 9, 15–16
  and architectural writing 105–6
  and astrology 58–9, 108–9
  Babylonian 25–6, 29
  compared with water clocks 27–8
  concave 25–6
  conical 26, 50
  cylindrical 26
  date-lines 89, 111–12
  day-curves 24*f*
  designs 93–4
  development of 18–19
  domestic 3
  dreams, symbolism in 107–8
  in Egypt 25
  fixed 26
  in funerary iconography 50, 102–3, 107n.78, 109–10
  Galen's specific mention of 63–4
  gnomon (shadow caster) 15, 23–5, 50–2, 54, 62, 93
  Greco-Roman style 1–2, 23–6
  hemispherical 26
  horizontal planar 26
  hour-lines 24*f*, 89, 111–12
  monumental 3, 62
  motifs incorporating 98–9
  and omens/portents 106–7
  pelecinum 26, 27*f*
  in Pergamon 62
  planar 93–4
  portable *see* portable sundials
  in Rome 62
  semi-circular 25–6, 95
  sloping 25–6
  social currency 87, 89–90, 92, 95, 100, 109n.86
  spherical 25–6, 93–4, 94*f*
  stone L-shaped 25–6
  as symbols *see* symbols, clocks as
  and tables 133

testability 72–3
trope of thinker and sundial 96–8
vertical planar 26
symbols, clocks as
  among Roman elite *see* Roman elite, clocks as symbols among
  of death and the afterlife 108–11
  in Galen's philosophy *see* Galen's clock-making paradigm
  of the human lifespan 100–8, 110–11
  of imperial power 87–92
  of mathematical ingenuity 93–6, 111
  of owners 108–9
  and philosophers 100–5
  of solar motion 45
Symons, A. 8–9

tables 39–40, 48, 58–9
  astronomical data 29, 132
  Hellenistic period 132
  mathematical 146
  scientific table-making 137
  star 19–20
  and sundials 133
  and water clocks 133
Talbert, R. 8–9
technology
  historians of 8
  user interaction with 6–7
  technological revolution 2
Telephus 175–8
telling the time
  in the Greco-Roman era 15–31, 93
  without clocks 17–19
temple priests 5–6
temporality
  desirability of temporal exactitude 135–55
  in Galen's fever case histories 120–6
  temporal indicators pegged to solar rhythms 125
  temporal markers of illness 53–4
  temporal trajectories of disease 2, 34–8, 40, 121
  temporal windows 41–2, 149
testability 71–4, 84
Theophilus 23
Thessalus of Tralles 59, 67n.31, 116–17
Thompson, E. P. 7
time
  see also clocks; hourly timekeeping; temporality; time-indications/time-indicators; timekeeping; time-reckoning; time-units
  body time 23, 185
  celestial 89–90

  civic 91
  "clock time" 23, 42–3, 185
  concept 101
  cosmic 91
  daily 17
  of day *see* time of day
  Eastern Standard Time 17
  hyper-indicated 17
  intra-day markers 121–2
  medicalized, standardization of *see* standardization
  "normative times" 47
  reifying 102
  river metaphor 101
  Roman civic 89–90
  sunrise and midday, between 17
  of year 17, 21–4, 131–2, 169, 172
time-indications/time-indicators 17, 185
  in medical context 33, 40–1
  short time 17–18, 20–1
timekeeping
  see also hourly timekeeping
  Hippocratic practices *see* Hippocrates of Cos; Hippocratic authors; Hippocratic Corpus; Hippocratics (Greek physicians)
  and Imperial Program, Roman period 87–92
  mealtimes *see* mealtimes
  qualitative indicators, based on 3
  and technological revolution 2
time of day 1, 131–2, 149
  see also hours, *hōra/hōrai*, *kairos* (opportune moment for medical intervention)
  announcing 49–50
  for bathing 166
  bloodletting 148–9
  to give treatment 38, 145–6, 149, 152, 154, 157
  mornings *see* mornings
  optimum, to give treatment *see kairos* (opportune moment for medical intervention)
  and summer 152–3
  of symptoms first appearing 145–6
  and winter 152–3
time-reckoning 17–19
  devices 17–18
  systems 18–19
time-units 17, 19–20
"Tomb of the Physician," Palazzina Capodaglio (Este, Italy)
  portable sundial found at 15, 16f, 48–9
Toomer, G. J. 29–30
"Torricelli's theorem" 28–9
Traversari, G. 98–9, 100n.43
twelve-hour day 21

twenty-four hour day 21
"Two-Sphere Model" 93

Ullmann, M. 124-5
urinary disorders 35
UŠ (fixed-length time-units) 20-1, 38-9
utility, concept of 9-10

Vachala, B. 35-6
Valerius Maximus 107n.77
　Memorable Deeds and Sayings 106-7
venesection *see* bloodletting
Venus 57-8
verifiability, concept of 9-10
Vitruvius 61-2, 64, 95-6
　On Architecture 94-5
von Lieven, A. 8-9
von Staden, H. 47
Vymazalová, H. 35-6

watches 3
water clocks 3-4, 7-8, 17-18, 27-30, 46, 61-2
　see also clocks; sundials
　in *Affections and Errors* (Galen) 9-10
　archaeological evidence 6, 9, 15-16
　and astrology 58-9, 109
　Babylonian 28-31
　compared with sundials 27-8
　designs 93-4
　Egyptian 28-31, 48
　"flower pot" outflow clocks 29, 47
　Galen's specific mention of 63-4
　Greco-Roman 27-8

inflow clocks 27-9, 64
medical 47-8
metaphor for human lifespan 101-2, 105-6
outflow clocks 27-9
　in Pergamon 62
　pointers 50-2
　in Rome 62
　as symbols of owners 108
　and tables 133
Webster, C. 144
week
　see also seven-day week
　medical 128-9
winter
　Greek sources 21-2
　health regimens 174-6
　solstice 24f, 78n.85, 88-9, 95, 168-9
　and time of day 152-3
Wittern, R. 137n.8
Wolkenhauer, A. 8-9

year
　number of days in 128-9, 182-3
　seasons of *see* seasons
　solar 89-90, 152
　time of 17, 21-4, 131-2, 169, 172
yellow bile 137-8, 151, 164-5
Young, M. W. 184

Zanker, P. 100n.42
zodiac 39-40, 133
　see also astrology; astronomy
　introduction of 39